Granite Genesis: In Situ Melting and Crustal Evolution

Granite Genesis: In Situ Melting and Crustal Evolution

by

Guo-Neng Chen and Rodney Grapes

Sun Yat-sen University,
Guangzhou, China

Springer

A C.I.P. catalogue record for this book is available from the Library of Congress

ISBN 978-90-481-7469-0
ISBN 978-1-4020-5891-2 (e-book)

Published by Springer,
P.O. Box 17, 3300 AA Dordrecht, The Netherlands.

www.springer.com

Printed on acid-free paper

Cover illustration:

Cartoon illustrating a geodynamic interpretation for the cycling evolution of continent rock material, emphasising the relationship between the two major material (energy) cycling processes in the continental crust and the mantle and formation of an intercrustal convecting granite magma layer.

CONTENTS

PREFACE

"We are never devoid of some hope that we shall eventually know everything. It may often be salutary, nevertheless, to recognize the remoteness of that time and to take stock of our ignorance".

N.L. Bowen–

1947 Presidential address to the Geological Society of America

There is a vast amount of published research on granite and we cannot claim to be familiar with it all. However, we are aware of the recent developments with respect to attempts at reaching a consensus on the problem of granite origin. With the concept of granitisation apparently dormant, if not dead, a paradigm of granite magmatism, i.e. melting, segregation, ascent and emplacement, has strengthened over the last 15 years. The combined research presented in this book describes a viable alternative mechanism of granite formation to this widely accepted intrusion model, namely, in-situ melting–intracrustal convection.

The core idea of the in situ melting origin of granite was formulated by GC, who worked as a member of one of the geological teams engaged in mineral prospecting and regional mapping in SE China between 1971 and 1977. It started from being unable to obtain satisfactory answers to the questions; if ore-minerals of hydrothermal deposits are not derived from granite magma, why are they found in their host rocks? If the ore-minerals are from granite magma, why are large granite masses barren of ore deposits? During seven year's of study at Sun Yat-sen University since 1977, GC spent most of his spare time in the university library trying to find an acceptable solution, but failed. GC gradually realised that perhaps what was generally believed about the origin of granite needed revision. So began an intensive 20-year study from which GC came to the conclusion that granite is produced and remains within the region of crustal melting. Collaboration on this idea began with RG in 2002 who was then at Freiburg University, Germany, and has been at Sun Yat-sen University since October 2005.

In the following pages, we demonstrate that thermal convection within a mid-upper crustal partial melting zone is essential for the formation of granite magma and that without convection partial melting of crustal rocks generates migmatites, not granite batholiths. Granite is layer-like on a crustal scale and the shape and size of individual granite bodies merely reflects the geometric relationship between the irregular upper part of a granite layer and the erosion surface, rather than the volume of intruded magma. Chemical and isotopic compositions of granites are considered to reflect processes operating within the magma system rather than unknown deep sources. Formation and crystallisation of a crustal magma layer results in reorganization and redistribution of elements within the crust, to form both granite and their related hydrothermal mineral deposits. The in situ melting–intracrustal convection model of granite origin integrates the two-knowledge systems related to continental geology and plate tectonics and explains related geological, geochemical and geophysical observations. The geochemical fields of elements established on the basis of both the in situ melting–intracrustal convection model and periodic table illustrates the harmony and unity between the microcosm and macrocosm of the natural world.

While the views advanced in this book differ from a number of traditional ideas in geology, the same evidence used to support these ideas is also adopted to strengthen our hypothesis. Our main aim is to offer another way of not only looking at granite but the Earth in the hope that this will stimulate further thought and study. As such, we hope that the book will be of interest to both professional and student Earth scientists.

Guo-Neng Chen and Rodney Grapes
Sun Yat-sen University
Guangzhou (Canton)
China
December 2006

ACKNOWLEDGEMENTS

GC is greatly appreciative of the many encouragements from academicians and in particular, Professors Zunyi Yang and Guoda Chen, during the course of this work. The understanding and dedication from GC's wife, Wang Dong, during 20 years of persistent study is warmly felt. Financial support from the sciences fund for the doctoral disciplines of universities from the Chinese Education Ministry, and from the sciences fund of Sun Yat-sen University, over the last 10 years has enabled this study to be completed. GC thanks Professor Robert Diffendal (University of Nebraska) for his helpful comments and assistance during a field excursion in the USA, and also thanks his colleagues, Professors Ke Zhang and Jingjian Cao for their help and cooperation. RG would like to thank his wife, Agnes, for her encouragement and help throughout the period of writing this book.

We also wish to acknowledge the help of Ph.D. students, Wenming Zhang, Yan Qiu, Zhuolun Peng, Xiaoming Lin, Haoming Ma, Rongsong Shao, Yanyong Liu; MSc students, Shansen Qui and Xianhe Zhang (Department of Earth Sciences, Sun Yat-sen University) and Christian Bratzdrum (Institute of Mineralogy and Geochemistry, Freiburg University, Germany). The assistance of the Earth Science librarians, Susanne Schuble (Freiburg University, Germany) and Gill Ruthven (Victoria University of Wellington, New Zealand), in locating references is gratefully acknowledged.

CHAPTER 1

INTRODUCTION

1.1. ROCK GENESIS AND ITS RELATIONSHIP TO GEOSYSTEMS

Rocks that form the outer part of the Earth, the crust, provide a record of constant change in the Earth's environment with time in relation to the movement and/or reaction of rock material during their formation. Thus, understanding the genesis of rocks is the foundation of the geological knowledge system. The nature, size, shape, properties and arrangement of constituent minerals gives rise to the diversity of rocks and their origin so that the challenge of geology is in determining their material source and mechanism of formation. Among the three kinds of rocks, igneous, sedimentary and metamorphic, the material source of metamorphic rocks need not be considered because its predecessor is also rock. Igneous rocks derived by crystallisation of liquids can be divided into mafic and felsic end members represented by gabbro/basalt and granite/rhyolite, respectively, with different sources.

1.1.1. Sedimentary Rocks and Continental Geology

The genesis of sedimentary rocks that form the outermost part of the solid earth was early clarified by Lyell (between 1830 and 1833; 1868) in his *Principles of Geology* and by the geosynclinal hypothesis of Dana (1873), i.e. sediments in basins are derived from weathering and erosion of rocks in mountains or other uplifted areas, lithification of the sediments during burial, return of the new sedimentary rocks to the Earth's surface by tectonic processes, their subjection to renewed weathering and erosion, and so on as shown in Fig. 1.1. The relationship of endogenic/exogenic forces of the Earth, and the sedimentary rock cycle, demonstrate environmental variation and related material transportation in terms of two different tectonic settings, areas of uplift and subsidence. Without knowledge of this relationship, most theories of continental geology would not exist, and it is the main reason why the geosynclinal hypothesis remained fundamental in geology for over a century.

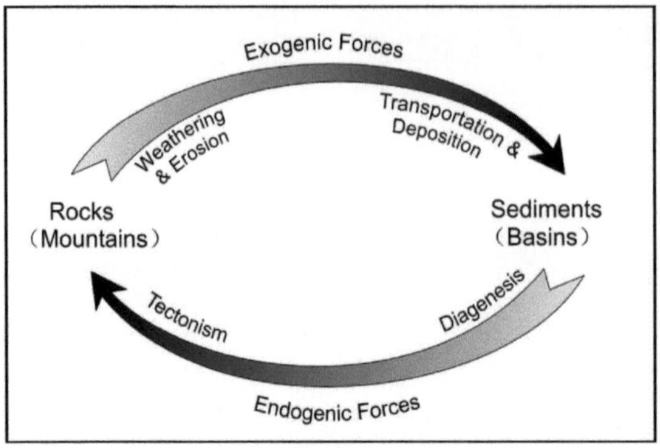

Figure 1.1. Genetic model of sedimentary rock formation and associated material cycle.

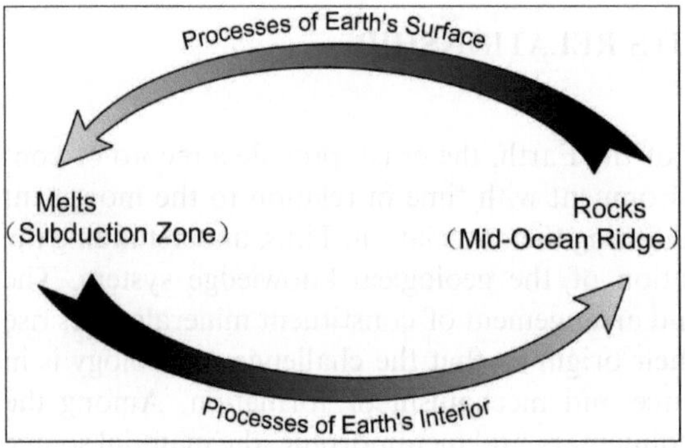

Figure 1.2. Genetic model of basalt and material cycling of the mantle.

1.1.2. Basaltic Rocks and Plate Tectonics

Basaltic rocks constitute the oceanic crust, and probably a significant part of the lower continental crust (Sima). From the work of Holmes who put forward the mantle convection hypothesis in the 1920's to that of Dietz (1961) and Hess (1962) who advanced the idea of ocean-floor spreading, the genesis of oceanic basalt and the corresponding material cycling of the mantle were successfully explained by geology, i.e. mafic magma from the mantle erupts from mid-ocean ridges to form the basaltic rocks of the ocean crust which moves away from the ridge axes, and is returned to the mantle by subduction (Fig. 1.2). Without this model, the plate tectonics hypothesis would not exist and it was probably the main reason why Wegener's hypothesis of continental drift was eventually accepted because it provided a mechanism for large-scale crustal movement.

1.1.3. 'Whence the Granites'

The genetic models of sedimentary and basaltic rocks explain environmental variables and material transport of the continents in relation to endo–exogenic forces, and energy/material flow in the Earth's interior in relation to tectonic plate

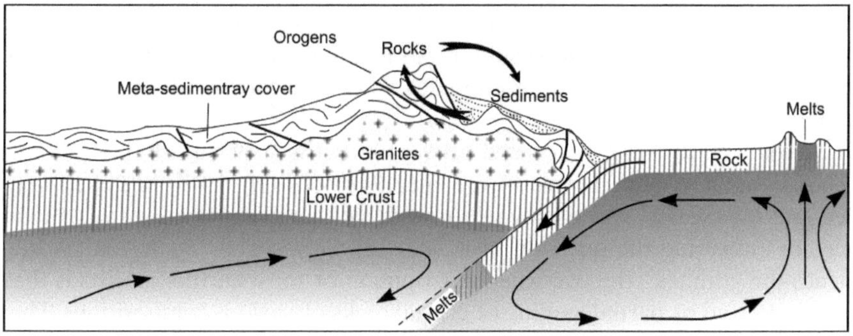

Figure 1.3. Cartoon showing spatial distribution of granite in the continental crust and relationship of the two material cycling processes known from geology.

movement. These models form the core theories of the two knowledge systems, i.e. *Continental Geology* and *Plate Tectonics*. However, as shown in Fig. 1.3, neither the formation process of sedimentary rock nor that of basalt produces granite that is considered to be distributed between the shallow crustal metasedimentary rocks and the lower continental crust.

Granite that belongs to the continent composes the fundamental part of the upper crust and is closely related to tectonism, metamorphism and mineralisation of the continental crust. The origin of granite, however, has been the subject of much debate since the eighteenth century when the science of geology was in its infancy. Accounts of the 'granite controversy' and the origin of granite can be found in Gilluly (1948), Pitcher (1993) and Young (2003) and will not be addressed here except to say that since the advent of plate tectonics in the 1960's, a framework has been provided in which the different interpretations of granite genesis can be re-explored and in many cases reconciled, although debate still continues. A misleading idea related to the basalt magma fractional crystallisation-differentiation scheme of Bowen (1914, 1922, 1948) is to combine the order of mineral crystallisation to the formation of basic to acid series igneous rocks. Only a small amount of residual granite liquid is generated by this process and this experimental evidence is clearly at odds with field evidence for the abundance of granite (Holmes 1926; Read 1957). While the mineral reaction series can be applied to the crystallisation of different magma compositions as recorded by their mineral textures, the first rocks formed are not necessarily mafic, and the last are not always felsic, because the rock crystallised from a melt depends on the composition of the melt and not the order of mineral crystallisation (Kennedy 1933). In this respect, the comment on Bowen's idea by Walton (1960) is relevant,

"Not that there was anything wrong with Bowen's chemistry or his application of it to the fractionation of basaltic magma; it was and will still remain a keystone of petrology. But there was a crucial tacit assumption involved in tying igneous theory so rigidly to a single model in which the evolution of most igneous rocks is dependent on the upward intrusion of basaltic magma into a level of the crust where it cools, crystallizes, and fractionates. The same chemistry can be applied to other models".

The granite controversy that raged between *granitisers* such as H.H. Read and *magmatists* such as N.L. Bowen in the late 1940's (Gilluly 1948) has effectively ended and most earth scientists would now agree that granites are of magmatic origin. In answering the question, *how does granite form?* or in Bowen's words 'Whence the granites?' the overwhelming opinion of most earth scientists is that granite is derived by partial melting of crustal rocks of various compositions. This idea essentially brings together the earlier competing explanations of granite genesis; magmatic (granites are igneous rocks resulting from the crystallisation of magma) and metamorphic (granites are the result of a dry or wet granitisation process that transformed sialic sedimentary rocks into granite), because granites are the result of ultra-metamorphism involving melting (anatexis) of crustal rocks. This explanation has important implications for the origin and chemical differentiation of the Earth's crust in relation to the source and evolution of thermal regimes, protolith composition, how much granite can be produced, over what time and at what temperature, the amount and source of available water needed, tectonic processes and plate tectonic settings—a truly holistic association.

1.2. GRANITES, MIGMATITES AND GRANITE PROBLEMS

1.2.1. Definitions

1.2.1.1. Granite

The maxim 'there are granites and granites' originally coined by H. H. Read remains just as true today as it did in 1933. In fact the number of 'granite types' has proliferated from at least 20 schemes that have been proposed to classify them (see Barbarin 1990, 1999 for summary, and Frost et al. 2001 for an appraisal of the more commonly used classification schemes). The most widely used classification schemes are geochemical and/or generic-alphabetical, i.e. S-, I-, M-, A-, and C-type granites (S = sedimentary source; I = igneous source; M = mantle source; A = anorogenic; C = charnockitic); calc-alkaline, alkaline, peralkaline, peraluminous, metaluminous granites; or are related to tectonic setting: 'orogenic' (oceanic and continental volcanic arc; continental collision), 'transitional' (post-orogenic uplift/collapse), and 'anorogenic' (continental rifting, hot spot, mid-ocean ridge, oceanic island) granites.

In this book the broad terms 'granitoid', 'granitic rocks' or simply 'granite' are used to mean quartz-bearing (>60 wt % SiO_2) plutonic igneous rocks in general. We adopt (or return to) where necessary a non-genetic classification of granitic rocks based on field and petrographic observation in that we use well-established rock names, e.g. granite, adamellite, granodiorite, tonalite, monzonite, diorite, together with qualifiers such as mineral (e.g. biotite-, muscovite-, two mica- etc.; quartz- in the case of syenite, monzonite and diorite), grain size (e.g. fine, coarse, medium, aplite, pegmatite), texture (e.g. porphyritic, granophyric, orbicular, etc), and colour (e.g. leuco-, meso-, melanocratic). Other textural/compositional terms, such as rapakivi granite, are in common usage and are retained because they are non-generic.

As quartz and feldspars (alkali feldspar and plagioclase) make up nearly 90 modal % of granitic rocks, the non-genetic IUGC classification scheme of Streckeisen (1976) can best be adopted to 'pigeonhole' them (Fig. 1.4A) . The classification scheme is, as with any such scheme, not without its faults, e.g. it does not include the mafic mineral component of granitic rocks, and the rock divisions may not always clearly distinguish natural populations, but it is straightforward and it is based on what can be seen. Our only modification is to retain the name 'adamellite' because it is entrenched in the literature. In the Streckeisen scheme the adamellite field forms part of a larger granite field. Minerals other than quartz and feldspars that may be easily distinguished in outcrop and which are used to identify variation in granitic rocks such as biotite, muscovite and hornblende (also fayalite, orthopyroxene, cordierite, garnet, tourmaline, topaz, ilmenite, magnetite-bearing varieties), can simply be attached to the Streckeisen classification name. In this way the granite name remains even if ideas on its origin and paleotectonic setting change and it is informative, e.g. porphyritic biotite granite or leucogranite. 'Trondhjemite' is another term in common usage, especially with respect to the trondjhemite-tonalite-granodiorite (TTG) association characteristic of Archean terranes. It is not represented in the Streckeisen scheme, but can be regarded as the leucocratic variety of tonalite, consisting of essentially plagioclase (oligoclase–andesine) and quartz with minor biotite and accessory Fe-Ti oxide. The relationship between modal mineralogy, including the common mafic minerals and muscovite, granitic rock type and colour index is shown in Fig. 1.4B.

We adopt a normative quartz, alkali-feldspar, plagioclase classification scheme because there are many more chemical analyses of granites than there are modal analyses and the normative scheme also allows characterisation of experimental melt (glass) compositions. The normative classification is approximated using the orthogonal plot of Streckeisen and Le Maitre (1979) (Fig. 1.5A) , where modal quartz is analogous to 100 Qz/(Qz + Or + Ab + An) and modal plagioclase/(alkali-feldspar + plagioclase) is analogous to 100 An/(Or + An) with normative Ab being ignored. Normative feldspar (An-Ab-Or) composition is also a convenient classification scheme for granitic rocks in that it allows distinction of the trondhjemitic variety (Fig. 1.5B).

1.2.1.2. Migmatite: terminology and classification

Migmatites, or mixed rocks, preserve a multitude of structures, mineral textures and compositions that reveal the process of rock melting (anatexis) that leads to the production of granite magma. In his classic work, Mehnert (1968) defines a

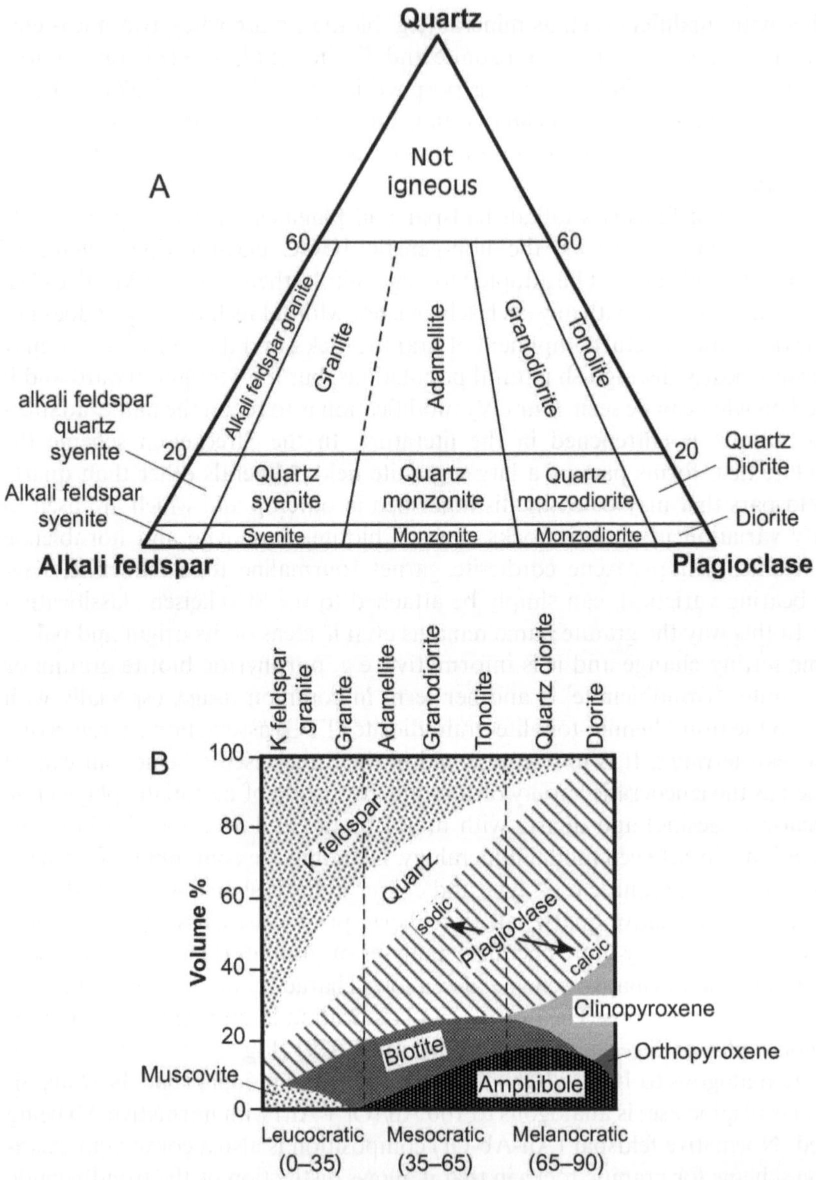

Figure 1.4. **A**. Modal classification of granitic rocks according to Streckeisen (1976). **B**. Modal and colour classification of granitic rocks that includes muscovite, biotite, hornblende and pyroxenes.

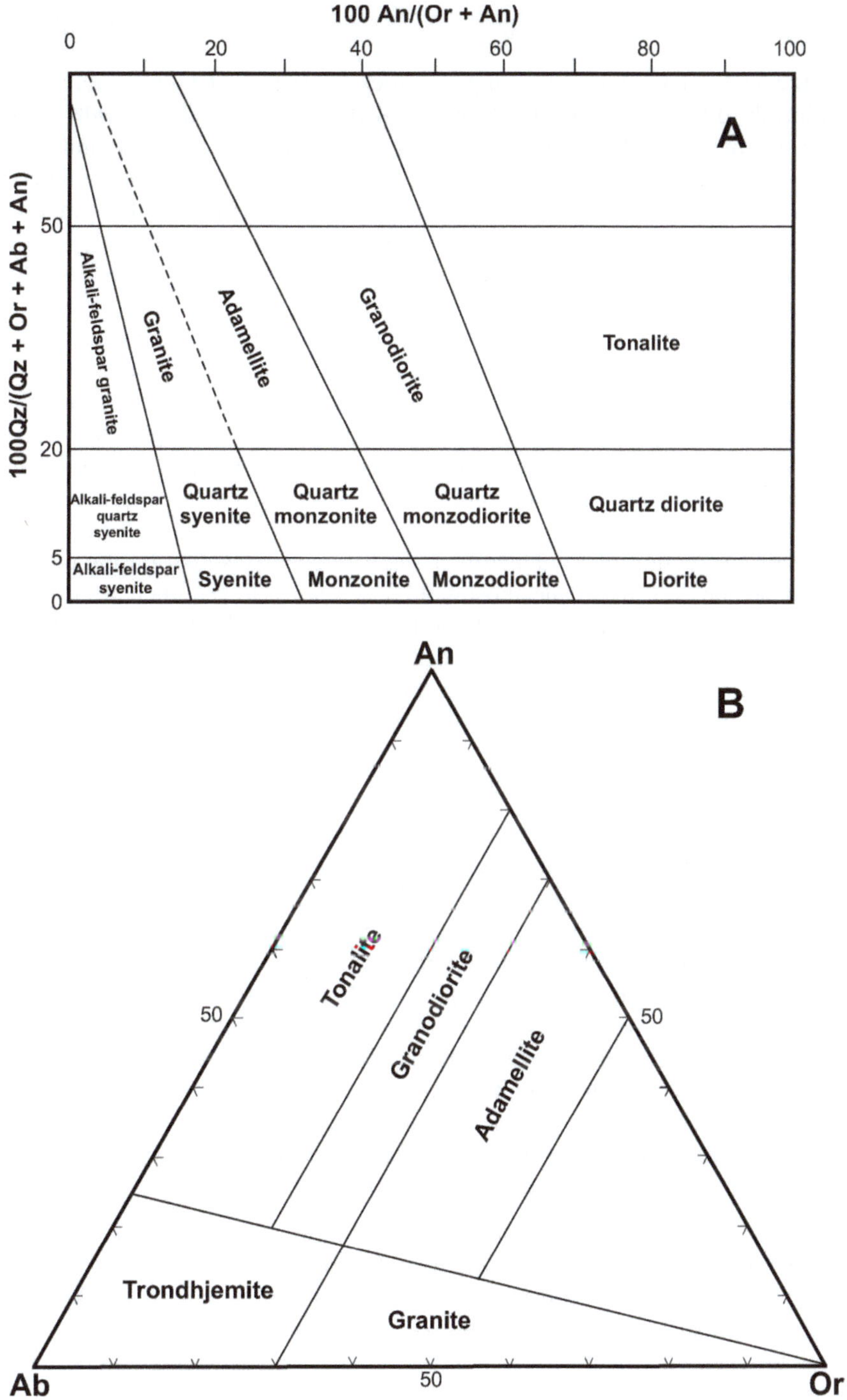

Figure 1.5. **A**. Normative classification of granitic rocks according to Streckeisen and LeMaitre (1979) with inclusion of an adamellite field. The field boundaries are best fits that separate chemically analysed rocks named according to the modal classification shown in Fig. 1.4A. **B**. Normative Ab-An-Or classification scheme of granites (after O'Connor 1965; Barker 1979).

migmatite as "a megascopically composite rock consisting of two or more petrographically different parts, one of which is the country rock generally in a more or less metamorphic stage, the other is of a pegmatitic, aplitic, granitic or generally plutonic appearance." Mehnert's definition is the basis for the latest nomenclature on migmatites proposed by the sub-commission for the nomenclature of metamorphic rocks (SCMR), (Wimmenauer and Bryhni 2002), who provide a revised definition of migmatite: "A composite silicate rock, pervasively heterogeneous on a meso- to megascopic scale. It typically consists of darker and lighter parts. The darker parts usually exhibit features of metamorphic rocks, while the lighter parts are of plutonic appearance. Whenever minerals other than silicates and quartz are substantially involved, it should be explicitly mentioned". Both definitions attempt to be non-genetic although use of the words metamorphic and plutonic underscores the common understanding that high grade (i.e. high-T) metamorphism and anatexis are prerequisites for migmatite formation.

The main distinction of a migmatite is the heterogeneous distribution of the light (leucosome) and dark (melanosome) parts relative to the unchanged portion. These are 'new' components of the rock, and are collectively termed the 'neosome'. The 'unmigmatised' parent rock (protolith) is called the 'paleosome'. Another dark-coloured component of a migmatite is the 'restite' from which a substantial amount of the more mobile components appear to have been extracted (Wimmenauer and Bryhni 2001). Additional terms that directly relate to migmatites are metatexis and diatexis.

1. Metatexis refers to the initial stage of anatexis where the paleosome is partly segregated into a more mobile part (metatect), and a non-mobilized (depleted) restite. If the migmatisation process has created a rock with obvious and discrete leucosomes, melanosomes, and intermediate coloured mesosomes, the migmatite can be called a 'metatexite'.
2. 'Diatexis' refers to an advanced stage of anatexis where the darker-coloured minerals are also involved in melting and the melt is not removed from its place of origin. The metamorphic paleosome has been largely disrupted, and the leucosomes and melanosomes are intimately combined in a *nebulitic* or *schlieren* texture. In such cases the migmatite can be termed a 'diatexite' (Ashworth 1985; Wimmenauer and Bryhni 2002). Clearly, flow has occurred and the diatexite could also be regarded as a 'dirty' or proto-granite magma.

In addition to quartz, feldspars, muscovite and biotite, migmatites derived from the partial melting of mica-bearing quartzofeldspathic rocks (the most common type of migmatite) contain cordierite, garnet, ± sillimanite, andalusite and kyanite and may be classified on this basis after evaluation of prograde and retrograde mineral textures. A *P-T* pseudosection for an average metapelite composition containing excess quartz and plagioclase in the MnNCKFMASH system is given in Fig. 1.6. The diagram shows stable assemblages coexisting with in situ–derived liquid (granitic melt) over a temperature range of 650–740°C that represent conditions of

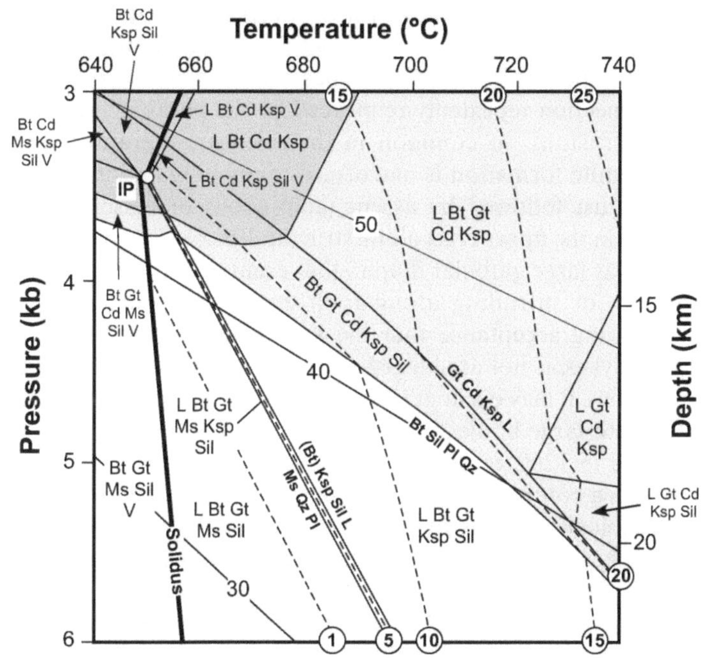

Figure 1.6. Suprasolidus *P-T* pseudosection in the MnNCKMASH system for the average pelite composition (normalised anhydrous mol % 73.64 SiO_2; 11.94 Al_2O_3; 7.23 FeO; 0.10 MnO; 3.95 MgO, 0.35 CaO; 1.41 Na_2O; 3.07 K_2O). Dashed lines at temperatures higher than the solidus and labelled 1 through to 25 indicate melt proportions in mol %. Dark grey area = Cd-only; light grey area = Cd + Gt; unshaded area = Gt-only. Linear geotherms 30, 40, 50°C/km are also shown. For chemical system and mineral abbreviations see Table 2.1 (after Fig. 4a of Johnson et al. 2003).

cordierite and garnet-bearing migmatite formation in the range 10–25 km depth. The areas of cordierite-only, cordierite + garnet and garnet-only depicted in Fig. 1.6 delineate fields in *P-T* space that reflect the formation of compositionally different migmatite terranes or of compositional variation in a single migmatite terrane. A field of low-temperature muscovite-bearing migmatite characterised by very low-melt fractions of ~1 mol % is also shown as part of the cordierite–garnet migmatite field. Temperature gradients >40°C/km will produce cordierite and cordierite– garnet-bearing migmatites while gradients <40°C/km will generate garnet-bearing migmatites at deeper crustal levels.

1.2.2. Granite Magma Intrusion and Its Problems

Geological studies of ancient orogens show that they contain large volumes of granite and geophysical studies of active orogenic zones suggest that part of the crust could be partially molten. Although common and often of similar appearance, granites reflect a complicated history of formation that has led to a diversity of opinions about their formation and continues to be a hotly debated topic in geology. Being at one time liquid, granite has become synonymous with intrusion, a genetic connection repeatedly reinforced by the phrases 'intrusive granites' or 'granite intrusions' so common in the scientific literature. The current paradigm of granite formation is one of crustal melting and melt segregation in the lower crust followed by ascent (continuous or episodic) via dykes with emplacement in the upper crust along structural/rheological 'traps', or rising through the crust as large globular diapirs that eventually freeze at some level to form a pluton, or intruding along fault zones to fill extensional openings. There is growing acceptance that the ascent and emplacement of granite magma, e.g. as dykes, is not as sluggish as previously thought and that provided flow is continuous, it may occur at timescales of less than 100,000 years, (Petford et al. 2000), with large bodies of granite (batholiths) for example of 1000 km^3, being formed in 1200 years from a 3 m wide, 1 km long (in plan) dyke, provided there was a continuous supply of magma from below (Clemens 2005). This is indeed a geologically catastrophic event! The age-old 'room problem' created by the idea that granite batholiths extended to unknown depths is explained by the notion that such bodies in fact have tabular sheet-like geometries (Chamberlin and Link 1927; McCaffey and Petford 1997; Crunden 1998) with thicknesses of only a few kilometres so that their 'emplacement' can more easily be explained by a combination of initially lateral (fault opening) and subsequent vertical (inflation by roof uplift and/or floor subsidence) movement. Diapirism as an alternative end member magma transport process is suggested by the roughly circular sections of many granite bodies from a few kilometers to one hundred kilometers in diameter and from analogy with the shape and finite strain field of salt diapiric intrusions (Bateman 1985). In the diapiric model, granite ascent from a ductile lower crust is much slower (10^5–10^9 years; Marsh 1982; Weinberg and Podladchikov 1994; Miller and Paterson 1999) than in dykes and there are more thermal and mechanical difficulties to overcome with transporting very large volumes of magma, especially into brittle upper crustal rocks. These ideas form the basis of the intrusion model of granite genesis (segregation, ascent, emplacement) (Fig. 1.7) and a vast array of field, geochemical, geophysical and experimental evidence combined with theoretical modelling has been used to support the model. Although theoretical models neither rule out diapirism nor fully support dyking as mechanisms to move granite magma upwards from its lower crustal source, the fact remains that the intrusion model requires that granite magmas be transported upward through progressively cooler crust to their emplacement site at a sufficiently fast rate to prevent them freezing.

Perhaps the weakest part of the intrusion hypothesis is in identifying a focusing site that enables granite melt to escape from its source area either by way of a conduit or in the form of a diapir. It has to be assumed that beneath every granite body there must be a deeper crustal source region. For example, Brown (2001)

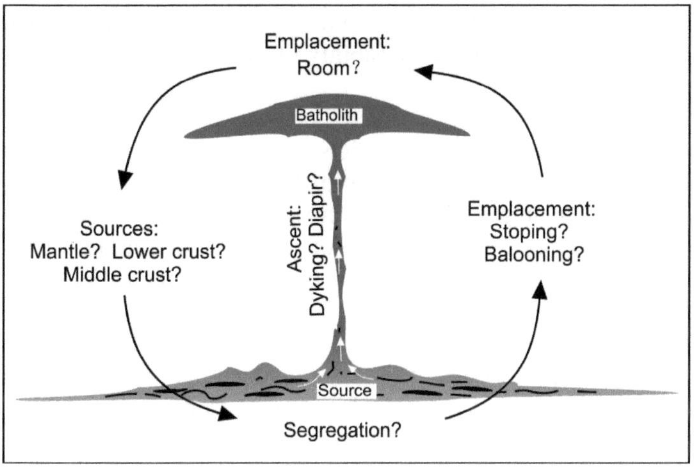

Figure 1.7. Schematic diagram to illustrate the intrusion model of granite genesis.

calculates that a horizontal semicircular half cone-shaped granite body with a diameter of ~30 km and a half height of 5 km could be formed by ~20 vol % fractional crystallisation of ~20 vol % melt segregated from an underlying ~15 km thick source! with a horizontal diameter of ~30 km, i.e. a lower crustal source volume for this one pluton of ~11,000 km^3. As pointed out by Brown, an implication is that granite bodies of this size might be spaced ~30 km apart. As many exposures of granite typically consist of more than one isolated occurrence, it also implies the existence of an extensive source region that must be related to a regional geotherm that enables crustal anatexis to occur.

Field evidence of dyke-like feeders for granite batholiths is rare, because their floors are seldom exposed. Obviously, if feeder dykes extend vertically for many kilometres as they are supposed to, a continuous link between granite batholiths and their source is unlikely ever to be found. Nevertheless, dykes are a requirement as well as a contention for the granite intrusion hypothesis. There are also problems with a rootless diapiric origin of granite where the surrounding rocks deform by viscous flow to allow the buoyant rise of a magma body. Although there is evidence from internal structures in granite bodies that suggest a diapiric origin (Pons et al. 1995) and modelling supports the possibility (Weinberg and Podladchikov 1994), the paucity of evidence such as rim synclines, near vertical cylindrical shear zones and strong thermal effects (Clemens 1988; Clemens and Mawer 1992) associated with granite bodies has raised doubts on the viability of such an ascent mechanism. There is no known structural/thermal evidence of an ascent trail for postulated diapiric granite bodies, so that diapirism, if it occurs, is probably restricted to deep crustal levels because of the exponential increase in crustal viscosity with decreasing depth that would be expected to halt diapiric ascent at about mid crustal levels (Clemens and Mawer 1992; although see Weinberg and Podladchikov 1994 for the possibility of granite diapirs reaching depths as shallow as 6 km).

'Ballooning' (Ramsey 1989) has been considered one way of solving the room problem for granite emplacement, and it predicts that the volume for pluton accommodation can be supplied by deformation of the thermal aureole during in situ radial inflation of a 'magma intrusion'. Inflation can occur by continuous feeding via dykes, or by way of the hotter 'tail' of a diapir continuing to rise whereas the main upper part has ceased its upward ascent. Structural patterns attributed to a ballooning pluton include a circular or elliptical shape in horizontal sections, planar fabric parallel to the contacts and more intensely developed at the border zones where it appears as a gneissic solid-state foliation, concentric zoning of rock facies, the central facies generally being more acidic in composition and later with respect to the marginal facies; a planar fabric parallel to the contacts in the host rocks due to pure shear increasing towards the igneous contact. However, all of these features can also be produced as a result of upward doming of part of an in situ melt layer as described below.

In a large number of cases where granite bodies show a discordant relationship with their host rocks (Fig. 1.8A), the space problem remains. If the granite is allochthonous, where is the pre-existing rock that formally occupied the area now occupied by the granite? Has it been passively engulfed (melted) or stoped by the granite magma? (Fig. 1.8B, C). Has the granite magma been tectonically emplaced along a fault and simply forced apart the country rocks? The stoping mechanism of magma ascent and emplacement first proposed by Daly (1914) has been invoked to explain the first possibility and implies thermal shattering of the host rocks induced by magma, invasion of fractures by the magma and sinking of the disrupted blocks. Sinking of largely unmelted host rocks in the magma could contribute to a more rapid rate of cooling so that the magma body would suffer an early thermal death and if so, one would expect to find abundant fragments of largely unmelted country rock frozen into the marginal areas of the granite.

Migration of granite magma along fault zones has been suggested for some orogenic settings e.g. Hutton et al. (1990); D'Lemos et al. (1992); Davison et al. (1995). This mechanism necessitates the formation of crustal scale fault/shear systems within which extensional jogs can be passively or forcibly filled with granite magma. Also, these active shear zones need to be situated above or perhaps within the zone of magma generation and extend for significant vertical distances through the crust. In many cases, however, granite bodies aligned along faults systems exhibit marginal shearing and truncation indicating that faulting was post-granite 'emplacement' or that the faults have been reactivated subsequent to granite emplacement. In other cases discordant granite contacts show no evidence of being faulted and in such cases it has been assumed that shear/fault zone avenues have been obscured by the effects of stoping and ballooning of the granite magma (D'Lemos et al. 1992). Statistical analysis of spatial relationships between faults and granite in the Armorican Massif, France

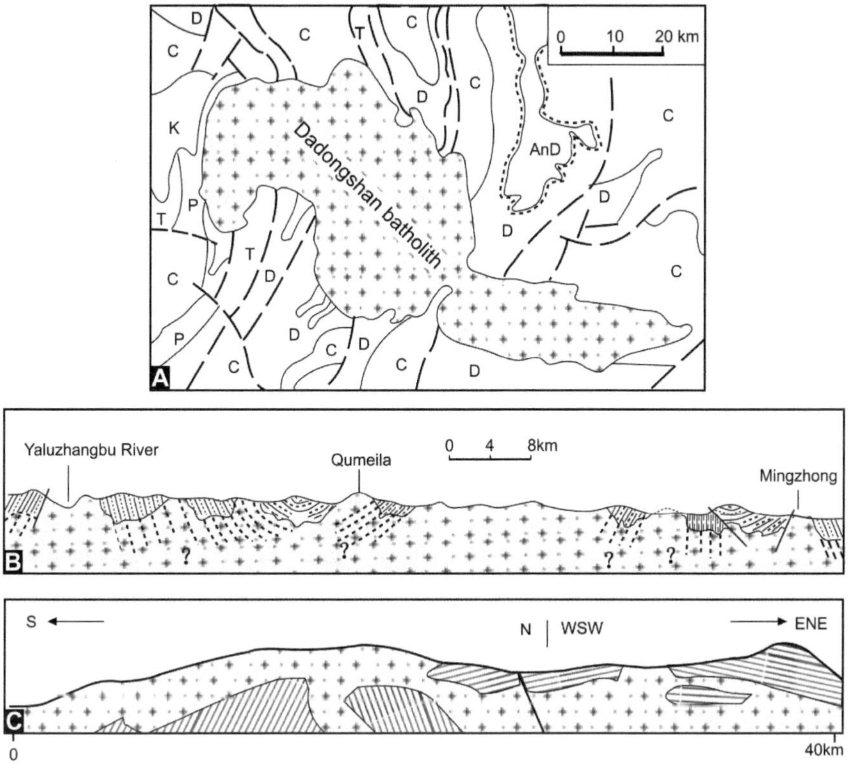

Figure 1.8. **A**. Simplified geologic map of the Dadongshan batholith, Guangdong Province, SE China, showing discordant contact relationship between batholith and structural trend of the country rocks. **B**. Cross section of the Gangdise batholith from the Yaluzhangbu River to Mingzhong, Tibet, showing crosscutting relationship between the granite contact and steeply-dipping metasediments with the suggestion that they have been melted (modified after SEGCA 1981). **C**. A 40 km long section of the Striegau granite, Silesia, Poland, showing variously orientated blocks of country rock within the granite (redrawn after Fig. 3 of Cloos 1923).

and southern Appalachians, USA, indicates a tendency for granite to occur away from faults, indicating that faults in these areas, and presumably in others, do not preferentially channel magma in general (Paterson and Schmidt 1999).

Magma transport is a mass transfer process where upward movement of magma is balanced and accommodated by movement of the country rocks. Thus, the prerequisites of granite intrusion are structural control and a suitable mechanism of episodic or continuous injection of magma. With these requirements and other exposure-related questions, it may be pertinent to ask whether it is necessary to separate granite melt from its source area at all. It is a question that also seems to have occupied the mind of James Hutton when he contemplated the magmatic origin of granite in 1790, asking "...whether it were not rather a body which had been originally stratified ... and afterwards consolidated by the fusion of these materials, or whether it were not rather a body transferred from sub-terraneous regions, and made to break and invade the strata." If granite magmas are not intruded by whatever means from a lower crustal metamorphic source, what is exposed must be granite together with the rocks of, or in the immediate vicinity of its source region, the relationships between them modified more often than not, by subsequent tectonic activity. This is the essence of 'in situ melting model' of granite genesis proposed here—that granite magma essentially remains within or close to its source area. The common association of granite and migmatite, e.g. in gneiss domes, and also granulite, provides unequivocal evidence of this, although granite bodies not associated with migmatite and surrounded by narrow metamorphic aureoles imposed upon relatively low-grade rocks, also require explanation. It is also clear that while low-melt (leucogranite) fractions can be efficiently transferred from migmatite to crystallise and sills, dykes and laccoliths in rocks immediately overlying the melting zone (e.g. Himalayas), this is not the case for the much larger volumes of granite, adamellite, granodiorite, tonalite, i.e. the 'stuff' of batholiths. In the following pages, we explore the alternative idea that granites have an in situ melting origin using the field, geochemical, geophysical, experimental and modelling evidence that has been marshalled in support of an intrusion origin.

CHAPTER 2

CRUSTAL MELTING: EXPERIMENTS AND CONDITIONS

2.1. INTRODUCTION

The formation of granite magma by a combination of lower *T* fluid-present melting and higher *T* fluid-absent dehydration melting of micas and amphibole in the Earth's crust largely depends on source rock composition, particularly whether the components necessary to produce a granite 'minimum-melt' composition (K_2O, Na_2O, Al_2O_3, SiO_2, H_2O) are available, the pressure at which melting occurs, temperature, type and amount of hydrous minerals present, and the presence or absence of free water. Water is continuously evolved during metamorphism from dehydration reactions. Whether this H_2O remains in the rock or not depends on its porosity, which in turn, is dependent on depth (lithostatic pressure). If the porosity of the rock is small, nearly all H_2O evolved by dehydration leaves the rock. Dihedral angle measurements for hydrous fluids in contact with silicates suggest that it might be difficult to completely remove a fluid phase from a rock undergoing progressive dehydration on metamorphism (Watson and Brenan 1987; Holness 1993; Laporte et al. 1997). Although deformation will enhance fluid escape, it seems possible that at the end of the dehydration event, some fluid will remain in the source rock along grain boundaries, as isolated pockets at grain corners or along grain edges as well as in fluid inclusions. Initial granite melts will be fluid-saturated but the amount of melt maybe small because the proportion of fluid phase present is also small. Therefore, unless there is an external source of fluid, fluid-absent melting will be the predominate mechanism for significant volumes of granite melt to form.

All granitic magmas contain dissolved water, e.g. up to ~6 wt % at 2 kb and ~10 wt % at 10 kb (Holtz and Johannes 1991). This water is derived from;

1. free water occurring along mineral grain boundaries, pores, fractures, fluid inclusions in minerals
2. structurally bound OH ions in phyllosilicates (as wt % H_2O^+) such as chlorite (10–12%), muscovite (~4%), biotite (3–4%), amphibole (~2%), or other less hydrous minerals such as epidote (1.5–2%), staurolite (~1%), cordierite (~1%).

The two sources of water give rise to two kinds of melting reaction;

1. water or fluid/vapour-present melting, H_2O-saturated melting (aH_2O = 1.0), or wet melting, where free water is available, e.g. anhydrous minerals + aqueous fluid/vapour = liquid anhydrous minerals + hydrous mineral + aqueous fluid/vapour = liquid
2. fluid/vapour-absent melting, H_2O-undersaturated (aH_2O = <1.0), or dehydration (incongruent) melting, where no free water is available, e.g. anhydrous minerals + hydrous mineral = liquid + anhydrous minerals.

2.2. MINERAL MELTING

(Mineral notations used below are listed in Table 2.1 together with nomenclature and composition).

Table 2.1. Mineral, composition, notation and oxide system notation

Andalusite	Al_2SiO_5	And
Anorthite	$CaAl_2Si_2O_8$	An
Albite	$NaAlSi_3O_8$	Ab
Apatite	$Ca_5(PO_4)_3(OH)$	Ap
Al-silicate (unspecified)	Al_2SiO_5	AS
Biotite	$K(FeMg)_3AlSi_3O_{10}(OH)_2$	Bt
Calcite	$CaCO_3$	Cc
Chlorite	$(MgFeAl)_{12}(SiAl)_8O_{20}(OH)_{16}$	Ch
Clinopyroxene (unspecified)	$Ca(MgFe)Si_2O_6$	Cpx
Cordierite	$(FeMg)_2Al_4Si_5O_{18}{\cdot}nH_2O$	Cd
Corundum	Al_2O_3	Co
Epidote	$Ca_2(Fe^{3+}Al)_3Si_3O_{12}(OH)$	Ep
Garnet (unspecified)	$(CaMgFeMn)_3(AlFe^{3+}Cr)_2Si_3O_{12}$	Gt
Hematite	Fe_2O_3	Hm
Hercynite	$FeAl_2O_4$	Hc
Hornblende (unspecified)	$(NaK)_{0-1}Ca_2(MgFeAl)_5(SiAl)_8O_{22}(OH)_2$	Hb
Ilmenite	$FeTiO_3$	Ilm
K-feldspar (unspecified)	$KAlSi_3O_8$	Ksp
Kyanite	Al_2SiO_5	Ky
Quartz (*β*)	SiO_2	Qz
Magnetite	$FeFe^{3+}{}_2O_4$	Mt
Muscovite	$K_2Al_4Si_8O_{20}(OH)_4$	Ms
Monazite	$(CeLaTh)PO_4$	Mz
Olivine (unspecified)	$(MgFe)_2SiO_4$	Ol
Orthopyroxene (unspecified)	$(MgFe)_2Si_2O_6$	Opx
Orthoamphibole (unspecified)	$(MgFe)_7Si_8O_{22}(OH)_2$	Oam
Osumulite	$(KNa)(FeMg)_2(AlFe)_2(SiAl)_{12}O_{30}$	Os
Plagioclase (unspecified)	Solid solution between Ab and An	Pl
Prehnite	$Ca_2Al_2Si_3O_{10}(OH)_2$	Pr
Pumpellyite	$Ca_4(MgFeAl)_5Si_60_{23}(OH)_3 2H_2O$	Pu
Pyrite	FeS_2	Py
Pyrrhotite	Fe_7S_8	Po

Table 2.1. Mineral, composition, notation and oxide system notation—Cont'd.

Rutile	TiO_2	Rt
Spinel (unspecified)	$FeMgAl_2O_4$	Sp
Staurolite	$(MgFe)_4Al_{18}Si_8O_{46}(OH)_2$	St
Stilpnomelane	$(KNaCa)_{0-1.4}(MgFeAl)_{5.9-8.2}Si_8O_{20}(OH)_4$ $(O,OH,H_2O)_{3.6-8.5}$	Stp
Sillimanite	Al_2SiO_5	Sil
Titanite	$CaTiSiO_4(OH,F)$	Tn
Tourmaline (unspecified)	$Na(MgFeAl)_3Al_6Si_6O_{18}(BO_3)_3(OH)_4$	Tm
Zircon	$ZrSiO_4$	Zr
(All Fe is Fe^{2+} unless otherwise indicated)		
Liquid	granitic	L
Vapour	H_2O	V

Oxide system notation

A	Al_2O_3	Mn	MnO
C	CaO	N	Na_2O
F	FeO	S	SiO_2
H	H_2O	Ti	TiO_2
K	K_2O		
M	MgO		

2.2.1. Topology of Melting

Topological relationships between subsolidus dehydration, vapour-saturated melting and dehydration melting of a rock containing a hydrous mineral such as muscovite, biotite and/or hornblende are shown in Fig. 2.1. Curve 1 represents a subsolidus dehydration reaction where a hydrous mineral-bearing assemblage reacts to form an anhydrous mineral assemblage + H_2O vapour. Curves 2 and 4 represent the H_2O-saturated solidus of the rock on the higher temperature side of which an H_2O-saturated melt (M_s) is formed. At pressures greater than the invariant point (I), curve 3 represents the vapour-absent solidus of the rock. This curve corresponds to a dehydration melting curve where a hydrous mineral (H) melts to form an anhydrous mineral assemblage (A) + H_2O-undersaturated melt (M_u). The line P_1 intersects invariant point (I) and represents the minimum pressure at which a hydrous mineral in a rock can coexist stably with melt. In the common case where a rock contains two or more hydrous minerals, these minerals react together with dehydration melting taking place over a multivariant *T–P* interval.

2.2.2. Muscovite Dehydration Melting

The most important muscovite dehydration melting reaction in sialic rocks is,

$$Ms + Qz + Ab = Ksp + AS + L$$

(Fig. 2.2) with addition of Ca (anorthite component in plagioclase) shifting the reaction to higher temperatures, e.g. from ~625 to 700°C at 5 kb (Thompson and

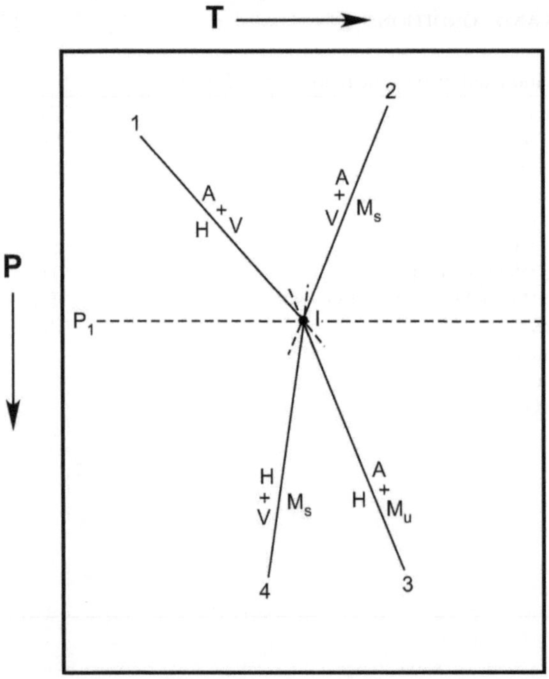

Figure 2.1. *P–T* diagram showing schematic relations between subsolidus dehydration, H_2O-saturated and dehydration melting reactions. **H** = hydrous mineral; **A** = anhydrous mineral; **V** = vapour (H_2O); **M** = melt with $\mathbf{M_s}$ and $\mathbf{M_u}$ = H_2O-saturated and H_2O-undersaturated, respectively; **I** = invariant point.

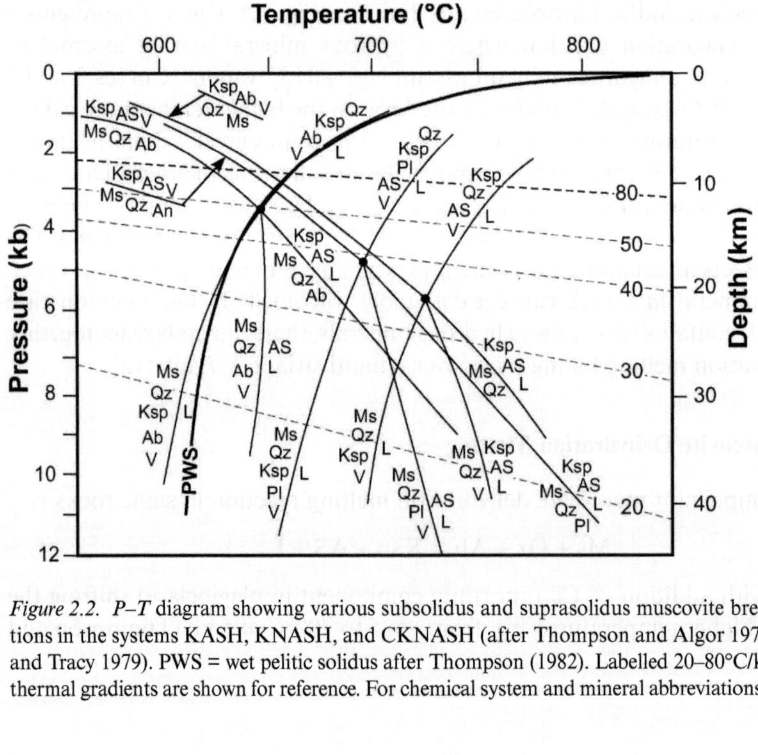

Figure 2.2. *P–T* diagram showing various subsolidus and suprasolidus muscovite breakdown reactions in the systems KASH, KNASH, and CKNASH (after Thompson and Algor 1977; Thompson and Tracy 1979). PWS = wet pelitic solidus after Thompson (1982). Labelled 20–80°C/km linear geothermal gradients are shown for reference. For chemical system and mineral abbreviations see Table 2.1.

Algor 1977). If muscovite contains some Fe, Mg and Ti, which is usually the case, then additional reaction products such as Fe-Mg spinel, cordierite and biotite may form (Brearley 1986; Grapes 1986; Rubie and Brearley 1987; Vielzeuf and Holloway 1988; Brearley and Rubie 1990; Patiño-Douce and Harris 1998). In general, however, muscovite dehydration melting occurs abruptly by a near univariant reaction because of its restricted compositional range.

2.2.3. Biotite Dehydration Melting

In the presence of quartz, plagioclase, ± Al-silicate, biotite will breakdown at higher temperatures than muscovite as shown from experimental studies summarised in Fig. 2.3. Model biotite breakdown reactions involve the formation of

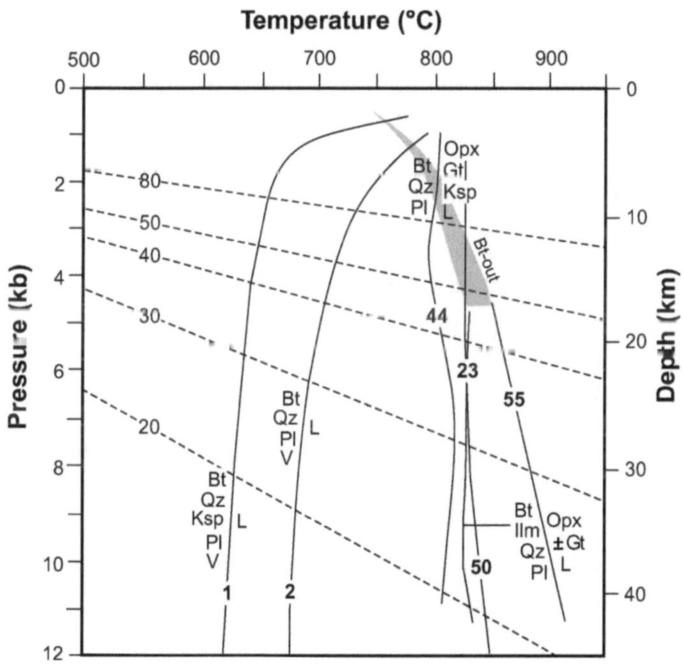

Figure 2.3. *P–T* diagram showing various biotite-bearing assemblage dehydration melting reactions. Reaction curves labelled **23** (Patiño-Douce and Beard 1995), **44** (Vielzeuf and Montel 1994), **50** (Singh and Johannes 1996) and **55** (Patiño-Douce and Beard 1995) = X_{Mg} content of biotite. Shaded area = range of biotite-out curves in greenschist–amphibolite grade greywacke compositions after Kifle (1992). Curves **1** after Luth et al. (1964), Piwinskii (1968) Johannes (1984), and **2** after Piwinskii (1968). PWS = wet pelitic solidus after Thompson (1982). Labelled 20–80°C/km linear geothermal gradients are shown for reference. For mineral abbreviations see Table 2.1.

one or more peritectic mafic minerals + liquid depending on protolith composition and pressure, e.g. the reactions,

$$Bt + Qz + Pl \pm AS = Ksp \pm Cd \pm Gt + L$$

$$Bt + Qz + Pl = Ksp + Opx \pm Cd \pm Gt + L$$

Additional breakdown products may include Fe-Mg spinel, Ti-magnetite and ilmenite, possible rutile, as well as new (more Ti-, Mg- and Al-rich) biotite (Grapes 1986; Brearley 1987a, b; Patiño-Douce and Johnston 1991) and orthoamphibole/subcalcic hornblende (Conrad et al. 1988; Kifle 1992). The position of the above reactions and other biotite-dehydration reaction curves in *P–T* space depends on the Mg/(Mg + Fe) (= X_{Mg}) and Ti-content of the biotite and X_{Mg} and TiO_2 of the host rock, with higher X_{Mg} and Ti shifting biotite-out reactions to higher temperatures (Vielzeuf and Montel 1994; Patiño-Douce and Beard 1995, 1996; Stephens et al. 1997; Pickering and Johnston 1998). In the case of biotite Ti, deprotonation (dehydrogenisation) resulting in loss of H by way of the substitution $TiO_2R_{-1}(OH)_{-2}$ will reduce the amount of hydroxyl in biotite (Henry et al. 2005), thereby extending its stability to higher temperatures. In addition, fluorine will also enhance biotite stability to higher temperature. Thus, in comparison to muscovite, dehydration melting of biotite can occur over a significantly greater range of temperature because of its compositional variation.

2.2.4. Hornblende Dehydration Melting

Dehydration melting reactions involving hornblende together with quartz and plagioclase, ± garnet, are shown in Fig. 2.4 and can be characterised as,

$$Hb + Qz \pm Gt = Opx + Cpx \pm Pl + L \text{ below 10 kb}$$

$$Hb + Pl + Qz = Gt + Cpx + L \text{ above 10 kb}$$

Ilmenite, titanite and spinel are possible additional phases where quartz is absent in the amphibolite protolith, e.g.

$$Hb + Pl = Cpx + Opx + Ca - Pl + Ilm + Tn \pm Sp + L$$

(Rushmer 1991) and hornblende is stabilised to higher temperatures. As with biotite, higher X_{Mg}, Ti-content and the presence of F and Cl will also stabilise amphibole to higher temperatures (Gilbert et al. 1982).

2.2.5. Biotite and Hornblende Melting In Granitic Rocks

Biotite- and hornblende-out curves determined for wet melting of granitic compositions are shown in Fig. 2.5. Biotite disappears over a wider *T*-range than hornblende (i.e. ~690–820°C verses 920–970°C, respectively, at 5 kb) reflecting its more variable composition.

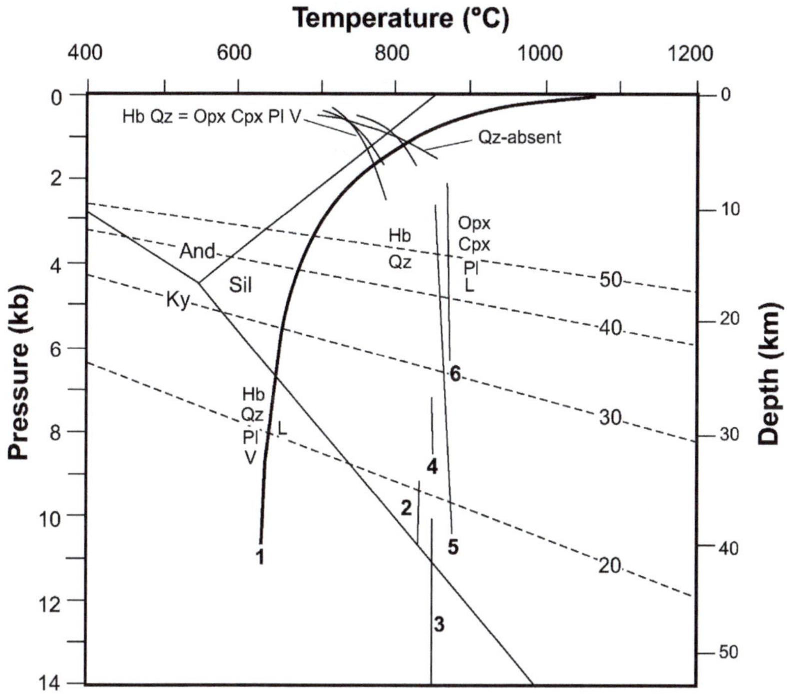

Figure 2.4. *P–T*–depth diagram showing hornblende dehydration melting reactions. Various low-*P* Hb + Qz and Hb with no Qz reaction curves from Choudhuri and Winkler (1967); Binns (1969); Spear (1981); **1**. Piwinskii (1968; tonalite 101); **2**. Conrad et al. (1988); **3**. Winther and Newton (1991); **4**. Rushmer (1991); **5**. Patiño-Douce and Beard (1995); **6**. Beard and Lofgren (1991). Al-Silicate stability fields after Pattison (1992). Labelled 20–50°C/km linear geothermal gradients are shown for comparison. For mineral abbreviations see Table 2.1.

2.2.6. Other Hydrous Minerals

Other hydrous silicates such as chlorite, staurolite and epidote (Fig. 2.6), could be important in generating granite melts. Although a large amount of water is released from the breakdown of chlorite, e.g.

$$\text{Ch} + \text{St} + \text{Ms} = \text{Bt} + \text{AS} + \text{V}$$

the reaction occurs at *T–P* conditions below the pelitic solidus in the NaKFMASH system (Spear et al. 1999). Only a dehydration reaction involving Mg-rich chlorite (clinochlore) to Fo + Cd + Sp + V in the absence of quartz or other minerals occurs at temperatures above the pelitic solidus (Cho and Fawcett 1986; Jenkins and Chernosky 1986). As this reaction is unlikely to occur in normal crustal rocks without the involvement of other minerals, chlorite dehydration probably does not directly contribute H_2O during melting.

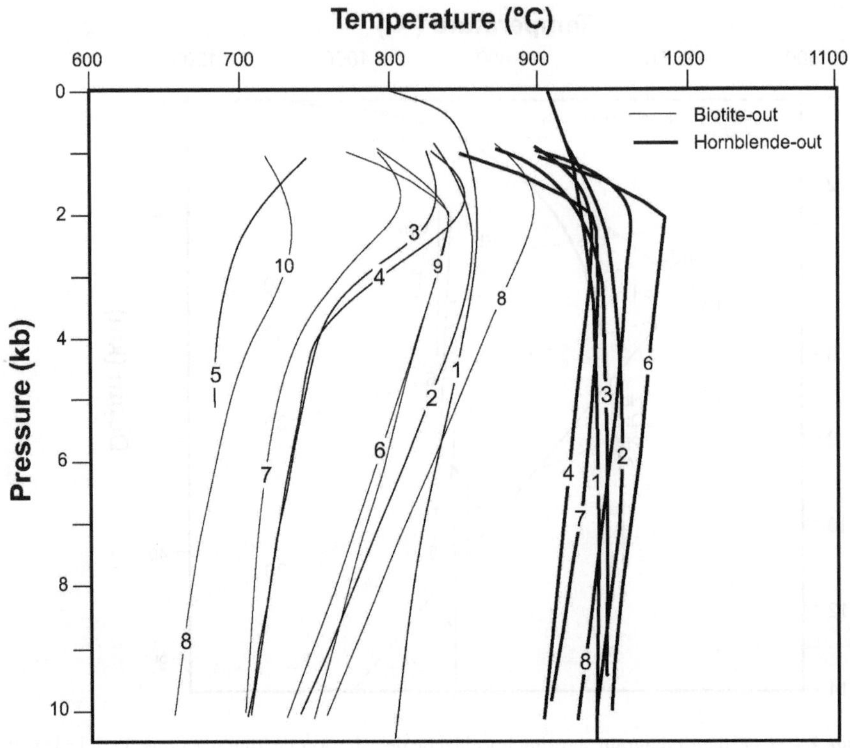

Figure 2.5. P–T diagram showing biotite- and hornblende-out reaction curves for various granitic compositions. **1**. Tonalite (101) (Stern and Wyllie 1975); **2**. Tonalite (101) (Piwinskii 1973a); **3**. Granodiorite (103) (Piwinskii 1973a); **4**. Granodiorite (102) (Piwinskii 1973a); **5**. Granite (104) (Piwinskii 1973a); **6**. Quartz diorite (DR510) (Piwinskii 1973b); **7**. Quartz diorite (DR126) (Piwinskii 1973b); **8.** Quartz diorite (CP2–1) (Piwinskii 1973b); **9**. Granodiorite (JSP6–2) (Piwinskii 1973b); **10**. Monzodiorite (MO-18) (Piwinskii 1973b).

Staurolite is typically found in aluminous Fe-rich rocks and dehydration melting by way of the reaction,

$$\text{St} + \text{Pl} + \text{Ms} + \text{Qz} = \text{Gt} + \text{Bt} + \text{AS} + \text{L}$$

(Spear et al. 1999; García-Casco et al. 2003) would produce a small amount of melt at temperatures below that of muscovite dehydration melting. In the NKFMASH system, the above reaction occurs between ~660 and 690°C and 6–10 kb (García-Casco et al. 2003) (Fig. 2.6).

Epidote is the first hydrous mineral responsible for fluid-absent melting (= solidus) above 10 kb in basaltic and andesitic bulk compositions (Vielzeuf and Schmidt 2001). Within the range of 5–15 wt % epidote in metabasaltic rocks, the amount of melt produced is small (typically < 10%) in comparison to that produced by hornblende decomposition (Schmidt and Poli 2004). The melting

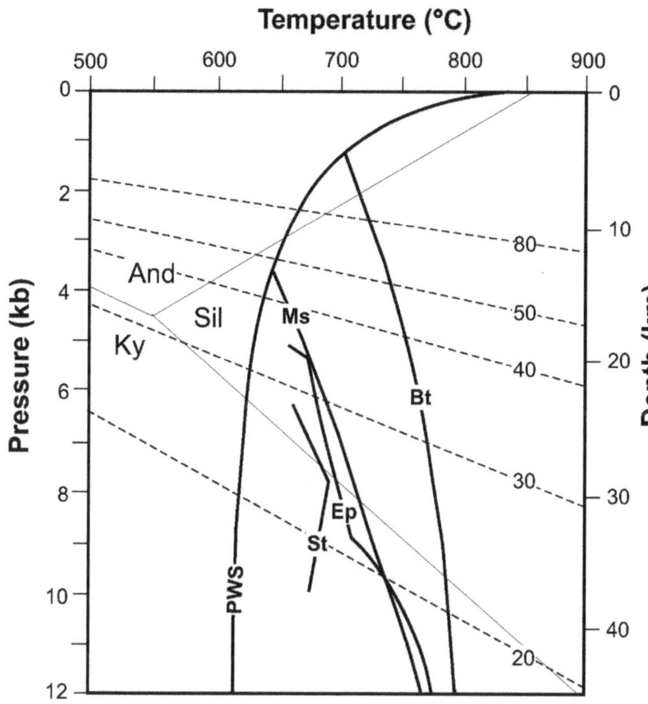

Figure 2.6. Stability curves for staurolite (after García-Casco et al. 2003) and epidote (in tonalite) (after Schmidt and Thompson 1996). PWS = Pelite wet solidus; Ms and Bt – muscovite and biotite-dehydration melting curves, respectively. Al-silicate stability fields after Pattison (1992). Labelled 20–80°C/km linear geothermal gradients are shown for reference.

curve of epidote in a tonalite composition is shown in Fig. 2.6. In volcanogenic metasedimentary protoliths, epidote is probably involved in a dehydration melting reaction at 10 kb such as,

$$Bt + Qz + Ep = Hb + Gt + Ksp + Pl + L$$

(Skjerlie and Johnston 1996).

2.2.7. Suprasolidus Decompression–Dehydration Reactions

Thompson (2001a, b) has drawn attention to the fact that during crustal decompression, e.g. of migmatite rocks, any hydrous minerals remaining after a deeper episode of dehydration melting may react with hydrous melt to release free H_2O and generate anhydrous restite assemblages. Such reactions are termed suprasolidus decompression dehydration reactions. They cause freezing of hydrous granitic melts and the release of H_2O. Examples of SDD reactions involving muscovite and biotite are shown in Fig. 2.7. The reactions have low dP/dT

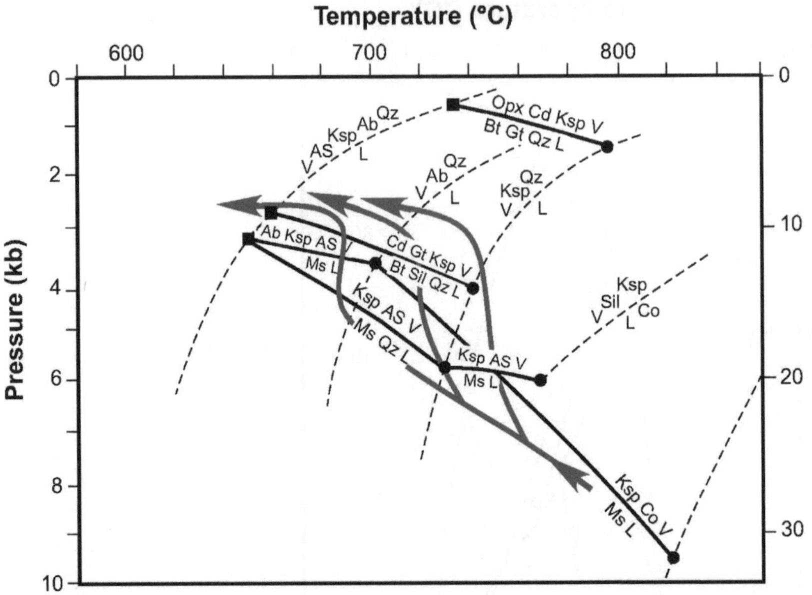

Figure 2.7. P–T–depth plot of suprasolidus decompression-dehydration reactions (SDDR) for muscovite and biotite (solid black lines with labelled reactions with invariant points shown by filled circles at temperatures above the solidus and filled squares on the solidus) that indicate freezing of hydrous melts and release of H_2O that may initiate a phase of renewed melting (after Fig. 1 of Thompson 2001b). For mineral abbreviations see Table 2.1.

slopes with hydrous mineral + hydrous melt on the high-pressure side so that dehydration to an anhydrous mineral assemblage + free H_2O will occur with decompression. The release of H_2O may induce a renewed episode of melting, e.g. in associated mica/amphibole-poor lithologies, older granites, that had not undergone melting at deeper levels due to lack of free H_2O.

Hypothetical decompression paths superimposed on Fig. 2.7 intersect several suprasolidus decompression–dehydration reactions (SDDR) curves (four muscovite and one biotite) between ~25 and 10 km depth and ~795–695°C producing vapour for additional melting to occur + anhydrous granulite assemblages of Ksp ± AS ± Ab ± Co (for muscovite) and Cd Gt Ksp (for biotite).

2.3. ROCK MELTING–EXPERIMENTAL EVIDENCE

2.3.1. Melt Compositions

Melt (glass) compositions have been determined from experimental melting of a wide range of rock types; 'mica-bearing metasedimentary' (peltic–psammitic phyllite/schist/gneiss/hornfels; artificial rock and artificial mineral mixes), 'intermediate' (dacite; volcanoclastic metasediment, artificial rock mixes), 'mafic' (greenschist, amphibolite, granulite; artificial mineral/oxide mixes), and 'granitoids' (tonalite, granodiorite, artificial mineral mixes). Experimental conditions, crystalline products and melt compositions relating to products of partial melting of the various starting rock compositions are given in Appendix 2 and glass compositions are plotted in terms of normative Qz-Or-Ab-An ratio and, for mafic compositions, Or-Ab-An granite classification schemes in Appendix 2 as Figs. A2a–d; B2a; C2a–c; and D2a. Analyses of bulk starting compositions are numbered according to the descriptions of compositions and are listed in tables included in Figs. A2a–d; B2a; C2a–d; and D2a. In the descriptions and tables, experimental rock compositions in the four main composition groups mentioned above are listed in order of decreasing silica.

The experimental results indicate that granite melt compositions of mica-bearing quartzofeldspathic rocks, mixed mica-bearing quartzofeldspathic and mafic rocks, dacite, quartz amphibolite and mafic rocks are controlled by *T*, *P*, fO_2, H_2O-present and H_2O-absent conditions. Melt compositions extend over nearly the entire compositional space of alkali-feldspar granite, granite, adamellite, granodiorite, tonalite, trondhjemite and quartz monzodiorite or monzodiorite with extensions into the quartz diorite, diorite and quartz monzonite fields as summarised in Fig. 2.8. Notably absent from all protoliths are melts representing

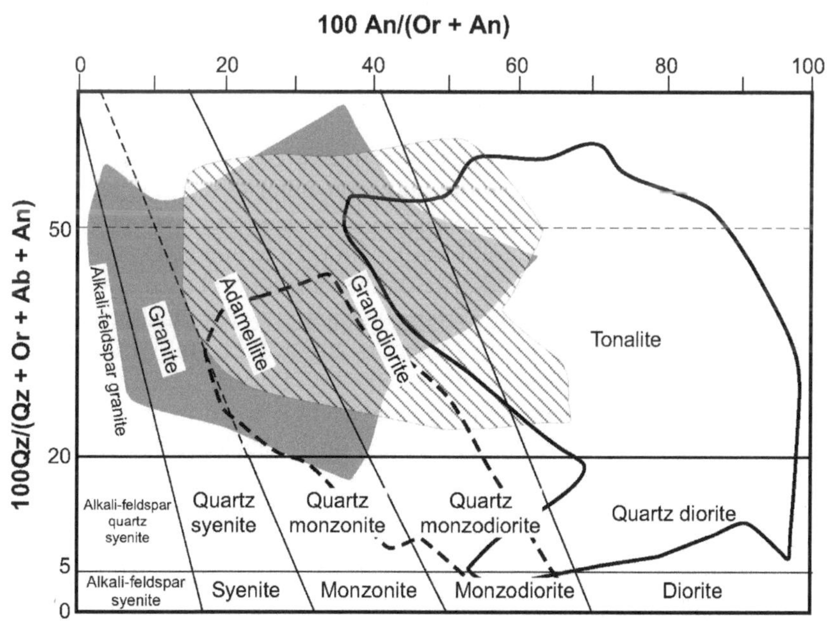

Figure 2.8. Summary diagram showing normative Qz, Or, Ab, An ratio composition fields of granite melts derived from melting of mica-bearing metasediments (grey-tone field); intermediate compositions (diagonal hatched field), mafic compositions (unshaded field enclosed by solid line; field enclosed by dashed line = K-rich mafic rocks). See Appendix 2 for details.

syenite. The experimental evidence indicates that it is possible to produce virtually the entire spectrum of granitic rock compositions by crystallisation of these melts (= magma) ignoring compositional variation resulting from in situ magmatic differentiation/fractionation.

Composition fields for granitic melts derived from mica-bearing quartzofeldspathic rocks (Samples A.1 to 21 in Appendix 2) mainly cluster in the fields of granite, adamellite, granodiorite with extension into the K-feldspar granite field (e.g. pelite rocks from Morton Pass; No.A.16) (Figs. A2a–d; Fig. 2.8). Compositions become granodioritic with increasing temperature reflecting progressive dissolution of plagioclase. With increasing temperature, melt SiO_2, Na_2O typically decrease and TiO_2, Al_2O_3, FeO (as total iron), MgO, CaO and K_2O generally increase.

Volcaniclastic metasediments composed of magmatic arc and continental detritus are typical of continental margins and form the protolith of granite magmas (adamellite, granodiorite, tonalite/trondhjemite with increasing temperature in Fig. B2a; Fig. 2.8) during orogenesis. In particular, the vapour-absent experimental results of Patiño-Douce (1995) indicate that fusion of equal amounts of biotite-bearing quartzofeldspathic rock and high-Al olivine tholeiite produces ~33 wt % H_2O-undersaturated granitic (adamellite) magma and mafic granulite residue ranging from noritic gabbro to pyroxene-garnet granulite. Melting of Al-rich pelitic and high Al-basalt compositions can generate adamellite with spinel or garnet norite restite.

Melt compositions from mafic protoliths (Samples C.1–20 in Appendix 2) are mainly tonalite, trondhjemite, and granodiorite (the TTG association), with extensions to quartz diorite, quartz monzodiorite and diorite (Fig. C2a–c; Fig. 2.8). High–K mafic rocks are exceptions in that they produce adamellite, quartz monzonite, quartz monzodiorite and monzodiorite melts. Again, with increasing temperature, melt compositions contain less SiO_2, Na_2O and are enriched in Al_2O_3, MgO, FeO (as total iron), MnO, TiO_2, and CaO.

Remelting of tonalite and granodiorite or their metamorphosed equivalents produces granodiorite–adamellite and adamellite–granite–alkali-feldspar granite melts, respectively (Samples D.1–4 in Appendix 2; Fig. D2a). In particular, the 4 kb melting experiments of Patiño-Douce (1997) on tonalite (quartz diorite in the normative classification scheme used here) and granodiorite produce high silica, low Ca, high Fe/Mg, K/Na and Ti/Mg melts that equate to so-called anorogenic-type (and rapakivi) granite.

Given the expected heterogeneous composition of protoliths undergoing anatexis in the crust, and ignoring the possibility of mafic magma mixing, variable granitic melt compositions could be produced within the region of anatexis, each with different geochemical/isotopic signatures. The granite or granite association end-product will therefore depend on the degree of homogenisation of the various 'magmas' due to mixing (convective) and also to subsequent in situ differentiation/fractionation on cooling.

2.3.2. Restite Compositions

Experimental generation of granitic melt compositions also involves production of comparable volumes of refractory solid residual assemblages. These can be characterised as aluminous granulite (Pl, Gt, Opx, Ilm and/or Pl, Cd, Bt, Sil, Sp, Ilm ± Co, ± Os, ± Sp, ± Sap) from metapsammitic-pelitic protoliths, intermediate granulite (Opx, Pl, ± Gt, ± Bt, ± Ksp, Ilm), and mafic granulite (Opx, Cpx and/or Hb, Pl, ± Oamp, ± Gt, ± Bt, ± Ksp) from intermediate igneous (tonalite, dacite)–mixed igneous/sedimentary, and mafic amphibolite protoliths, respectively. In many restite assemblages Qz may be present (unmelted, reconstituted relics) and accessories are typically Ap, Mz, ± Aln, ± Tn, Zr and sulphides. Rapakivi granites are commonly associated with anorthosite–norite–mangerite–charnockite complexes (Anderson 1983; Haapala and Rämö 1992) that could represent the restite (mainly Opx + Ca-Pl) of low-pressure melting of granodiorite–tonalite.

2.3.3. Rock Solidi

Wet and dry rock solidi for mica-bearing quartzofeldspathic metasedimentary, mixed sialic/mafic and dacite, basaltic, and granitic compositions are plotted in terms of *T–P* in Figs. 2.9–2.12. Wet mica-bearing quartzofeldspathic metasedimentary solidi vary over a temperature range of 610–670°C at 4 kb, and higher temperature dry solidi range ~740–860°C at 6 kb (Fig. 2.9). Wet and dry solidi of mafic amphibolite compositions occur over a greater temperature range, e.g. 690–780°C/6 kb for wet solidi and 820–1025°C/6 kb for dry solidi (Fig. 2.11). Wet solidi of granitic rocks range between 615 and 720°C/6 kb (Fig. 2.12).

2.3.4. Melt Fraction

Water-saturated melting typically results in only a small volume of melt unless an external fluid flux is substantial, and melting will stop once available free water is consumed. For example, if 10 wt % H_2O is required to saturate a granite melt generated at 5 kb, and if the rock being melted has a porosity of 0.1 vol %, then only <0.1% melt generation would be predicted (Clemens and Vielzeuf 1987). As long as free water is available melting will continue until one of the reactant phases is consumed. In the case of the granitic assemblage Qz Ksp Ab, melting is at the eutectic and Ksp will be the first phase to be consumed. Heating beyond this will continue to produce a small amount of melt (<2 mol %) as long as some water is available until the temperature of the muscovite dehydration reaction is reached.

With dehydration melting involving water supplied by the incongruent breakdown of OH-bearing minerals, the volume of melt produced reflects modal amounts of these minerals in the protolith undergoing melting. In general, muscovite breakdown will yield only small quantities of melt (unless the rocks are

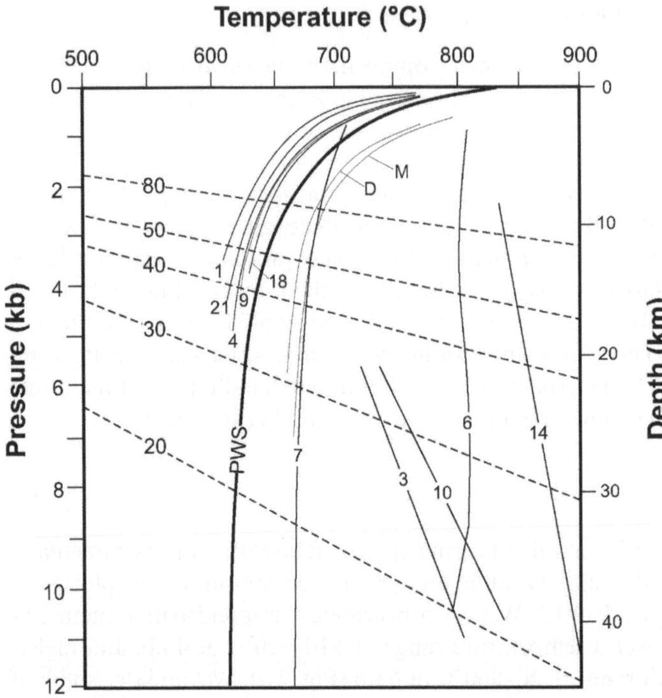

Figure 2.9. Wet and dry solidi for mica-bearing quartzofeldspathic rocks. Numbers refer to bulk compositions described in A of Appendix 2. M = metatexite (Qz Pl, Bt, Hb); D = diatexite (Qz, Pl, Bt) wet solidi from Büsch et al. (1974). PWS = pelite wet solidus after Thompson (1982). Dashed lines labelled 20–80°C/km = simplified linear geothermal gradients.

particularly muscovite-rich) and melt is produced over a very narrow temperature interval of only a few degrees. This is the main reaction responsible for the production of metatexite migmatites and leucogranite melts. For a muscovite-rich schist and melting between Qz Ms Na-Plag, muscovite and a large portion of the sodic plagioclase are first consumed. According to Harris and Inger (1992), for each wt % of muscovite consumed about two units of melt are formed, i.e. 20 vol % muscovite can provide a melt fraction of ~0.4. Biotite breakdown occurs at a higher temperature range than that of the muscovite dehydration reaction because of solid solution in the biotite as described above and produces significantly greater amounts of granitic melt. Because many amphibolite-grade quartzofeldspathic lithologies are biotite-rich, up to ~50 vol % granitic melt may form a diatexite. The effect of bulk rock composition on biotite melting in terms of Ti and Mg content is illustrated in Fig. 2.13 which shows that the upper limit of biotite stability in a TiO_2-bearing metagreywacke composition is up to ~80°C higher than in Ti-free compositions. A large pulse of melt production occurs over a narrow temperature range of as little as 15°C

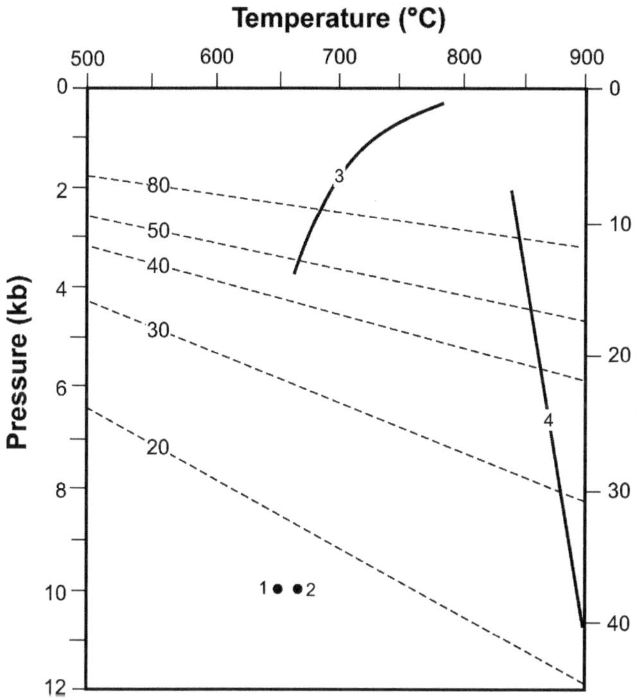

Figure 2.10. Wet and dry solidi for dacite (1), volcanoclastic greywacke (2,3) and quartz-rich amphibolite (4). Numbers refer to bulk compositions described in B of Appendix 2. Dashed lines labelled 20–80°C/km = simplified linear geothermal gradients.

between 800 and 860°C depending on bulk rock Mg content in the latter, whereas in the former, temperatures of ~950°C must be attained to produce melt fractions ~60%. At higher temperatures, hornblende-rich mafic rocks are also potentially highly fertile granitic magma sources and melt volume is correlated with the breakdown of hornblende. For example, Rushmer (1991) reports that the vol % melt increases from 3–5% near the solidus (800°C) of an amphibolite of island arc tholeiite affinity, to ~10–15% when hornblende starts to melt at 825°C, and up to ~40% when hornblende finally disappears at 950°C.

Amounts of granite melt produced by vapour-present and vapour-absent dehydration reactions with increasing temperature at different pressures in different rock types are shown in Figs. 2.14–2.17. In vapour-absent mica-bearing quartzofeldspathic metasedimentary rocks at between 820 and 900°C over a wide range of *P*, the amount of melt produced can reach a range of 60–70% as a result of biotite breakdown (Fig. 2.14A, B). Contours of melt fraction (mol %) in the model pelite composition, show that 50% melting occurs at ~800°C/2kb and at ~900°C/8kb and at temperatures below 800°C at pressures >4kb the

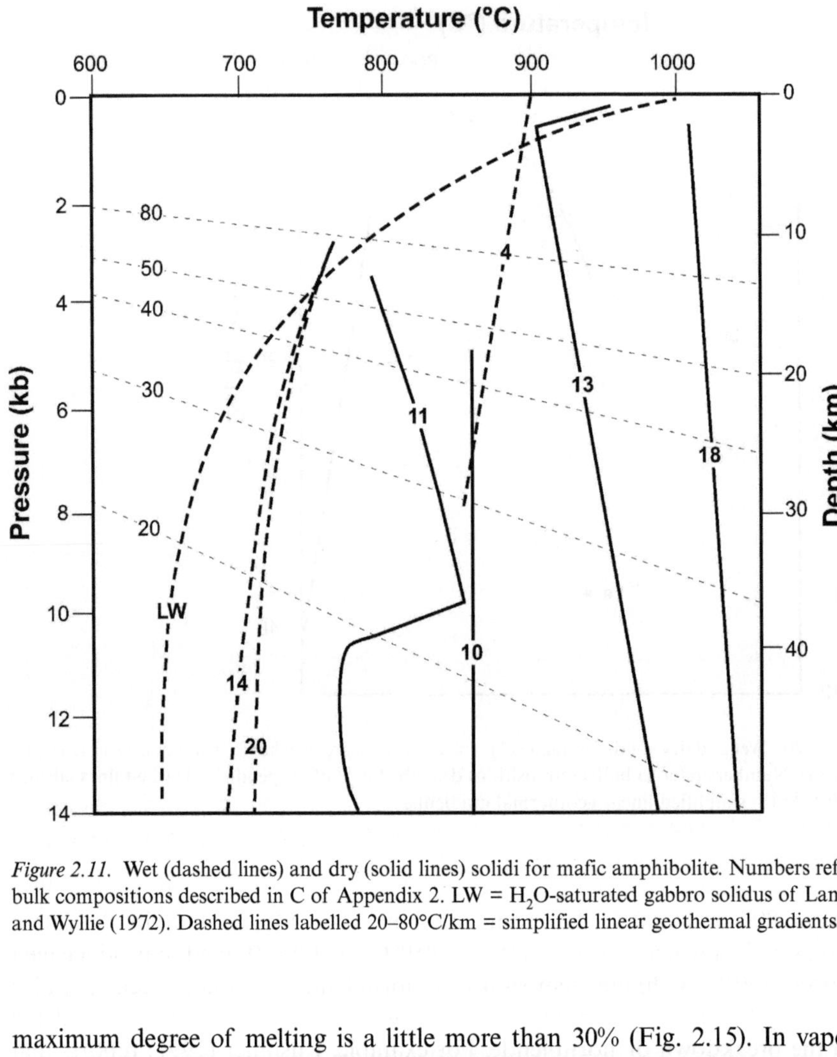

Figure 2.11. Wet (dashed lines) and dry (solid lines) solidi for mafic amphibolite. Numbers refer to bulk compositions described in C of Appendix 2. LW = H_2O-saturated gabbro solidus of Lambert and Wyllie (1972). Dashed lines labelled 20–80°C/km = simplified linear geothermal gradients.

maximum degree of melting is a little more than 30% (Fig. 2.15). In vapour-absent melting of mafic amphibolite compositions < 10–50% melt is generated at 850°C/3–10 kb and at 950°C/1–6.5 kb between 20 and 60% melting occurs (Fig. 2.16). With H_2O-saturated melting of granite compositions, significant amounts of melt of up to ~80% can be produced between 690 and 850°C with melt volume increasing with increasing P at any given T (Fig. 2.17).

2.3.5. Conclusion

Given the expected heterogeneous composition of protoliths undergoing anatexis in the crust, and ignoring the possibility of mafic magma mixing, variable granitic melt compositions could be produced within the region of anatexis.

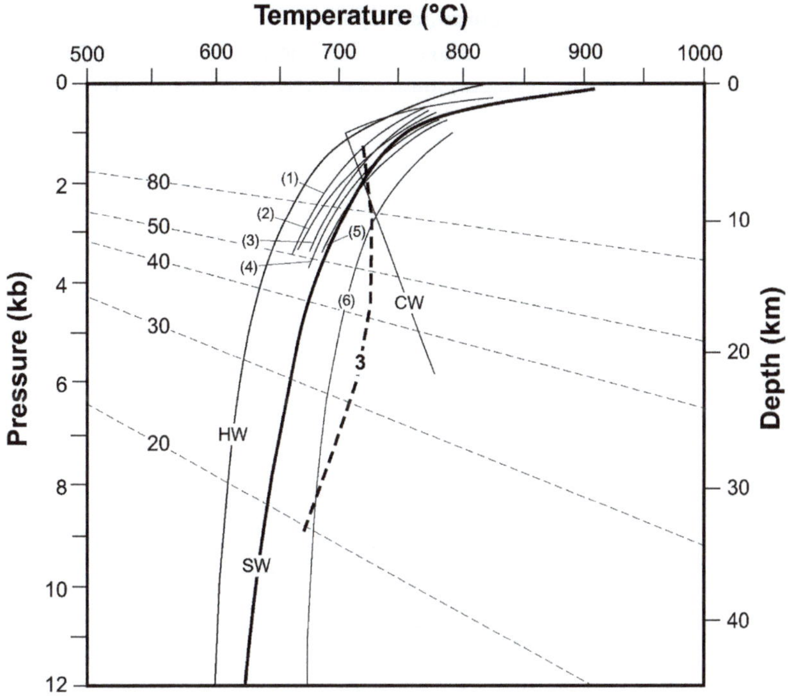

Figure 2.12. Wet solidi for granitic rocks. *Bracketed* numbers refer to granodiorite **1**,**2**,**3**, after Robertson and Wyllie (1971), Piwinskii (1968), Piwinskii and Wyllie (1968), respectively; **4** and **5**,**6** after Piwinskii (1968) and Piwinskii and Wyllie (1968), respectively. *Unbracketed* number **3** refer to bulk composition described in Appendix 2.D. **HW** = muscovite granite after Huang and Wyllie (1973, 1981); **SW** = Tonalite after Stern and Wyllie (1975); **CW** = peraluminous granite after Clemens and Wall (1981). Dashed lines labelled 20–80°C/km = simplified linear geothermal gradients.

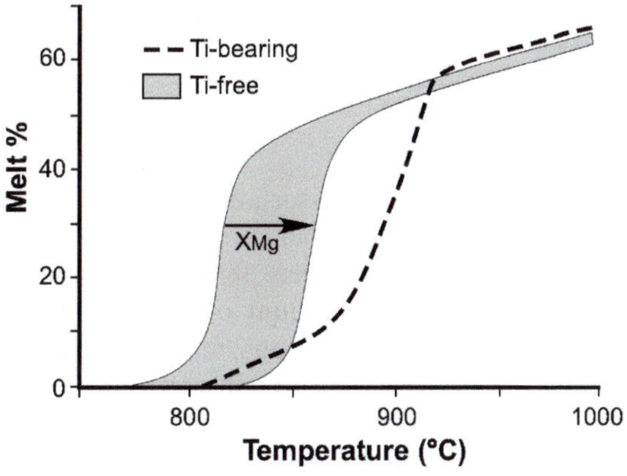

Figure 2.13. Diagram showing fluid-absent melting of biotite in Ti-bearing and Ti-free meta-greywacke compositions with variable *X*Mg at 5 kb (after Fig. 16 of Stevens et al. 1997).

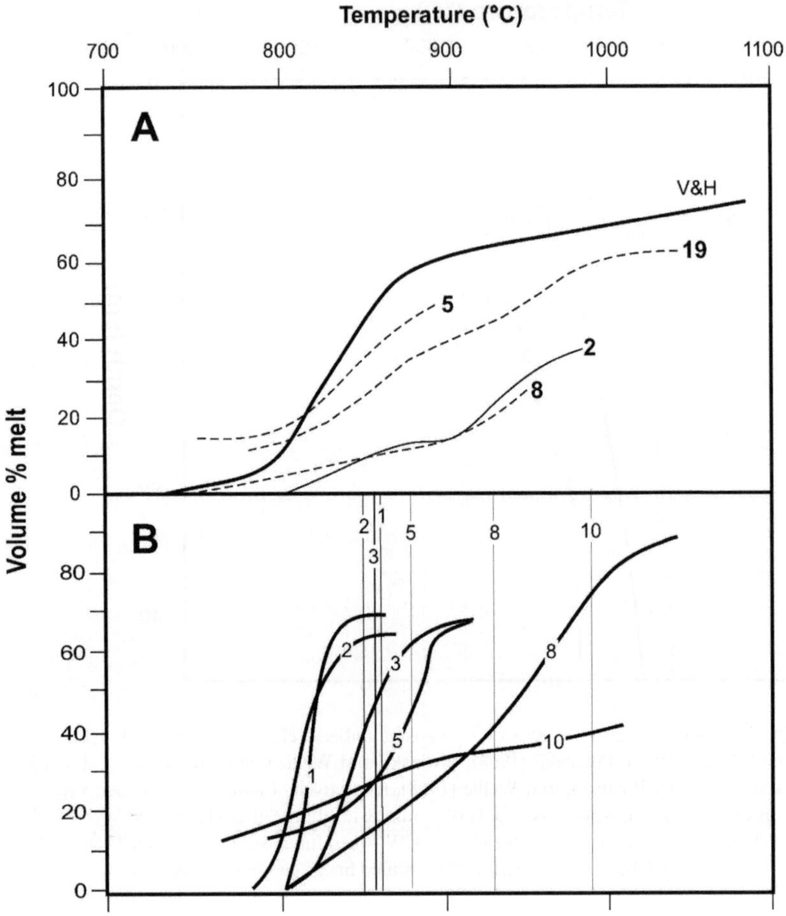

Figure 2.14. Melt vol % verses *T* plot for dry melting of various mica-bearing quartzofeldspathic compositions. **A**. At 10 kb; Numbers refer to rock compositions described in A of Appendix 2. V & H = Vielzeuf and Holloway 1988. **B**. At different *P* (labelled) for metagreywacke No.A.6 in Appendix 2. Vertical lines (*P*-labelled) = Bt-out temperature (redrawn from Fig. 5 of Montel and Vielzeuf 1997).

The granite or granite association end-product will therefore depend on a multitude of interrelated variables such as *P*, maximum *T* reached during anatexis and how long the rocks were able to remain at temperatures substantially above the solidus, fO_2, the amount of H_2O available and its source, the degree of homogenisation of the various 'magmas' produced from mixed protoliths due to mixing (convective), and to subsequent in situ differentiation/fractionation on cooling. Similarly, residual granulite assemblages of melting will also vary depending on protolith composition at similar *T*, *P*, etc.

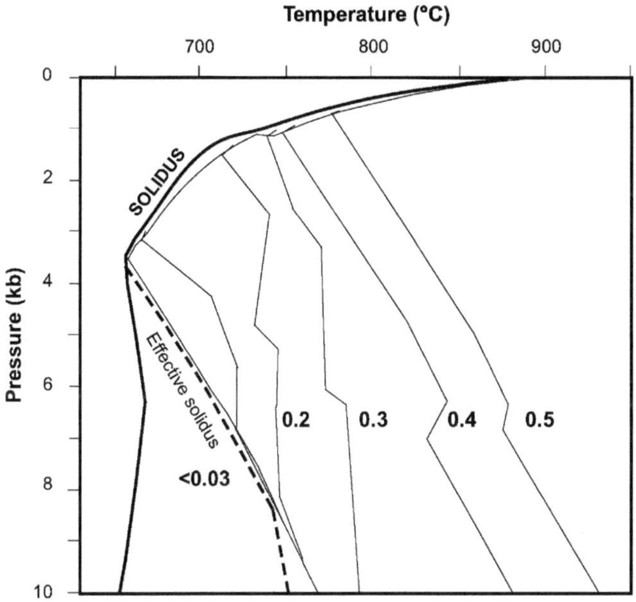

Figure 2.15. *P–T* pseudosection of model pelite composition in the system NCKFASH with excess Qz and just enough H_2O to saturate immediate subsolidus mineral assemblage, showing mol % melt contours (after Powell and White 2001). See Table 2.1 for chemical system nomenclature.

2.4. STRUCTURE AND COMPOSITION OF THE CRUST

The thickness of the Earth's crust, i.e. from the surface to the Mohorovicic (Moho) discontinuity, varies with tectonic setting, being thinner (~27 km) in rift and forearc environments and thicker (>50 km) for Mesozoic–Cenozoic orogenic belts (e.g. Andes, Himalayas and Tibet, European Alps), with an average continental crust thickness of ~40 km (Rudnick and Fountain 1995). Relative thicknesses of seismically-defined upper, middle and lower crustal sections from different tectonic settings are shown in Fig. 2.18. Studies of deep crustal xenoliths, remote sensing and surface heat flow indicates that in general the crust becomes more mafic with depth and that there is a rapid decrease in heat-producing elements downwards from the surface (Fig. 2.18) that reflects a decrease in average heat production from 1.65 to 0.19 $\mu W/m^3$ (Rudnick and Gao 2003).

Normative ratios of bulk upper, middle and lower crustal compositions from data compiled by Rudnick and Gao (2003) are plotted in Fig. 2.19 and indicate that they range from granodiorite–quartz monzodiorite (upper and middle crust) to quartz diorite and diorite (lower crust) with an average continental crust composition of quartz diorite. Direct evidence of the composition of the middle–lower crust is from exposures of high-grade amphibolite–granulite facies

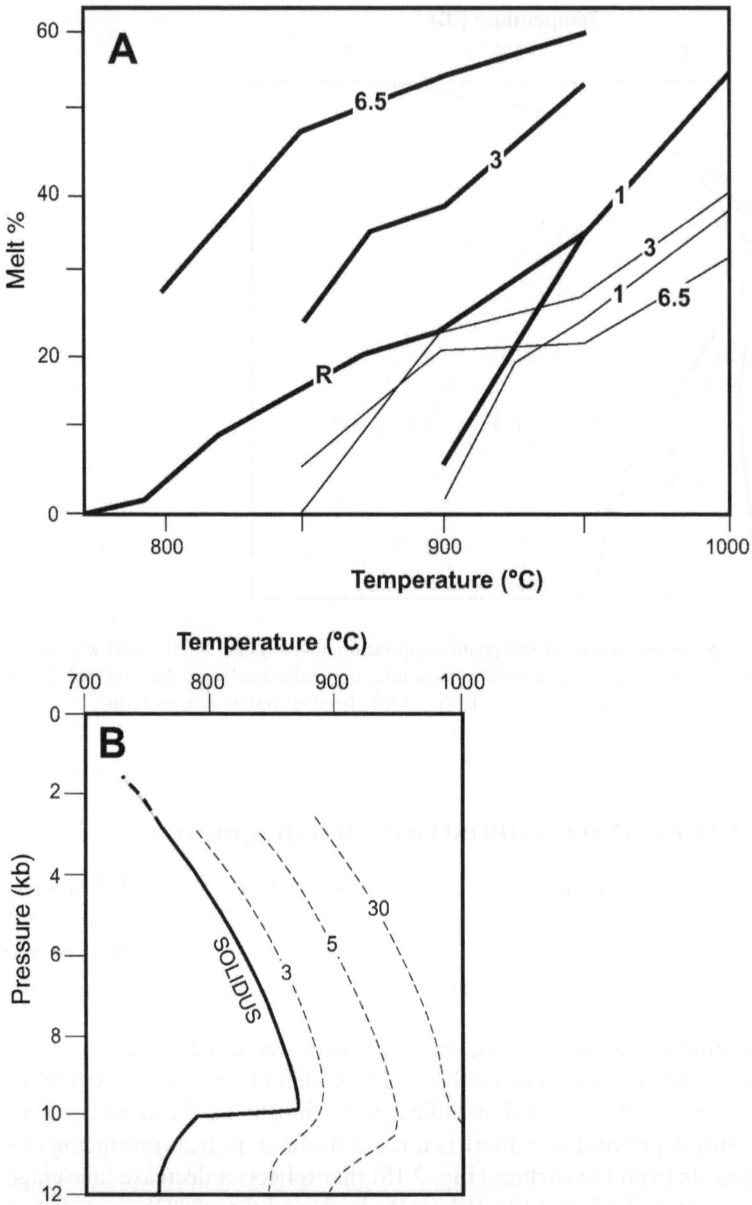

Figure 2.16. Melt vol % versus *T* plot for wet (thick lines) and dry (thin lines) melting of amphibolite compositions. **A.** At different *P* (labelled), amphibolite composition No.D.4 in Appendix 2 (Beard and Lofgren 1991). R = amphibolite composition No.D.5 in Appendix 2 (Rushmer 1991). **B.** *T–P* plot showing form of solidus and melt % contours for dry melting of amphibolite No.C.11 in Appendix 2 (redrawn from Fig. 6. of López and Castro 2001).

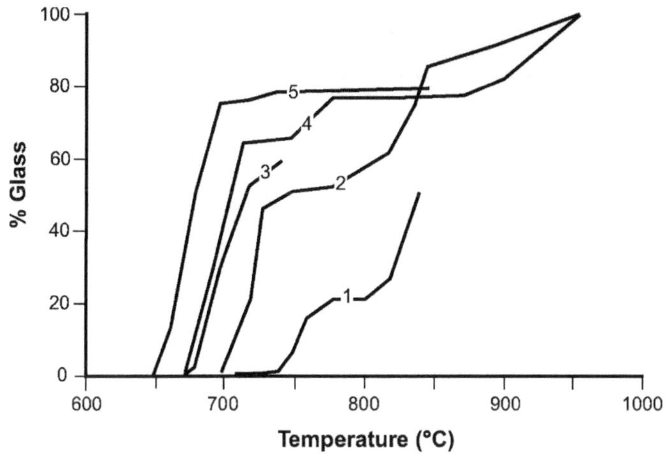

Figure 2.17. Melt % versus T plots for wet melting of granodiorite at various pressures (labelled). Data from Piwinskii (1973a).

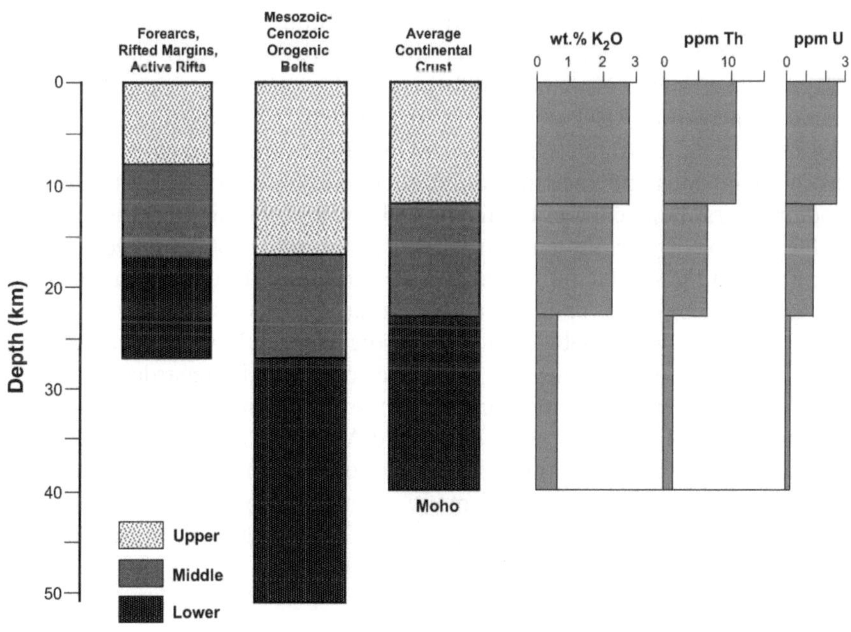

Figure 2.18. Typical crustal thicknesses for tectonic settings and average continental crust together with average wt % K_2O, ppm Th and U values for upper, middle and lower crust (data from Rudnick and Gao 2003).

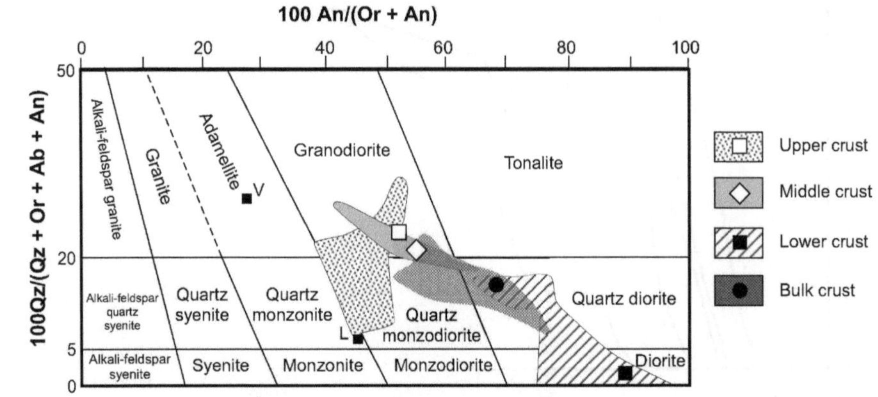

Figure 2.19. Normative Qz-Or-Ab-An ratio plot showing composition fields and recommended compositions (open and filled symbols) for upper, middle and lower crust. Data from Tables 1, 3, 5, 7, 8, 9, 10 and 11 of Rudnick and Gao (2005). Note that two proposed lower crustal compositions, **L** (Liu et al. 2001) and **V** (Villaseca et al. 1999) are anomalous in that they plot well away from the composition field of other lower crustal estimates.

metamorphic terranes of 'cratonic areas', e.g. Vredefort dome (South Africa), Kapuskasing Uplift (Canadian Shield), Wutai-Jining Zone (North China); Musgrave (Central Australia), Ivrea-Verbano zone (western Alps, Italy); 'continental arcs', e.g. Sierra Nevada (California), and 'oceanic arcs', e.g. Kohistan (Pakistan), Talkeetna (Alaska), where maximum depths of exposure range from 13 to 45 km. In these examples, mid-crustal lithologies are represented by a wide spectrum of igneous and gneissic granitic rocks (tonolite, trondhjemite, diorite, granodiorite, granite), pelitic-psammitic, metabasic and calc-silicate schist/gneiss/granulite, marble, quartzite, and also anorthosite and gabbro. Lower crustal lithologies contain similar felsic-mafic gneiss/granulite lithologies but usually with a greater abundance of mafic and ultramafic rocks, especially in oceanic arc sections. There is still debate on whether the exposed lower crustal sections are typical or not and whether the granulitic rocks can be regarded as a true residue (restite) of crustal melting or represent exhumed metamorphosed mafic underplate material (Clemens 1990; Zandt and Ammon 1995) or both. The proposed diorite–quartz diorite lower crust bulk composition (Fig. 2.19) suggests the possibilities of hybridisation of mafic magma during assimilation of the crust (Patiño-Douce 1995) or large scale imperfect segregation of granite melt and granulite restite (Rudnick and Presper 1990).

2.5. WATER IN THE CRUST

In addition to fertile protoliths described above, the two most important requirements for crustal melting are water and heat. The presence of water in excess of that in rock-forming hydrous silicates exerts a profound influence on the melting of crustal rocks because it controls the degree of melting to be expected at a given temperature. Also the hydrous nature of granite magmas implies that partial melting which gives rise to these magmas must involve transfer of H_2O from the protolith ± fluid to the melt. It appears inescapable that initiation of melting in rocks, i.e. the formation of granitic melt (minimum melt) by fusion of anhydrous quartz and feldspars, must also involve a free fluid phase. If all the available free water at the melting site is dissolved in the melt, melting stops and it requires a temperature increase to release additional, internally-derived water from the breakdown of any hydrous minerals present.

"Water is the dominant low-density fluid, the dominant solvent, the dominant chemical transport agent, the dominant silicate-reaction catalyst and the dominant catalyst of rock deformation in the crust" (Fyfe et al. 1978). Free water occurs in fractures, pores and as films, perhaps only a monolayer thick, along mineral grain boundaries (free, absorbed or non-hydroxyl water) and in fluid inclusions in minerals. It can occur as a liquid, vapour or supercritical fluid depending on temperature and pressure. In continental crystalline crust at 4–5 km depth the temperature of these deep fluids ranges between 100 and 200°C (Stober and Bucher 1999) and deeper, e.g. 9.1 km (KTB, Germany; Möller et al. 1997) and 12.5 km (Kola, Russia; Kolzlovsky 1984) temperatures are 240 and 270°C respectively. At greater depths there is a gradual transition to so-called higher temperature hydrothermal fluids (Barnes 1995).

The structural water-content (H_2O^+) of rocks, i.e. that held in hydrous silicates, generally decreases with depth as a result of dehydration reactions during prograde metamorphism. An example of structural water (H_2O^+) contents of micaceous metasedimentary rocks representing a transition from metagreywacke sandstone/siltstone to their schistose equivalents representing a depth interval of about 25 km is illustrated in Fig. 2.20. The plot mainly reflects the modal variation of muscovite and biotite (increasing biotite with respect to muscovite with increasing depth = increasing metamorphic grade, although chlorite also occurs in the lowest grade rocks) and indicates the amount of water available if temperatures were high enough to cause dehydration melting of muscovite and biotite at any particular depth. A similar downward decrease in bulk H_2O-content of various lithologies is indicated from the superdeep Kola hole (GR-3) bottoming at 12 km depth, Russia (Kremenetsky 1991) (Fig. 2.21). Open (fractures) and closed (within mineral) porosity increase with depth to values of >2%, but with open porosity values decreasing to <0.5% in the lower 1000 m of the borehole (Fig. 2.21).

Petrological (Rubie 1986; Frost and Bucher 1994; Yardley and Valley 1994, 1997) and geophysical (Wannamaker 1994; Warner 2004) arguments suggest that the lower crust probably contains little or no pervasive free water and therefore is effectively dry or at least contains ~0.1% or less free water. However, water could be present in the lower crust as scattered domains each composed of an interconnected network of fluid-filled fractures (Connolly and Podladchikov 2004). Alternative models in terms of variation of fluid pressure with depth within the crust are illustrated in Fig. 2.22 and may be applicable to areas of

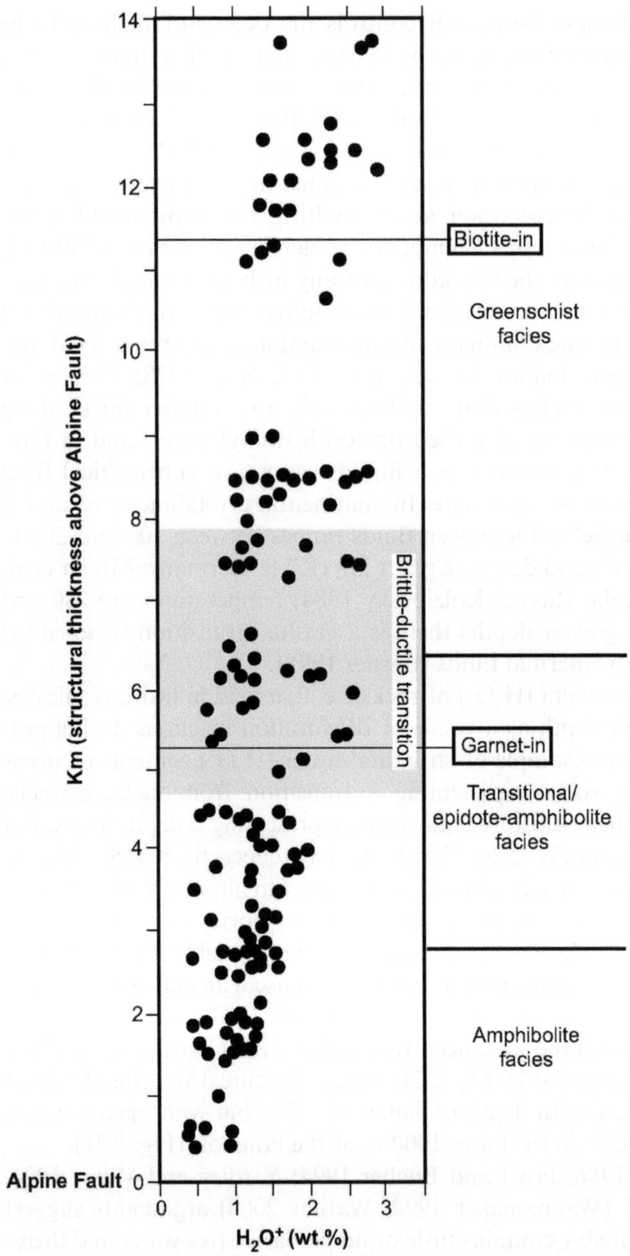

Figure 2.20. Plot of structural water (wt % H_2O^+) in mica-bearing quartzofeldspathic rocks verses structural distance above the Alpine Fault, central part of the Southern Alps, New Zealand (Grapes, unpublished data). Positions of the biotite-in and garnet-in isograds after Grapes (1995). Position of brittle–ductile transition interval from Little et al. (2002). The crustal section has been exposed by uplift on the Alpine Fault to form the Southern Alps.

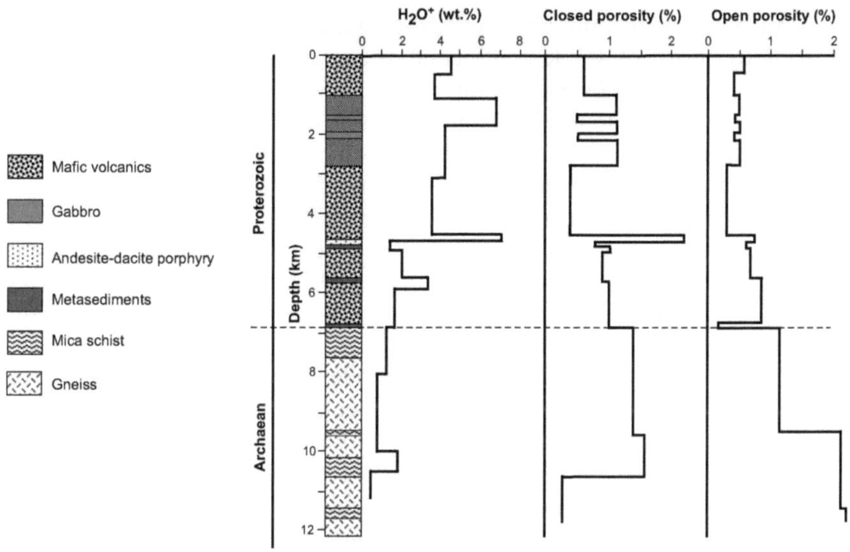

Figure 2.21. Plot of wt % H_2O^+, % closed and open porosity versus depth in the 12 km superdeep Kola borehole (SG-3), Russia (data from Kremenetsky 1991).

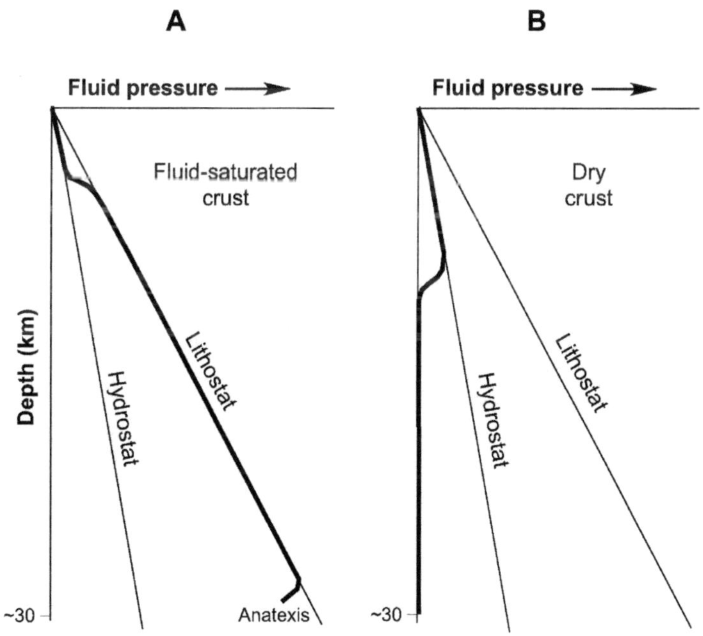

Figure 2.22. Models for variation of fluid pressure with crustal depth. **A**. Fluid-saturated crust where fluid pressures increase from hydrostatic values in the upper crust to lithostatic values with increasing depth. **B**. Dry crust in which fluid occurs at hydrostatic pressure in the upper crust and is absent due to reaction in the lower crust, where fluid pressures are buffered to very low values by mineral equilibria (redrawn from Fig. 1 of Yardley and Valley 1994).

active subsidence and metamorphism (fluid-saturated lower crust, A), and essentially dry lower crust of stable continental areas consisting of old, high-grade rocks (B). As pointed out by Yardley and Valley (1997), in (A) the lower crust will be a high-pressure fluid source while in (B) it is a fluid sink. Downward propagation of the brittle–ductile transition during compression with crustal thickening could effectively carry upper crustal fluids into the middle and lower crust on a timescale that allows compaction to pressurize the fluids. This idea is supported by evidence of progressive pressurization of fluid inclusions inferred to represent originally shallow fluids (Gleeson et al. 2000), and by evidence of the involvement of meteoric fluids in the metamorphism of lower crustal rocks (Wickham and Taylor 1985; Upton et al. 1995).

In contrast to the lower crust, there is an abundance of free water in the upper crust that is a reflection of a fracture network connected with the surface and which decreases with depth (Fig. 2.23). The ultimate depth extent of fracture permeability in the crust is not known, but in stable continental areas earthquake focal depths that are thought to mark the lower limit of the brittle zone that can reach 40 km (Smith et al. 1989; Slemmons et al. 1991). Deep drilling results from Russia (Kremenetsky and Ovchinnikov 1986; Borevsky et al. 1995) and the German KTB (Emmermann and Lauterjung 1997, and companion papers) have shown frequent fluid entries under nearly hydrostatic pressure over the entire drilling depth (up to 12 km). In the case of the Saatly deep borehole, Russia, large scale downward fluid flow in fractures continuing beyond borehole depth (8.3 km) caused loss of circulation that stopped further drilling (Borevsky

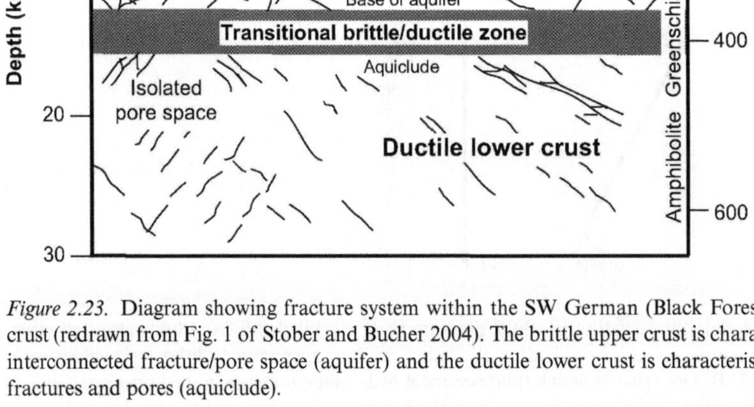

Figure 2.23. Diagram showing fracture system within the SW German (Black Forest) continental crust (redrawn from Fig. 1 of Stober and Bucher 2004). The brittle upper crust is characterised by an interconnected fracture/pore space (aquifer) and the ductile lower crust is characterised by isolated fractures and pores (aquiclude).

et al. 1995). These data support the idea of an overall drawdown of fluid into basement rocks. Eventually, such gravity-driven fluid flow will reach depths (>12 km) where temperatures, e.g. >300°C, will be high enough for hydration reactions to form phyllosilicate minerals (with ~4–12 wt % H_2O). This will effectively use up much of the free water and result in a 'drying out of the rocks.' At greater depth, rocks undergoing metamorphism involving dehydration reactions will release large amounts of water (Fyfe et al. 1978; Baumgartner and Ferry 1991), but on cooling any free water remaining in the rocks will be largely consumed during retrograde reactions so that the metamorphosed rocks will also effectively become dry (Yardley 1981).

The very different hydrological regimes of the upper and lower crust reflect different rock strengths. In the upper crust, the rocks have sufficient strength to maintain a network of fractures through which fluids can circulate independently of the stress state of the rocks (Etheridge et al. 1984). Consequently, fluid pressure in the upper crust will follow a hydrostatic gradient that is independent of lithostatic pressure (Fig. 2.22A). The hydrostatic pressure regime continues down to a critical depth representing a percolation threshold that is determined by rock composition and structure, temperature and stress that can be approximated by the brittle–ductile transition (Gavrilenko and Gueguen 1989). For rocks below the brittle–ductile transition that behave plastically on relative short timescales, porosity will only be maintained if the pores are filled with fluid at or close to lithostatic pressure. The anomaly posed by the conditions required to maintain high-fluid pressure simultaneously with lower rock strength of the ductile region can be explained by a model proposed by Connolly and Podladchikov (2004) whereby fluids are concentrated in self-propagating hydraulic domains which behave as weak zones within stronger adjacent fluid-poor rocks. Because the mean stress gradient in a weak inclusion depends on its orientation with respect to far-field stress, the direction of fluid flow in such domains is sensitive to tectonic forcing. Under a lithostatic pressure gradient, fluids will tend to migrate upwards so that, in contrast to the upper crust, lower crustal fluids can be regarded as ephemeral 'one-pass systems' (Walther and Orville 1982; Wood and Walther 1986).

In compressive tectonic settings, an inverted pressure gradient below the brittle–ductile transition may develop by relaxation of the yield stress developed in the brittle portion of the crust (Stüwe et al. 1993; Stüwe and Sandiford 1994; Petrini and Podladchikov 2000). At depths where this inverted pressure gradient is less than the hydrostatic gradient of the interstitial fluid, and if the fluid is subject to rock confining pressure, it will migrate downward to a depth at which the rock pressure gradient is equal to the hydrostatic gradient of the fluid. This defines a level of tectonically-induced neutral buoyancy that also acts as a barrier to upward fluid flow within vertically oriented structural features that are the most favorable for facilitating fluid expulsion. Elementary analysis based on the depth of the seismogenic zone and experimental rheological constraints indicates that tectonically-induced buoyancy would cause fluids to accumulate into a long-term horizontal, stagnant, although overpressured, ~1 km-thick horizon 2–4 km below the brittle–ductile transition. Because the crust is heterogeneous there could be multiple depth levels at which tectonically-induced fluid accumulation and stagnation might occur. The possibility of ponded water below the brittle–ductile transition may explain anomalous mid-crustal reflectivity and it has important implications for the wet melting of crustal rocks. It may be no coincidence that thermal modelling (see below) often predicts anatexis occurring at mid-crustal depths in the vicinity of the brittle–ductile transition.

2.6. CRUSTAL HEAT AND PARTIAL MELTING

2.6.1. Introduction

The geotherm in average continental crust is inferred to be no higher than about 500°C at a Moho depth of 35 km implying a gradient of ~15°C/km, and as such the wet solidus temperatures for any 'fertile' rock composition over the entire thickness of continental crust will not be attained (Fig. 2.24). The wet solidus of the model pelitic composition (PWS) is about 10°C lower than that of the wet granite solidus (Tuttle and Bowen 1958), and for melting to occur at 20 km depth it would require a geotherm of about 30°C/km. For crustal anatexis to occur at shallower levels, an anomalously high-thermal gradient in excess of about 40°C/km is required. In addition, as shown above, temperatures need to be high enough to significantly overstep crustal rock solidi in order for a significant

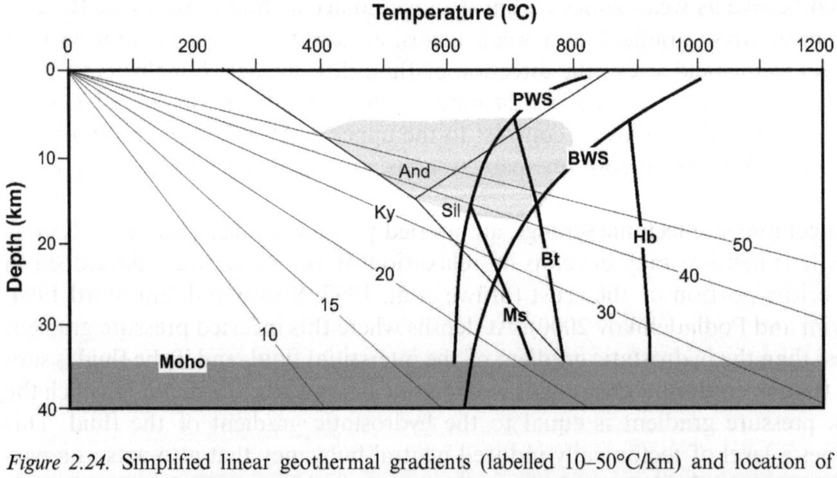

Figure 2.24. Simplified linear geothermal gradients (labelled 10–50°C/km) and location of wet pelite (PWS), basalt (BWS) solidi, and muscovite (Ms), biotite (Bt), hornblende (Hb) dehydration melting curves in a 35 km thick crust. PWS after Thompson (1982); BWS after Wyllie (1977); Burnham (1997); Ms after Huang and Wyllie (1981); Bt extrapolated from Le Breton and Thompson (1988) and Vielzeuf and Holloway (1988); Hb after Rushmer (1991). Al-silicate stability fields after Pattison (1992). Grey shaded area = field of high-*T*/low-*P* metamorphism.

melt fraction, i.e. >25%, to form. For example, at 20 km depth, Fig. 2.24 shows that dehydration melting would begin at 630°C for muscovite, 740°C for biotite and 900°C for hornblende, and depending on the shape of the melt fraction verses temperature curve, it would take an additional temperature increase in the range of 20–60°C for the melt fraction to reach ~25 vol %. With the probability of very low porosity in the lower crust, at 35 km depth the dry solidus for basaltic rock at ~1100°C would only be intersected by linear geotherms > 30°C. Clearly, with such geotherms, most of the lower crust would be largely molten, although only if it consisted of 'fertile' mica-bearing metasedimentary and/or granitic rocks. As there is no geophysical evidence for the lower crust being molten, it implies that either suitably fertile rocks do not occur in the lower crust, e.g. they are composed of refractory granulite, or that the geotherm is cooler at such depths.

In post-Archean orogenic terranes the thermal structure is a product of internal heat production, advection, and basal crustal heat flux. Normal orogenic evolution is driven by subduction followed by collision leading to crustal thickening and decreased convergence rate, resulting in self-heating of the thickened orogen. In continental arcs such as the North and South American Cordillera, heat advected with mantle-derived magmas is also potentially important. The physical causes of 'higher than normal' geothermal gradients that are associated with high-*T*/low–medium-*P* metamorphism and anatexis have received considerable attention over the last 20 years. Modelling, both general and in relation to specific areas, has shown that a combination of several, typically interrelated processes can strongly perturb the geothermal gradient by displacing isotherms toward the surface to create a thermal antiform generating temperatures that exceed the wet solidi of various rock compositions and induce dehydration melting of micas and amphibole (Gerbi et al. 2006).

1. Thickening of radiogenic heat-producing rocks (England and Thompson 1984, 1986; Thompson and England 1984; Chamberlain and Sonder 1990; Patiño Douce et al. 1990; De Yoreo et al. 1991; Thompson and Connolly 1995; Huerta et al. 1998; Jamieson et al. 1998; Sandiford and Hand 1998; Sandiford et al. 1998; Gerdes et al. 2000).
2. Intracrustal shear heating (Harrison et al. 1997; Leloup et al. 1999; Nabelek and Liu 1999, 2004).
3. Lithospheric extension due to rifting and gravitational collapse (Wickham and Oxburgh 1985; Dewey 1988).
4. Delamination and convective removal of thickened mantle lithosphere and thermal boundary layer, respectively (England 1993), and slab breakoff (Davies and von Blanckenburg 1995).
5. Ascent of a mantle plume (Watson and McKenzie 1991), a mantle-related transient heat pulse (Bodorkos et al. 2002), and injection of mafic magma into the continental crust (Shaw 1980; Spera 1980; Lux et al. 1986; Huppert and Sparks 1988a, b; Fountain et al. 1989).
6. Near isothermal decompression at elevated temperature (Teyssier and Whitney 2002).

2.6.2. Thickened Crust

Subduction followed by collision leads to crustal thickening and decreased convergence rate allowing self heating of the thickened orogen in which anticlockwise *T–P* paths (pressure increasing downward) are characteristic. Generalised thermal modelling of crustal thickening has been addressed by England and Thompson (1984, 1986), Thompson and England (1984), and Zen (1988) and indicates that partial melting of the crust can occur in response to thermal relaxation of overthickened crust. In the case of crustal thickening by 'instantaneous' overthrusting during collision the normal geotherm is transformed into a sawtooth pattern as shown in Fig. 2.25A for a hypothetical (although unlikely) wet mica-bearing quartzofeldspathic (pelitic) crust end member composition as modelled in one dimension (vertical) by Thompson and Connolly (1995) that maximizes the importance of melting. Greenschist to amphibolite facies subsolidus dehydration reactions are indicated by steep positive d*T*/d*P* lines labelled R1–4 with R1 and R2 releasing 1 wt % H_2O and R3 and R4 releasing 0.5 wt % H_2O at higher temperatures. Higher temperature (granulite facies) suprasolidus muscovite, biotite and hornblende dehydration melting reactions are labelled

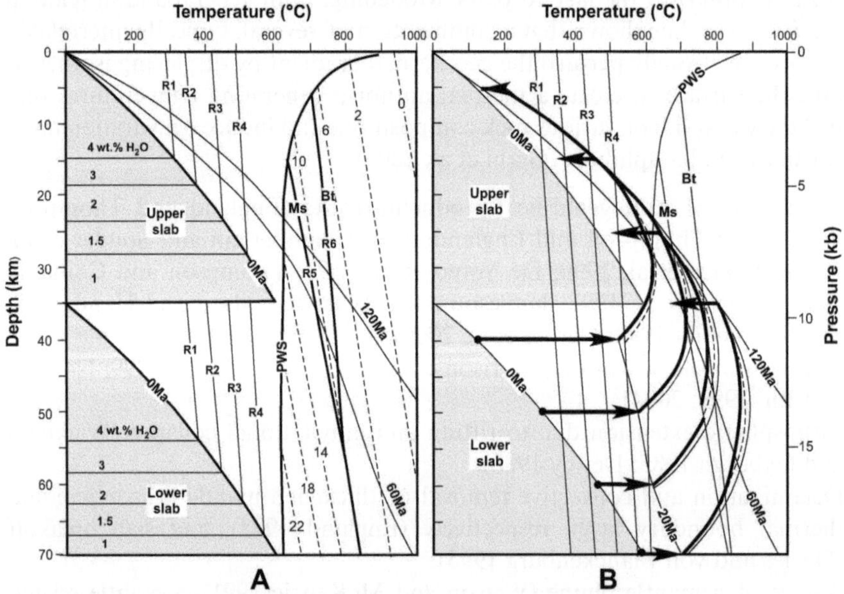

Figure 2.25. Model temperature, depth/pressure, time relationships and uplift paths of thickened crust of average pelitic composition in a continental collision setting. Geotherms labelled in Ma. In B, incorporation of latent heat of melting (solid curves) results in a 20–40°C lowering of *P–T–t* paths. For other symbols/curves refer to Fig. 2.24. See text. (**A**. Combined from Figs. 1a, b of Thompson and Connolly 1995; **B**. after Fig. 3 of Thompson and Connolly 1995).

R6, 7 and 8 respectively. In the case considered, an initially 35 km thick crust is doubled during continental collision and it takes 120 Ma after the thickening event for a maximum temperature geotherm to be attained by upward migration of the wet pelitic solidus temperature isotherm at a rate of ~70 m/Ma and the 800°C isotherm at ~130 m/Ma. Assuming an incubation period of 20 Ma after thickening followed by linear uplift that removes the 35 km thick upper slab in 100 Ma, *P–T-t* uplift paths of lower slab rocks from depths of 40, 50, 60 and 70 km are shown in Fig. 2.25B. The uplift paths of rocks initially at depths of 40 and 50 km just intersect the wet pelitic solidus (~30 km/50 Ma), muscovite (~35 km/60 Ma) and biotite (~40 km/80 Ma) dehydration melting curves at low angles and thus will generate only the small amounts of melt in each case. Only rocks from depths >60 km will undergo extensive melting as they attain temperatures significantly above that of the biotite-dehydration melting curve. At 120 Ma, the deepest rocks (70 km) reach a final depth of 35 km where they are still ~50°C above the biotite melting curve. With melt remaining in situ, the uplifted crustal section will largely comprise a migmatite to granite/granulite restite sequence. Migmatite with ~1 vol % leucogranite will occur between 5 and 14 km depth underlain by migmatite with ~7 vol % leucogranite down to a depth of ~24.5 km. Between 24.5 and 35 km, the hypothetical pelitic crust will contain ~18 vol % biotite granite and granulite restite (Fig. 2.26). Figure 2.26 also indicates that melt will be retained for up ~30 Ma following cessation of uplift at 120 Ma so that there should be sufficient time to grow a tabular-like layer of granite magma (see Chapter 3). For 'hotter' initial geotherms obviously larger amounts of melt would be produced.

2.6.3. Burial of High-Radiogenic Rocks

Crustal thickening is invariably associated with radiogenic heat production that can result in geothermal gradients of >30°C and raise temperatures high enough to cause anatexis at mid–upper crustal levels. Internal heat production depends on the concentration of heat-producing radioactive isotopes of K, Th and U in crustal rocks and is typically in the range of 0.5–3 $\mu W/m^3$ but can be significantly higher for rocks that contain localised concentrations of heat-producing elements (Sandiford et al. 1998). A number of researchers have proposed that crustal rocks enriched in heat-producing elements can elevate geotherms to temperatures high enough for anatexis to occur, e.g. the Arcadean thermal high of New England, USA, that consists of thousands of square kilometres of high-temperature (sillimanite-bearing), low-pressure metamorphic rocks and associated synmetamorphic granite (Chamberlain and Sonder 1990). Calculations by Chamberlain and Sonder (1990) indicate that temperatures could have been high enough (>700°C) to produce abundant granitic melt at relatively shallow depth. Similarly, Sandiford and Hand (1998) have shown that distributions of heat production and measured surface heat flow production in several Proterozoic terranes in Australia can generate conditions for anatectic conditions at 10–20 km

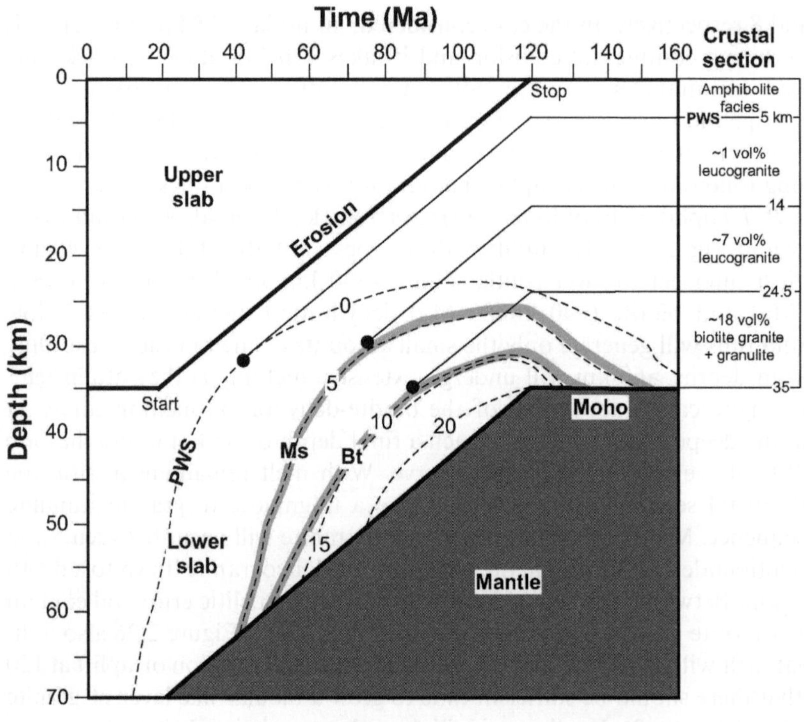

Figure 2.26. Depth/time evolution of the model pelitic anatexis-uplift depicted in Fig. 32B for standard geotherm conditions (thermal conductivity [K] = 2.25 W/m/K; mantle (reduced) heat flow [Q_m] = 30 mW/m²; radiogeanic heat production [A_o] = 2.33 μW/m³). The 35 km thick upper slab is thinned by linear erosion over 100 Ma. Wt % in situ granite melts are shown by dashed contours at 5, 10, 15 and 20%. Zero % melt line = wet pelitic solidus. The thick grey lines along the 5 and 10% contours represent the muscovite and biotite-dehydration melting reactions that are 'smeared' because their reaction rates are controlled by heat flow and fluid flow. Solid circles represent maximum height of the partially melted rock column. They occur as tangents to the melt contours and therefore at greater depths than the shallowest level reached by the contour lines because the crust is thinning. Thus, the 0% solid circle will end up 5 km below the surface at 120 Ma when uplift stops. Those for muscovite and biotite-dehydration melting will finish at 14 and 24.5 km below the surface, respectively, resulting in the migmatite–granite crustal section shown on the right (after Fig. 4 of Thompson and Connolly 1995 and Fig. 2 of Thompson 1999).

depth. As the conditions for melting come from within the crust, it is therefore possible to produce a 'perched' granitic magma layer at relatively shallow crustal depths by subsidence and burial of rocks containing suitably high concentrations of K, Th and U. Granites produced by this process might be expected to occur as extensive 'blobs' rather than as the linear belts typically associated with subduction zones or crustal extension.

Patiño-Douce et al. (1990) have demonstrated the possibility of melting at mid-crustal depths in regions with normal heat flow and with thickening

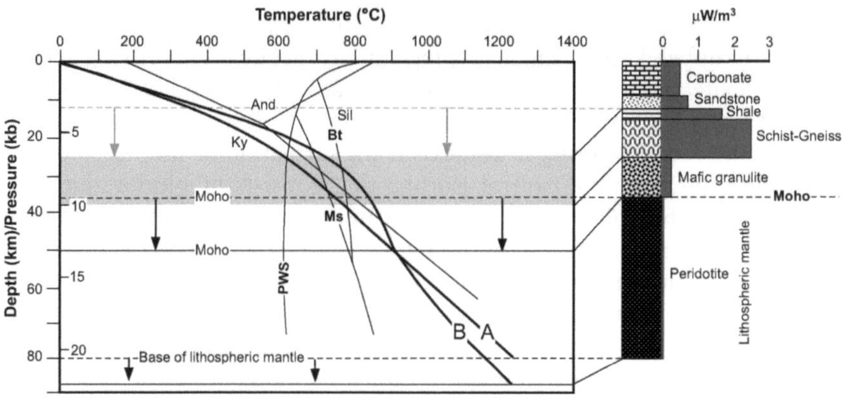

Figure 2.27. Modelled burial-*T-t* relationships of partial melt layer formation in radiogenic shale-schist-gneiss rocks, Seiver Hinterland, western USA (after Patiño-Douce et al. 1990). Original crustal section together with radiogenic heat production values (μW/m³) given on right. Radiogenic rocks originally at a depth of between ~12 and 25 km (original geotherm **A**) are buried by crustal thickening to 25–38 km depth (grey-shaded zone) where they undergo partial melting (geotherm **B** at 45 Ma after thickening). For other symbols and curves see Fig. 2.24. See text.

ratios <<2.0 using the Seiver hinterland of the North American Cordillera as an example, where crustal thickening to ~60 km is estimated to have occurred in the late Mesozoic. Their one dimensional *T–P–t* model for mid-crustal melting is illustrated in Fig. 2.27 and involves crustal thickening over a period of 30 Ma by thrust sheet stacking from an initial 36 km to 54.5 km due to ~100 km of shortening. The lower crust accommodates this deformation by ductile contraction and the lithospheric mantle is undeformed with the Moho behaving as a detachment zone. Erosion starts at the same time as deformation and occurs at a rate of 0.5 km/Ma. Melting begins ~25 Ma after the beginning of deformation within a mid-crustal fertile interval initially at a depth between 12 and 26 km comprised of shale overlying metapelitic–greywacke schist and gneiss that have a heat production rate of 16.8 and 25.0 mW/m^2, respectively. Although the surface heat flow is 60 mW/m^2, the high heat-production layer results in higher temperatures in the middle crust that would be the case, if heat production occurs in a near-surface layer or is homogeneously distributed throughout the entire crust. By the end of the orogenic event at 30 Ma, only a small amount of melt (~5%) is generated by muscovite dehydration melting involving 7 km of crust resulting in the formation of migmatite. Higher granitic melt fractions of 30–40% are only generated ~10 Ma after the end of deformation over a vertical interval of ~2 km when the rising geotherm attains temperatures of biotite-dehydration melting near the base of the fertile layer at the depth of ~37 km and a temperature of ~820°C. Figure 2.27 shows thermal conditions at 45 Ma after the beginning of deformation (curve B) and 15 Ma after its cessation during which time 7.5 km of uplift has occurred. With reference to the muscovite/biotite-dehydration melting curves and the 45 Ma geotherm (B), the 14 km thick layer of partly melted rock consists of an upper 3 km of migmatite (essentially metatexite) overlying a 9 km layer of diatexite/granite magma occurring over a depth interval of 25–36 km.

Another example combining crustal thickening and a layer of high radiogenic heat-producing rocks resulting in anatexis is the formation of the ~10,000 km^2 South Bohemian batholith, central Europe that consists of biotite and two-mica granite. The batholith is part of the Variscan Orogen, a well-studied Late Devonian–Early Carboniferous collision zone between the Armorican terrane (Bohemia, Saxo-Thuringia) and North Gondwana (Moldanubia) that is characterised by large volumes of granite and associated high-*T*/low-*P* metamorphic rocks. Results of a two-dimensional thermal–kinematic model of crustal anatexis to form the South Bohemian batholith by Gerdes et al. (2000) are shown in Fig. 2.28 at 10 Ma intervals from 350 Ma (10 Ma after onset of collision involving 220 km of overthrusting at a rate of 8 mm/year). The diagrams show the development of a migmatite layer in radiogenic metapsammitic–pelitic rocks followed by increased melt production at temperatures >850°C involving biotite-dehydration melting to form an extensive perched partial melt layer with between 35 and 55% granite melt in the middle crust (30–25 km depth). The model predictions are in agreement with the dated time lag between an amphibolite facies temperature maxima at ~335 Ma and, although not shown in the model, inferred development of granite magma layer up to a depth of ~19 km that crystallised between ~330 and 320 Ma. At this stage, the modelling indicates that about 16–18 km of exhumation (at an assumed erosion rate of 0.5 mm/year) would have occurred. Further thinning of the crust from a thickness of 40–45 km to ~32 km was completed before the Permian at which time the top of the granite/migmatite would have reached depths of between 11and 6 km.

2.6.4. Shear Heating

Crustal scale shear heating is a plausible mechanism to induce anatexis at temperatures up to and above those of muscovite and biotite-dehydration melting from mid to upper crustal levels (i.e. ~25–15 km depth). Shear (or strain) heating requires that rocks retain sufficient strength to heat up while deforming at geologically slow rates. Although this is not the case for granitic rocks that become weak below the brittle–ductile transition, foliated rocks such as mica schist retain their strength (~35 MPa for a strain rate of 10^{-15} s^{-1}; ~65 MPa for a strain rate of 10^{-7}/s) at high temperatures near to and beyond the solidus (Fig. 10 of Nabelek and Liu 2004). A problem with shear heating is the time over which it can be maintained. Once melting begins, it is expected that the contribution of shear heating to raising the geotherm would drop as shear stress within the shear zone becomes negligible and in so doing limit the amount of melting. However, within the volume of deforming rocks, variable rates of movement on individual thrust planes within the shear zone and variable shear strengths of interlayered rocks of different compositions (e.g. metaquartzite, pelite, psammite, amphibolite etc.), implies that melting will occur at different

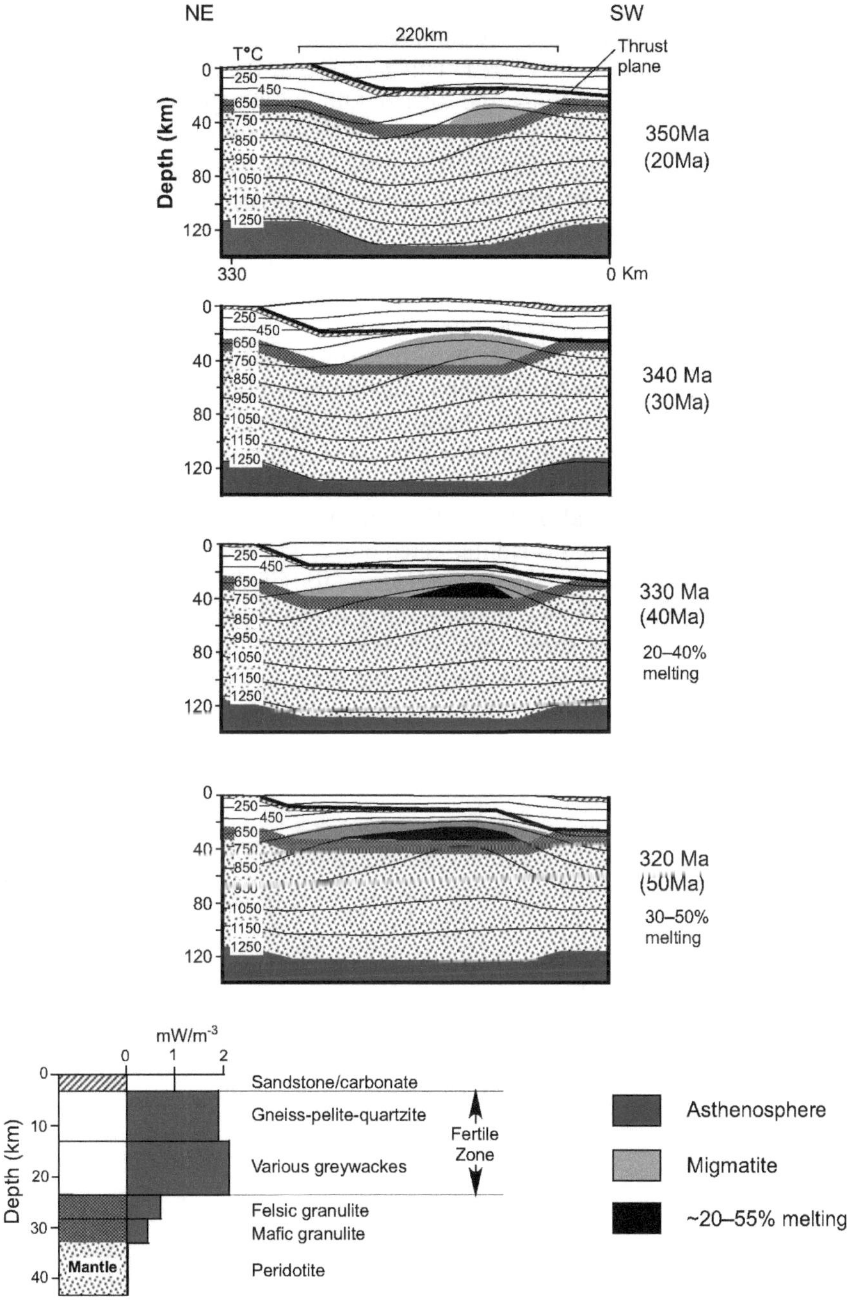

Figure 2.28. Modelled burial-*T*-*t* relationships of partial melt layer and granite formation in radiogenic gneissic pelite–greywacke rocks, Bohemian massif, central Europe, shown at 20, 30, 40 and 50 Ma after crustal thickening (simplified after Gerdes et al. 2000). Original crustal section with radiogenic heat values (μW/m³) shown at bottom. See text.

times and in different areas within the shear zone. As the active part of the shear zone weakens, shear stress becomes amplified in other areas of the deforming zone where melting has not occurred. This means that the amount and rate of melting will vary both vertically and horizontally within the shear zone as it evolves to form a partial melt layer. Therefore, the overall positive heat contribution to the geotherm within the shear zone will either be maintained at a certain level for a finite time depending on the proportion of heterogeneous protolith that has melted, or it will increase in unmelted parts of the zone with continuous shearing.

A number of shear heating models for generating melting conditions in thickened continental crust for different rates of thrusting, depth of shear zone and radiogenic heat-production profiles have been proposed by Nabelek and Liu (2004). One of these models that show maximum overlap of evolving geotherms, pelitic solidus and mica dehydration melting curves is given in Fig. 2.29. In this case, the duration of thrusting within a 4 km thick shear zone centred at 25 km depth is assumed to be 50 Ma at a rate of 4 cm/yr. Modelled evolving geotherms at 10 Ma intervals indicates that for the shallowest depths of partial melting, the wet pelitic solidus is intersected at ~20 km depth (2 km above the top of the shear zone), muscovite in rocks at the top of the shear zone start to melt at 20 Ma, and at 30 Ma the beginning of biotite-dehydration melting occurs in rocks below the shear zone at ~29 km depth. At 50 Ma the biotite melting curve is intersected at 25 km within the shear zone.

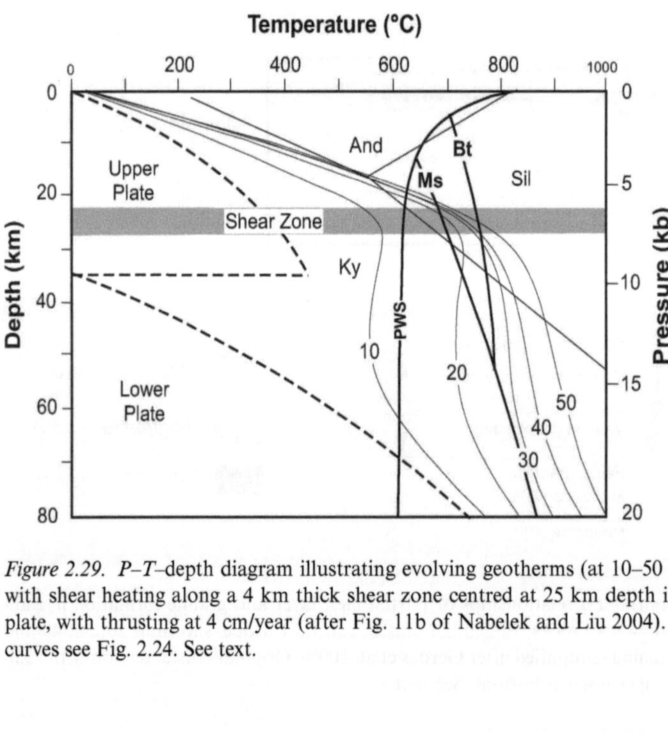

Figure 2.29. *P–T*–depth diagram illustrating evolving geotherms (at 10–50 Ma) in thickened crust with shear heating along a 4 km thick shear zone centred at 25 km depth in the upper, overthrust plate, with thrusting at 4 cm/year (after Fig. 11b of Nabelek and Liu 2004). For other symbols and curves see Fig. 2.24. See text.

It can also be noted from Fig. 2.29 that at 20 Ma after thrusting the top of the partial melted crust, i.e. rocks at temperatures > the wet pelite solidus, lies at ~32 km depth. After 60 Ma of thrusting, the top of the partial melt zone has thickened upward to a depth of ~17 km and would consist of an upper 15 km of migmatite (largely metatexite + leucogranite) underlain by diatexite + biotite granite that would extend down to the lower boundary of suitably fertile lithologies that may be present.

Recent collision orogens, e.g. the Himalayas (Le Fort et al. 1987; Zhu and Shi 1990; Harrison et al. 1997, 1998; Nabelek and Liu 2004), and ancient orogenic belts, e.g. the Black Hills, South Dakota, USA (Nabelek and Liu 1999; Nabelek et al. 2001) and Maine, USA (Brown and Solar 1998a, b; Solar et al. 1998), are examples where the association between major shear zones and leucogranite has been proposed.

2.6.5. Extension and Removal of Lithospheric Mantle

Extension of the lithospheric mantle is associated with thinning/rifting of the continental crust and has been proposed as a mechanism for generating conditions of high-*T*/low–medium-*P* metamorphism (Lachenbruch and Sass 1978; Wickham and Oxburgh 1985; Sandiford and Powell 1986). With the lithosphere–asthenosphere boundary represented by the 1200°C isotherm, a steepening of the lithosphere geotherm is an inevitable consequence of replacement of lithosphere by asthenosphere. Primary mechanisms invoked for the replacement of all or part of the lithospheric mantle by asthenosphere involve convective thinning (Houseman et al 1981; Molnar et al. 1998) and mantle delamination that results in the crust being directly underlain by hot asthenospheric mantle (Bird 1979; Loosveld and Etheridge 1990). The process is predicted to cause significant heating of the lower crust with the thermal anomaly propagating upwards into the middle and upper crust. The resultant high heat flow associated with regions of crustal extension, e.g. large areas such as the Basin and Range Province, USA (1.3 $\times 10^6$ km^2), or narrower rift/graben features such as the Pyrenees, France, the Lachlan Fold Belt, SE Australia, or the present day Taupo Volcanic Zone, New Zealand, can produce temperatures of 700–900°C at ~30 km depth or shallower, resulting in granulite facies metamorphism with biotite and hornblende dehydration melting reactions giving rise to granite melts.

Examples of lithospheric thinning are shown in Fig. 2.30, involving a 40 km thick crust and a mantle lithosphere/asthenosphere boundary at 120 km depth where the temperature is 1200°C. In (A), the temperature at the base of the crust is ~500°C and no melting occurs. In (B) with ~50% crustal thinning (i.e. to a thickness of 20 km) due to axially symmetrical extension of mantle lithosphere by pure shear, the base of the crust remains at ~500°C indicating decompression without convective heating. In (C), crustal thinning to ~30 km results from axially symmetrical extension of mantle lithosphere through non-uniform shear resulting in basal crustal temperatures of ~600–700°C indicating that melting of

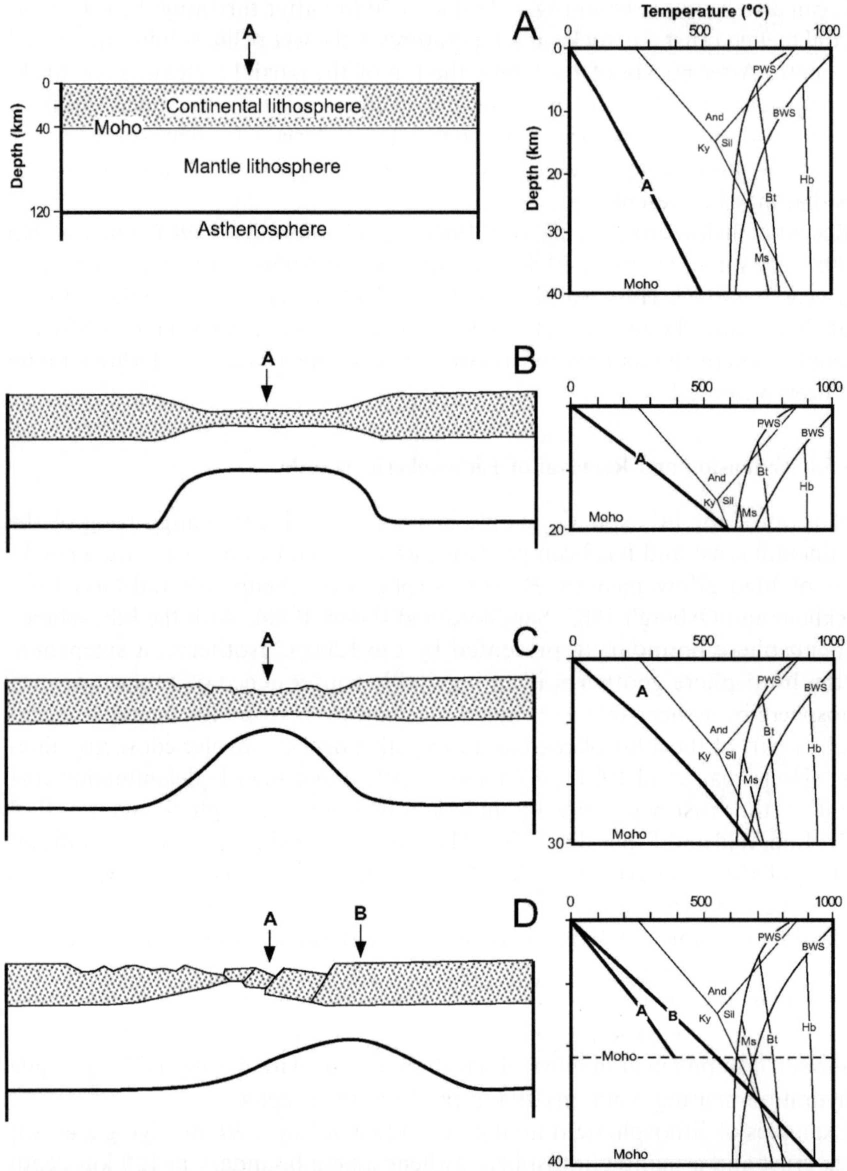

Figure 2.30. Lithospheric thinning and its effect on the crustal geotherm below positions A, and B in relation to crustal melting where PWS = pelite wet solidus; Ms and Bt = muscovite and biotite-dehydration melting, respectively. Al-silicate stability fields after Pattison (1992). Modified after Fig. 1 of Sandiford and Powell (1986). See text.

suitable lithologies could occur. In this case, as the mantle lithosphere thins more rapidly than the crust, the geotherm is the result of both decompression and heating. In (D), asymmetric extension is accommodated in the crust by simple shear along a detachment zone, below which extensional strain is distributed in a ductile region of pure shear. The most perturbed geotherm occurs where the mantle lithosphere is thinnest (below position B), rather than where the crust is thinnest, and melting of crustal rocks will occur within the 20–30 km depth interval. Clearly, in all the above cases, and particularly (B) and (C), mantle melting and underplating of the crust by basaltic magma is also required to further perturb geotherms in order for extensive anatexis to occur at mid–upper crustal depths.

In the case of complete delamination of the lithospheric mantle, Fig. 2.31 depicts an example where delamination is associated with asthenospheric convection continuing for a period of 20 Ma. This results in partial melting of

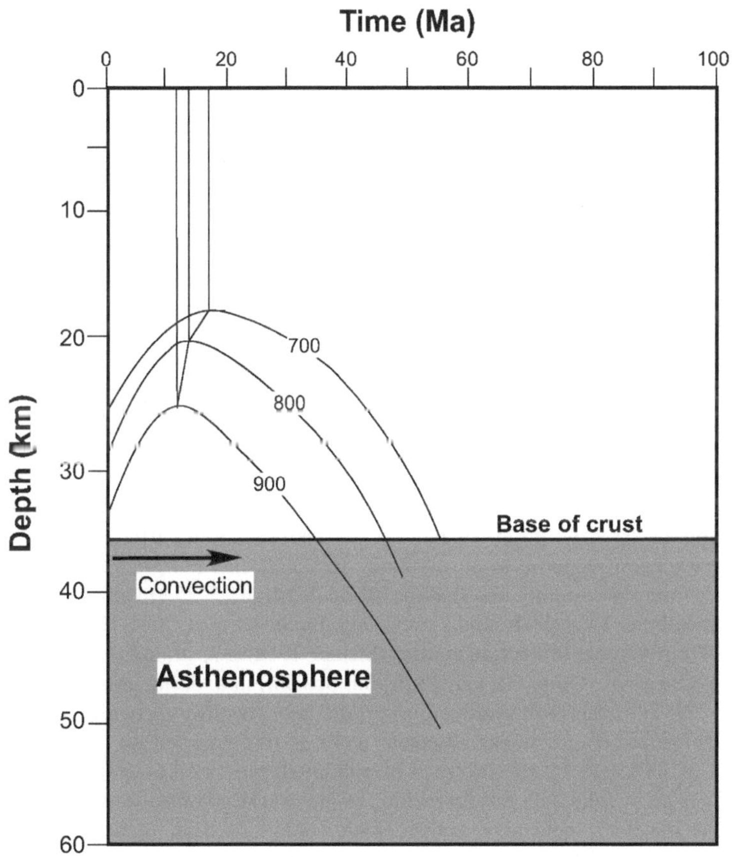

Figure 2.31. Depth/time plot of three isotherms (700, 800, 900°C) representing steps of melting in 35 km thick crust by 'hot mode' delamination of mantle lithosphere and where asthenosphere convection continues for 20 Ma (T_a = 1300°C) (after Fig. 3b of Thompson 1999).

appropriate lithologies at minimum depths of ~25 km (900°C = amphibole dehydration melting) at 11 Ma, ~21 km (800°C = biotite-dehydration melting) at 13 Ma, and 18 km (700°C = muscovite dehydration melting) at 17 Ma following delamination, respectively. After 55 Ma the temperature at the base of the crust cools to 700°C.

Perhaps the most important geophysical feature below **active** rifts zones is anomalously low seismic velocities and the anomalous attenuation of seismic waves that suggest very high temperatures and partial melting. This is manifested by high heat flow and in many cases with magmatic and geothermal activity, e.g. East African Rift; Salton Sea area, southern California; Rio Grande Rift, New Mexico; Central Volcanic Region, New Zealand; Larderello, Italy.

The central volcanic region (CVR) of the North Island of New Zealand, represents a back-arc basin formed within continental lithosphere related to NW subduction of the Pacific Plate off the east coast of the North Island (Stern 1985, 1987; Cole 1990) (Fig. 2.32). The region is one of active extension, volcanism and abnormally high heat flow estimated at 4000–5000 MW (Bibby et al. 1995). The currently active part of the CVR is the Taupo volcanic zone (TVZ) that is characterised by abundant rhyolitic volcanism including thick sequences of ignimbrites erupted from calderas within the last 1.6 Ma (Wilson et al. 1995), with present day andesitic volcanism occurring from Mt. Ruapehu in the South to White Island in the North (Fig. 2.32). The area contains more than 20 high-T geothermal systems that together discharge more than 4000 MW of heat (Bibby et al. 1995). Mesozoic greywacke crops out to the East (Axial Ranges) and West of the CVR and TVZ (Fig. 2.32), and forms the downfaulted basement of the TVZ, being overlain by > 2 km of volcanic and volcaniclastic rocks. Magnetotelluric (Ogawa et al. 1999; Ingham 2005) and shear wave velocity (Sherburn et al. 2003; Bannister et al. 2004) data indicate the presence of high-conductivity/low-velocity zones at depths of 10–15 km consistent with the presence of partially melted rock with lateral extents of ~10–20 km. Long wavelength magnetic anomalies parallel to the NE-SW trend of the TVZ underlie nearly all of the < 0.7 Ma ignimbrite eruption calderas at between ~4 and 7 km depth suggest the existence of granite magma that has cooled to below its Curie temperature of 575°C (Soengkono 1995) (Fig. 2.33). We interpret this data to indicate the downward cooling of an regionally extensive granite magma layer developed from partial melting of the basement greywacke in response to an originally very high-thermal gradient due to mantle lithospheric thinning resulting in a shallow Moho depth of 15–25 km and possibly mantle melting beneath the TVZ (Stern and Davey 1987; Bannister et al. 2004) (Fig. 2.32). The occurrence of vitrified schistose (migmatite) (Steiner 1958) and granulite xenoliths (with Qz, Pl, Ksp, Cpx, Opx, Omp, Gt, Cd, Ti-Bt, Sil, Sp, Ilm) in andesite lavas (Steiner 1958; Graham 1987) and voluminous eruption of rhyolite provide direct evidence of partial melting of greywacke basement rocks. If this occurred up to shallow depth, i.e. ~5 km as indicated by the presence of solidified, but still hot granitic rock (Soengkono 1995), experimental data for wet melting of greywacke compositions East and West of the TVZ (Kifle 1992) indicates that their solidi would have been intersected by a geotherm of between 134 and 120°C/km, respectively (Fig. 2.34). In Fig. 2.34, glass composition fields from experimental melting of eastern and western greywacke (Conrad et al. 1988; Kifle 1992) are shown in comparison with TVZ rhyolites in terms of normative ratios. Ignoring the possibility of in situ crystal fractionation, it is evident from this plot that the rhyolites (and granite) could have been derived from partial melting of a mixture

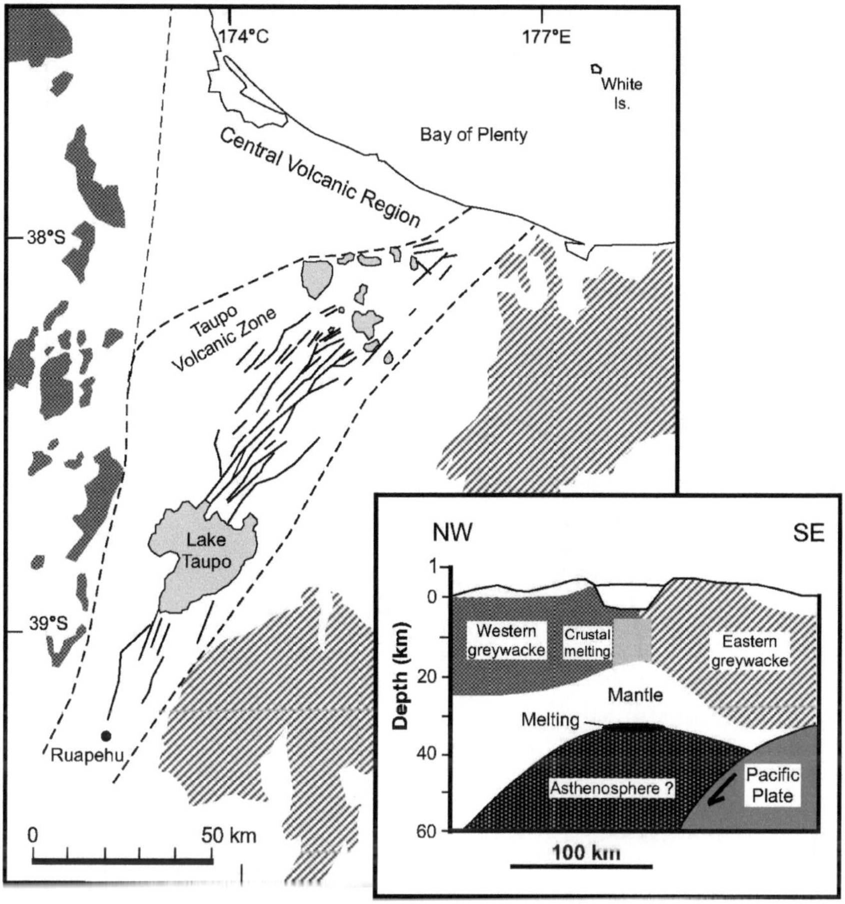

Figure 2.32. Simplified geological map of the central volcanic region and currently active Taupo volcanic zone, central North Island of New Zealand. The zone of rifting (indicated by NE–SW-trending normal faults) separates western (volcanogenic) and eastern (granitogenic) greywacke basement that underlies the Taupo volcanic zone. Inset shows crust–mantle–asthenosphere section across the central part of the central volcanic region, westward subduction of Pacific oceanic plate, thinned mantle lithosphere and inferred zone of greywacke melting beneath the rift zone.

of both greywacke compositions with a greater volume of western greywacke which contains a large volcaniclastic component. Restite phases (newly-formed and relic) from experimental melting of these greywackes are as follows: Qz, Pl, Bt, Opx, Cpx, Omp, Hb, Gt, Ilm, Ap (western greywacke) and Qz, Pl, Cd, Os, Opx, Oamp, Sil, Sp, Mt (eastern greywacke), i.e. the mineral phases present in granulite xenoliths, and the rhyolites contain 1–>50 vol % phenocrysts of Pl, Opx, Cpx, Hb, Omp, Ti-Bt, Ilm, Mt, together with resorbed Qz.

Another example, the Larderello geothermal field, Italy (Fig. 2.35) , occurs in an area that has been undergoing post-collisional extension since the early

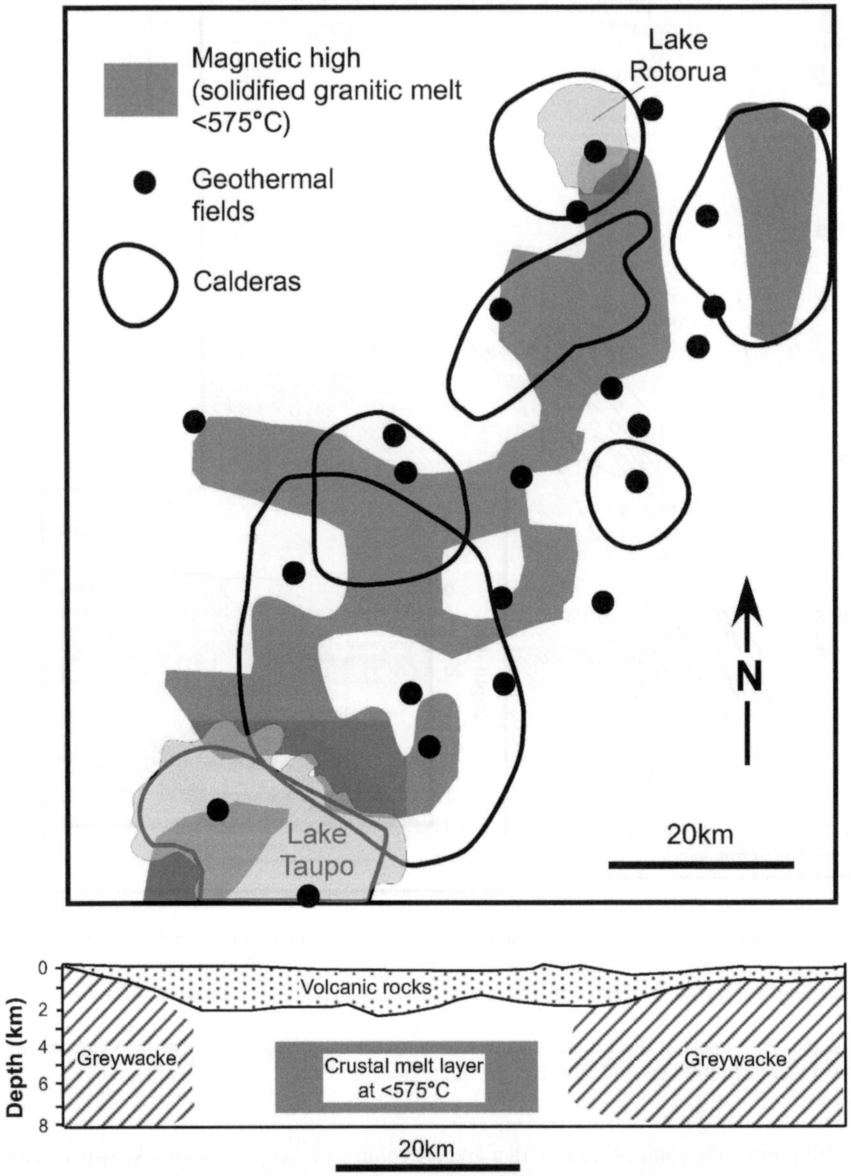

Figure 2.33. Map showing distribution of crustal magnetic highs, main geothermal areas and outlines of caldera eruption centres between Lakes Taupo and Rotorua, Taupo volcanic zone, central North Island of New Zealand (modified after Fig. 8 of Soengkono 1995). For location see Fig. 2.32. Below: Interpretive E-W crustal section across the Taupo Volcanic Zone.

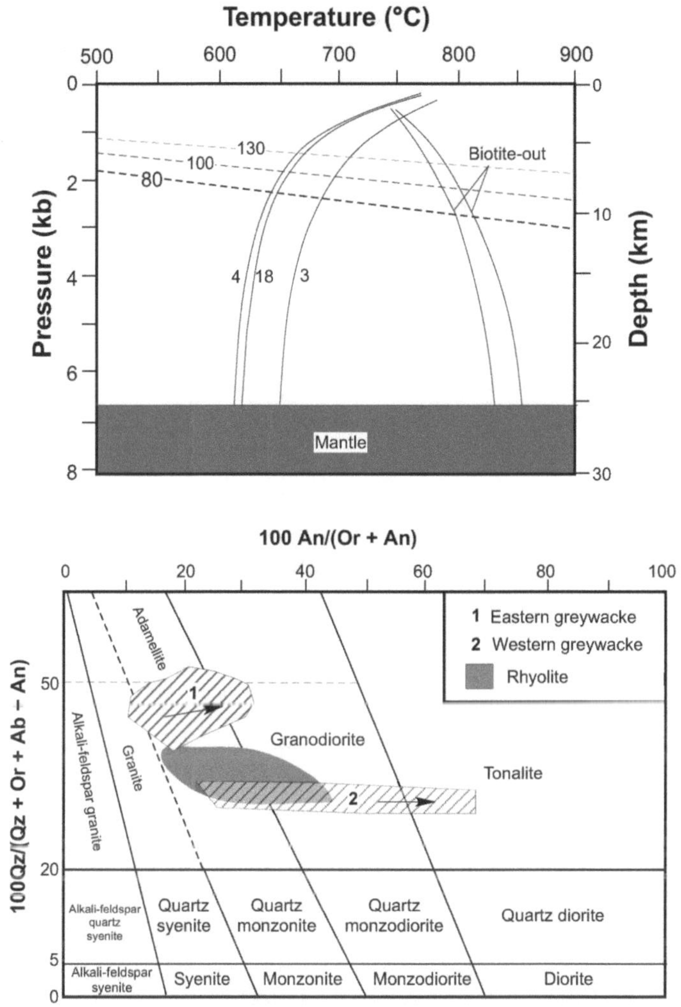

Figure 2.34. Above. P–T–depth diagram showing wet solidus curves for eastern (Nos.A.4 and18 in Appendix 2) and western (No.B.3 in Appendix 2) greywacke compositions bordering the central volcanic region, central North Island, New Zealand (see Fig. 32). Biotite-out curves refer to compositions A.4 and 18. See text. *Below*. Normative Qz, Ab, Or, An ratio diagram of granite nomenclature showing field of experimental glass compositions for eastern (Nos.A.4,18) and western (No.B.2) greywacke compositions in comparison to field of rhyolite compositions from Taupo volcanic zone (after Steiner 1958). Arrows indicate increasing *T*. See text.

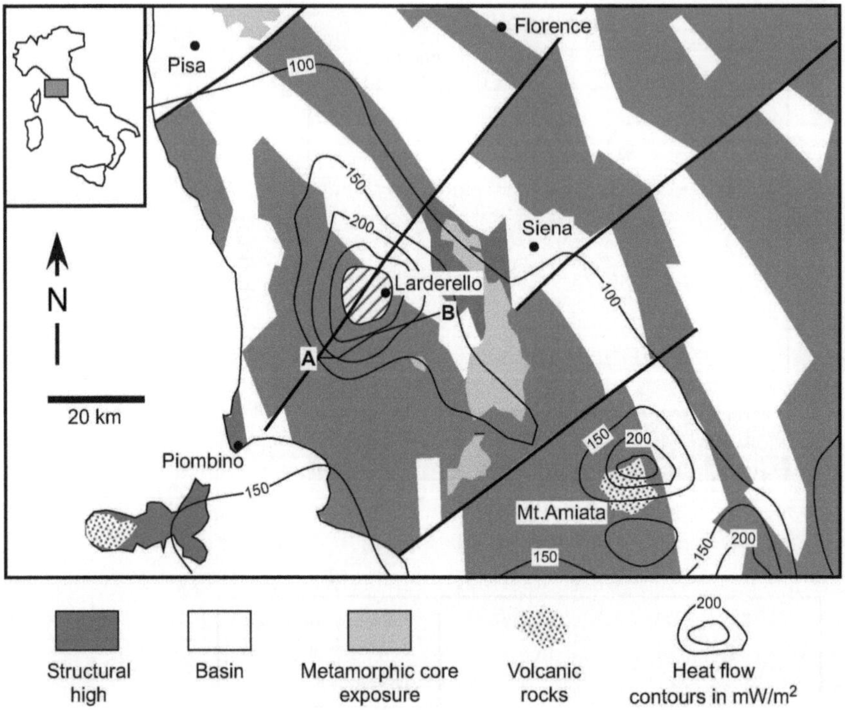

Figure 2.35. Sketch map of the southern Tuscany region, Italy, showing the structure and location of the Larderello geothermal area and heat flow contours (after Figs. 1 and 2 of Cameli et al. 1993).

Miocene resulting in uplift and a shallowing of the Moho to ~25 km. The area is characterised by a positive heat flow anomaly with a geothermal gradient of ~100°C/km, a Bouguer gravity minima, a marked negative magnetic anomaly together with slow cooling rates of basement metamorphic minerals, that implies the presence of cooling granite at shallow depth (Fig. 2.35) (Puxeddu 1984: Cameli et al. 1993). Exploration wells have confirmed the presence of hydrothermally altered, fractured granite, and contact metamorphism (And, Cd) has affected overlying Hercynian basement gneiss, schist and phyllite (Gianelli and Ruggieri 2000). A seismic reflector (known as the K-horizon) at between 4 and 8 km depth is considered to represent the change from brittle to ductile behaviour and the presence of hydrothermal fluids at temperatures in the range of 400–450°C (Barielli et al. 2000) (Fig. 2.36). Near the Larderello thermal high, the reflector horizon appears to coincide with the top of the granite; further away it lies below the granite contact (Cameli et al. 1993; Barielli et al. 2000). The extensive negative gravity anomaly over the region suggests that at depths of between 8 and 12 km the temperature of the granite should be ~800°C, and therefore partly molten. The granite is part of a regionally extensive layer

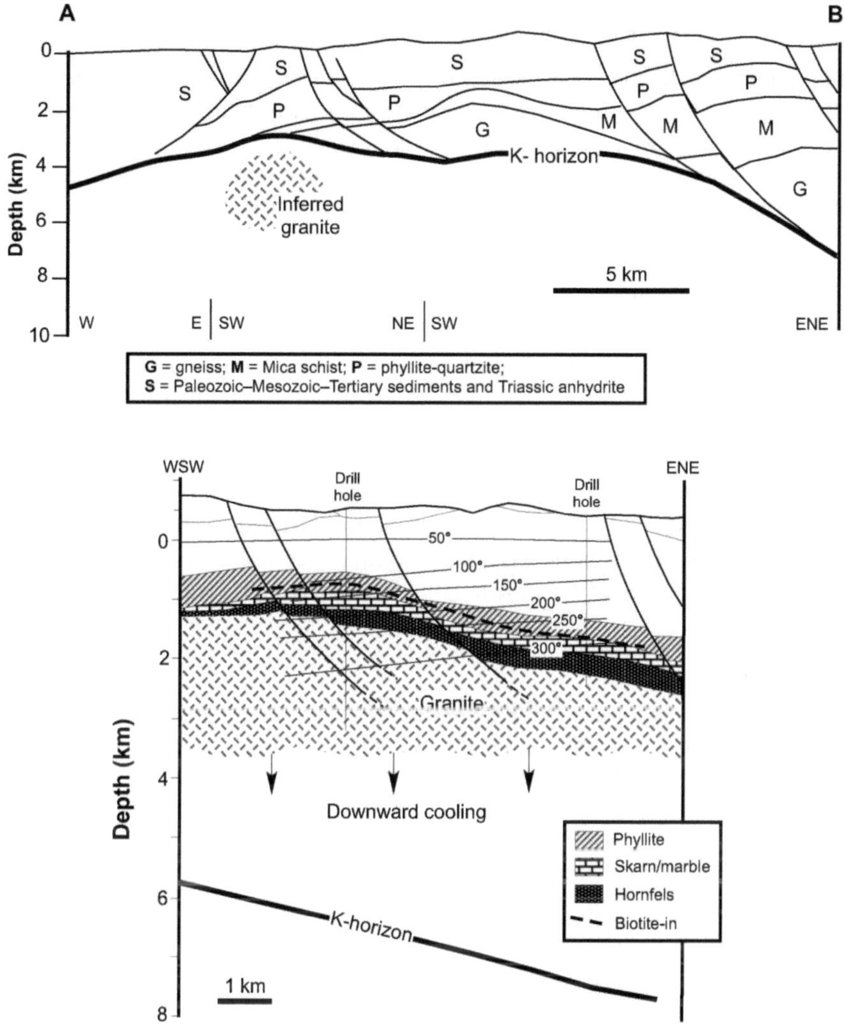

Figure 2.36. Sections across and close to the Larderello geothermal field, southern Tuscany, Italy. Upper section location is shown in Fig. 2.35 and is after Fig. 4 of Cameli et al. (1993). Lower section after Barielli et al. (2000; section No.2). See text.

beneath SW Tuscany, including the Tyrrhenian Sea, resulting from anatexis (up to 10 km depth) of Hercynian gneiss–schist basement rocks that overlie granite beneath the Larderello area. Volcanic rocks of the Tuscan magmatic province contain abundant xenoliths with Pl, Ksp, Qz, Cd, Sil, Gt and Zr that are considered to be relics of partial melting and are the same as the main and accessory constituents of gneiss and schist of the Hercynian basement.

2.6.6. Intrusion of Mafic Magma

As the amount of fertile crust needed to produce granite magmas is extremely small in comparison to the volume of the mantle, the mantle and derivative basalt magma may well be an important source of heat for crustal melting. Lithospheric thinning and upwelling of hot asthenosphere as described above are also conducive to mantle melting and intrusion of mafic magma to increase crustal temperatures (Fig. 2.37). Intrusion of mafic magma at ~1200°C into crustal rocks can cause rapid (10^2–10^3 years) melting (Huppert and Sparks 1988a, b; Bergantz 1989) compared to the timescales and length scales of an orogenic event e.g. 10^5–10^6 years (Thompson and Connolly 1995). Huppert and Sparks (1988a) point out that if emplacement of basaltic magma adds heat to the crust at a faster rate than heat can be extracted by conduction along the geotherm, then the crust will heat up resulting in a steepening of the geotherm. However, according to Thompson (1999), in order to raise temperatures to 800°C (biotite dehydration melting) at 20 km depth, it would need enormous intrusion thicknesses of 10, 25 and 60 km to be emplaced at depths of 25, 30 and 35 km, respectively, of which there is no geological or geophysical evidence. Indeed, intrusion of large bodies of basaltic magma into the crust typically evidenced by

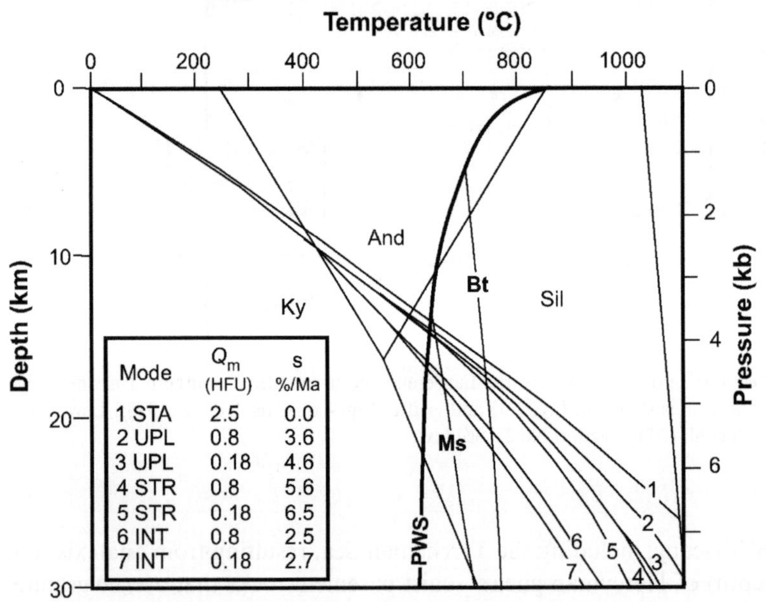

Figure 2.37. Temperature profiles calculated from measured heat flow for average Battle Mountain High conditions (after Fig. 11 of De Yoreo et al. 1991). In the table, STA = static; UPL = mafic magma underplating; STR = stretching; INT = mafic magma intrusion. Q_m = mantle (reduced) heat flow; 1 HFU = 41.8 mW/m^2; s = strain rate in %/Ma. For other symbols/curves refer to Fig. 2.24.

peridotite–gabbroic diorite plutons, e.g. Ivrea, Italy, have only a limited thermal effect on the country rocks (Barboza et al. 2000) and there is little evidence of substantial crustal melting associated with such intrusions, that may have been fairly rapid. On the other hand, the intrusion of swarms of basaltic dykes or sills, overlapping in space and time, could be a more effective way to cause crustal melting (Petford and Gallagher 2001).

The effects of mafic sills intruded into fusible protoliths have been discussed by Huppert and Sparks (1988a) and Fountain et al. (1989) with the general conclusion that heat from the intrusion can rapidly produce a melt layer with a thickness of about two-thirds the thickness of the intrusion in crustal rocks already at elevated temperature, i.e. 500°C or greater. As stated above, the melting process would be enhanced if mafic magma was sequentially intruded instead of at one time. Thickness/time relationships for crustal melting by basaltic sills are shown in Fig. 2.38. The diagram indicates that 500 m thickness of basalt at 1200°C intruded as multiple sills into fertile crustal rocks at 500°C will, at the time the basalt is 60% crystallized and convection has stopped, produce a 294 m thick granitic melt layer at 934°C that will have undergone 50% crystallisation in only about 210 years after intrusion. If the country rock temperature is 500°C, e.g. at ~15 km and 20 km depths for hot and warm geotherms respectively, then contact temperatures of a 1200°C sill will be ~850°C, i.e. above biotite-dehydration melting, and significant melting can occur. Once a partial melt layer has been formed,

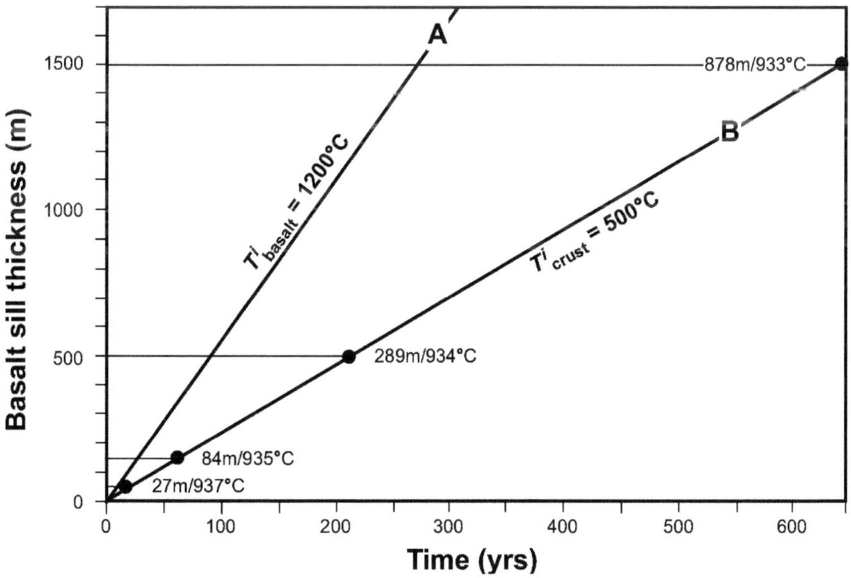

Figure 2.38. Thickness/time relations for melting of crustal rocks (initially at 500°C) by basaltic sills (initially at 1200°C), (after Fig. 5 of Thompson 1999). See text.

Huppert and Sparks (1988a) suggest that it would be difficult or impossible for subsequent basalt (ascending as dykes) to intrude through it and magma would pond beneath and within the original sill to extend or even increase the intensity of the heating event of the already hot crustal rocks.

A present day example of crustal melting by mafic magma intrusion may be the area beneath Soccoro in the Rio Grande Rift, USA, where a strong seismic reflector with a characteristic horizontal dimension of ~70 km is located at a mid-crustal depth of about 19 km. This has been interpreted as indicating the presence of basalt magma with a current volume of ~104 km^3. It is suggested by Fialko et al. (2001) that uplift of a few millimeter per year over the last ~90 years could be mainly due to a volume increase at a rate of 0.1 km^3/year caused by in situ melting of quartzofeldspathic crustal rocks above the basalt. An ancient example of crustal melting cause by intrusion of mafic magma is the well-known high-*T*/low-*P* metamorphic Buchan area of NE Scotland. Droop et al. (2003) demonstrate syn-metamorphic intrusion of mafic magma into warm (~550°C) pelitic–semi-pelitic rocks at depths of 15–18 km resulted in peak-metamorphic temperatures of ~900°C and extensive (~60%) fusion via biotite dehydration melting. It is considered that gravity-driven segregation of granitic melt from a quartz-free restite of Ksp, Cd, Opx and Sp in the migmatite host rocks occurred in the absence of regional tectonic deformation giving rise to contemporaneous two-mica granite.

A potentially important mechanism of crustal melting in a continental margin setting involves channelised upward intrusion of mafic magma into the crust below a volcanic arc, e.g. the Andes (Laube and Springer 1998). The mafic conduits thermally equilibrate with the surrounding crust and water released from cooling magma enhances crustal anatexis to form a region of partial melting as illustrated in Fig. 2.39 within the confines of the conduit zone. Seismic investigation of the northern Izu-Bonin-Mariana (IBM) arc suggests the existence of such a mid-crustal layer of felsic igneous rocks between ~7 and 12 km overlying a mafic lower crust (Suyehiro et al. 1996). In the Tanzawa Mountains of central Honshu, Japan, an exposure of Miocene tonalite with subordinate gabbro is considered to represent the mid-crust of the ~22 km thick intraoceanic IBM arc that collided with central Honshu in mid-Miocene time (Kawate and Arima 1998). It seems likely that the tonalite and a pyroxenitic restite is the result of in situ melting (~45%) of prehnite–pumpellyite basaltic–high Mg-andesitic volcanic and volcaniclastic arc rocks that occur at a depth of ~6 km.

2.6.7. Crustal Thinning and 'Diapiric' Decompression

Exhumation of deep crustal rocks by near-isothermal decompression may result from partial melting that weakens the crust facilitating orogenic collapse and crustal thinning, or from erosion or tectonic processes that trigger anatexis (Zeitler et al. 2001). Many high-grade kyanite-bearing rocks (e.g. gneiss domes) show evidence of near isothermal decompression of >12 km at high temperature (>700°C) coupled with dehydration melting. Decompression mechanisms that

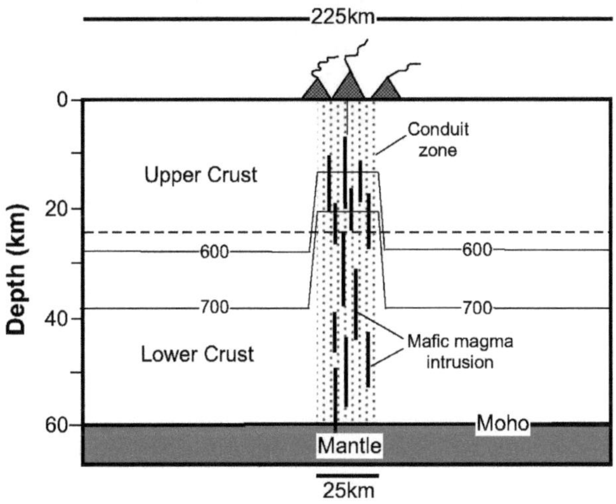

Figure 2.39. Arc (andesitic) magma intrusion and perturbation of isotherms in conduit zone beneath the active (e.g. Andean) arc (adapted from Fig. 4 of Laube and Springer 1998). See text.

involve removal of upper crust by erosion, crustal thinning and diapirism have been proposed by Teyssier and Whitney (2002) to explain the presence of large volumes of granite in migmatite terranes as shown in Fig. 4.40A, B. The initial condition for the conceptual models of Teyssier and Whitney (2002) is an orogen after crustal thickening with a rigid upper crust and a deep crust that had reached high temperature due to thermal relaxation (Fig. 2.40A). A horizontally laminated lower crust implies that lateral flow causing crustal thinning occurs by bulk pure shear resulting in collapse and therefore decompression. This will be relatively slow if restricted to the lower crust (Fig. 2.40D) and faster if the upper crust is extended, thinned or rapidly eroded (Fig. 2.40C). For a given rate of lateral flow in a 50 km thick crust and average displacement rates based on plate velocities of 10–100 km/Ma, thinning is substantially faster (50–5 km/Ma) in a narrow (100 km-wide) orogen than in a wide (1000 km-wide) orogen (5–0.5 km/Ma). However, decompression results in minimal melting as the uplift curve only just intersects the biotite-dehydration melting curve.

Figure 2.40D shows the example of thick, hot, dominantly solid crust that rises buoyantly and decompresses in the case where lower density rocks underlie higher density rocks. This would be the situation if a partial melt layer was formed, e.g. Andes, Tibet, and could trigger diapirism. The magnitude of diapiric decompression would be significant (>12 km) because the material motion is essentially vertical (Fig. 2.40D) and self-sustaining because of partial melting causing a decrease in viscosity enhancing upward flow and essentially isothermal decompression. Sinking regions marginal to the rising diapiric areas undergo a pressure increase with the result that the end product could be a series of migmatite–granite-cored domes (the deformed partial melt layer) at mid to upper crustal depths and higher pressure granulitic rocks residing at deeper

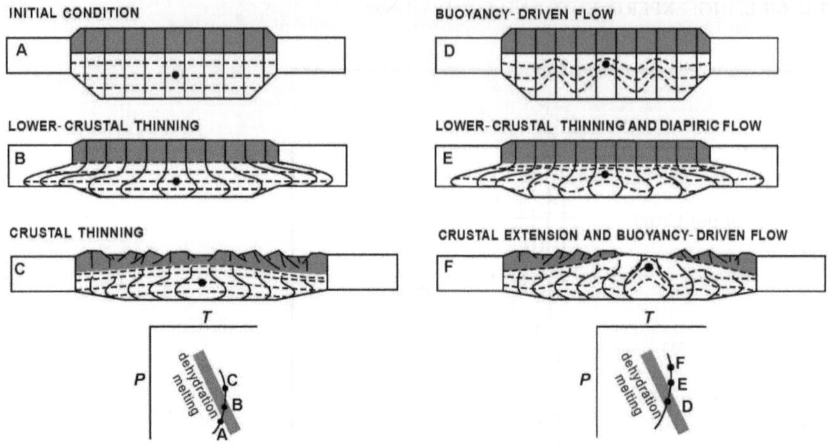

Figure 2.40. Models for crustal flow, partial melting and decompression with reference to displacement of filled circle point as shown in lower *P–T* diagrams in relation to biotite-dehydration melting (after Fig. 2 of Teyssier and Whitney 2002). See text.

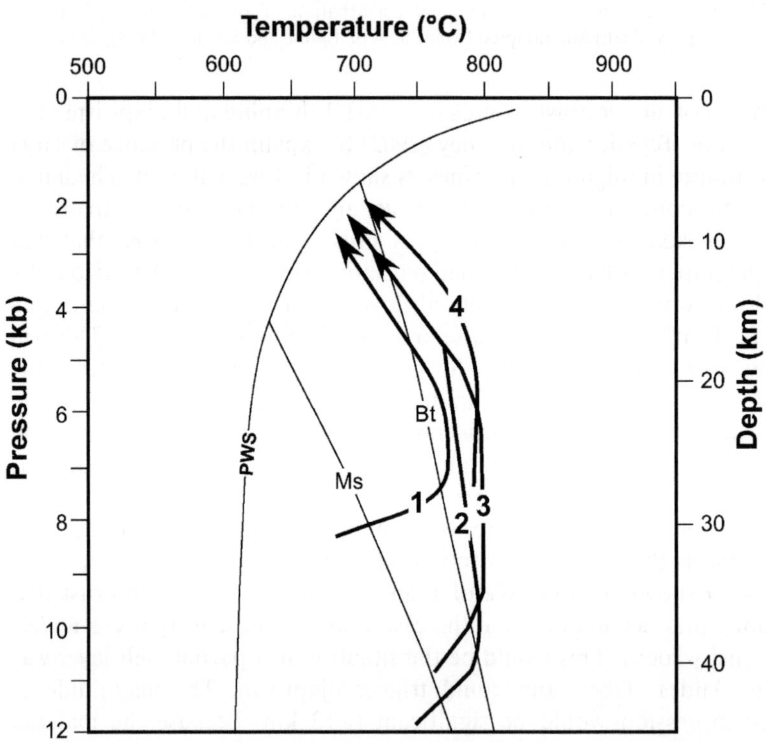

Figure 2.41. *P–T* paths for migmatite-cored gneiss domes. **1**. Thor-Odin dome, British Columbia, Canada (Norlander et al. 2002); **2**. South Brittany, France (Audren and Triboulet 1993); **3**. Velay dome, Massif Central, France (Montel et al. 1992); **4**. Tormes dome, Iberia, Spain (Escuder et al. 2000). PWS = pelitic wet solidus; Ms and Bt = muscovite and biotite-dehydration melting curves, respectively (modified after Fig. 1C of Teyssier and Whitney 2002).

levels. The granite melt is retained in situ except for small scale leucogranite segregations. Bouyancy-driven diapirism (Fig. 2.40D) may be coupled with thinning of the thickened crust (Fig. 2.40E), and would be more important if the upper crust is removed by erosion, tectonic denudation or extension (Fig. 2.40F). These cases involving diapirism produce maximum amounts of melting during decompression because significant temperature overstepping of the biotite-dehydration melting curve occurs.

Examples of *T–P* decompression paths of migmatite–granite-cored gneiss domes are shown in Fig. 2.41. Decompression is probably explained in terms of diapiric uprise of rheologically unstable zones of anatexis in orogens, and could be an important process in the thermal, mechanical and chemical evolution of some continental areas (Brun et al. 1994; Vanderhaeghe et al. 1999; Teyssier and Whitney 2002) (Fig. 2.42). Exhumation (relative vertical motion of rocks with

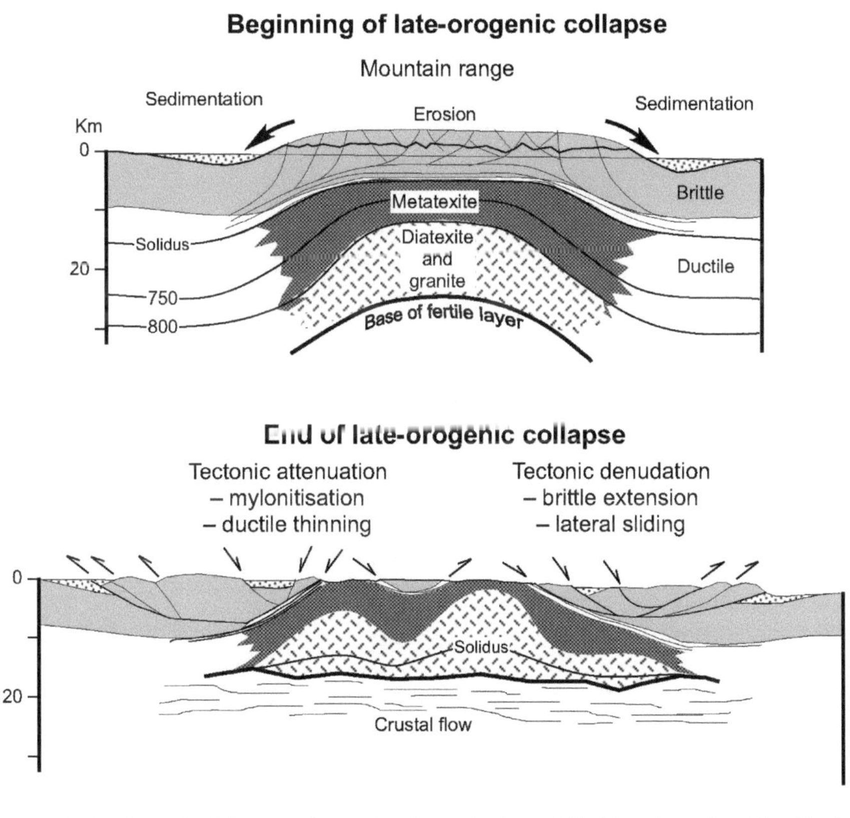

Figure 2.42. Generalised diagrams illustrating the mechanism of diapiric exhumation (after Fig. 10 of Vanderhaeghe et al. 1999). Two examples that could be explained by diapiric exhumation are the migmatite core complexes of the Velay Dome, French Variscides and the Thor-Odin dome area, Canadian Cordillera, described in Chapter 4 where thermobarometric, geochronological, structural and sedimentological data imply that 15–20 km of exhumation has occurred (Vanderhaeghe et al. (1999)).

respect to the Earth's surface) of these granite–migmatite rocks involves removal of the upper crust by erosion and tectonic processes. Erosion can account for rapid unroofing such as proposed for the Nanga Parbat syntaxis, western Himalayas (Zeitler et al. 2001), and it requires evidence of rapid uplift and related deposition. In most cases, however, erosion probably plays a secondary role to tectonic processes of exhumation.

CHAPTER 3

IN SITU MELTING AND INTRACRUSTAL CONVECTION: GRANITE MAGMA LAYERS

3.1. INTRODUCTION

With respect to the origin of granite, two models, i.e. the traditional intrusion model, and the in situ melting model presented in this book, emphasise that granite magma is predominantly the result of partial melting of crustal rocks. The magma intrusion model requires magma to escape from a deep source (area of anatexis) and be emplaced within upper crustal rocks. The in situ melting model regards the formation of granite as a result of entropy variation within the system from order (protolith) to disorder (magma) to new order (granite). The areal extent of a partial melting region in the crust depends on the geometry of the isotherms and its thickness depends on the duration of heating. Although temperatures increase with depth, the lower boundary of a crustal-scale partial melt layer is considered to be determined by the deepest level of the fertile protolith undergoing melting so that they can be characterised as 'rootless' or perched layers.

3.1.1. Geophysical Evidence for Crustal Melting

Intracrustal low velocity layers in young and old orogenic belts strongly suggest that extensive partial melt layers within the mid-upper crust are a reality, e.g. Himalayas, (see below), Andes (see below), Basin and Range Province of western USA (Eaton 1980), North (Liu 1985) and SE China (Ma 2006), mid-European Variscan orogen (Prodehl and Giese 1990), the area south of the North German Basin in the European Geotraverse (Wedepohl 1996 in Johannes and Holz 1996) (Fig. 3.1). Examples of extensive partial melt layers beneath the two highest plateaus on Earth, the Himalayas-Tibet and the Andes, are described below.

3.1.1.1. Himalayas and Tibetan Plateau

The Tibetan plateau is associated with a pronounced satellite magnetic low that coincides closely with the topographic edges of the plateau and suggests that the

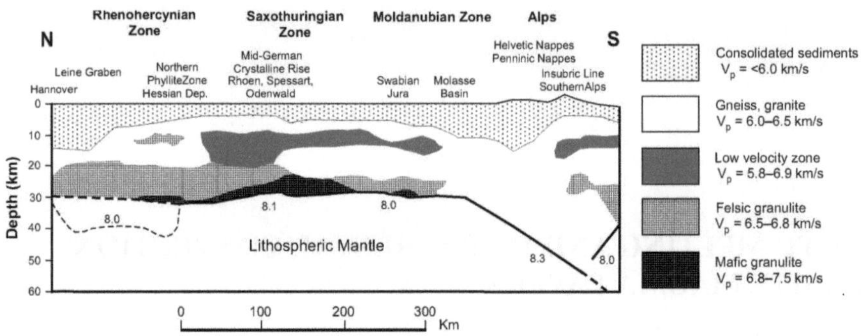

Figure 3.1. Part of a seismic–geological section of the continental crust along the European Geotraverse (EGT) from Hannover (Germany) to the Insubric Line of the southern Alps (Switzerland) (after Fig. 1 of Johannes and Holz 1996 from Fig. 1a of Wedepohl 1996, unpublished data).

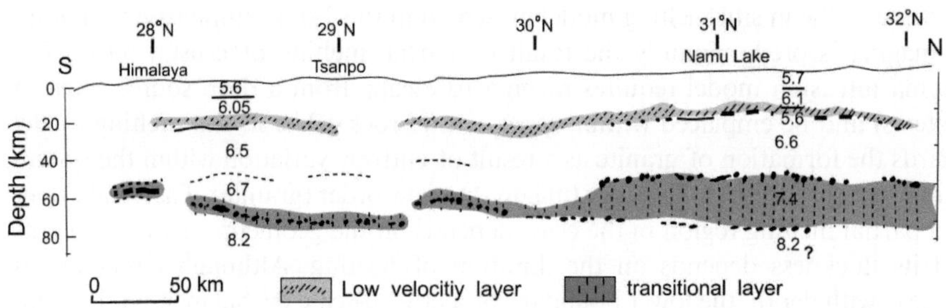

Figure 3.2. N–S cross section showing crustal velocity structure of the Tibet plateau and location of intra-crustal low-velocity zone (after Xiao et al. 1988). The lower 'transition zone' is interpreted to be cooling mafic magma.

crust is hot with the Curie isotherm (~575°C) residing in the upper crust at a depth of ~15 km (Alsdorf and Nelson 1999). A low-velocity zone occurs beneath Tibet at a depth of 20–30 km as shown in Fig. 3.2, and crustal melting at temperatures in the range of ~600–650°C is likely to have occurred across the plateau at depths between ~16 and 18 km under an upper crustal geotherm of ~37°C (Kind et al. 1996; Nelson et al. 1996; Owens and Zandt 1997). Geophysical investigations have also revealed structural analogies between the High Himalayan range and the northern Tibetan crust. In particular, seismic bright spots imaged by the INDEPTH project at 15–18 km depth have generally been interpreted as the culmination of melting to form a partial melt layer (Liu 1984; Xiao et al. 1988; Brown et al. 1996; Chen et al. 1996; Kind et al. 1996; Nelson et al. 1996; Li et al. 2003; Xie et al. 2004; Unsworth et al. 2005).

3.1.1.2. The Andes

The Andes is the largest subduction-controlled mountain belt on Earth with a length of about 7,500 km and in the central part between 16° and 25°S is 800 km wide with elevations greater than 6,000 m. Compressive deformation and uplift of the central Andes occurred in the Miocene and Pliocene and overlaps with a major dacitic ignimbrite eruption phase from caldera complexes, the largest of which is the Altiplano-Puna volcanic complex with a volume of 10^4 km^3 of ignimbrite. The onset of volcanic activity in the Late Miocene (ignimbrite flare-up) reflects a marked change in the thermal state of the crust that is thought to persist today in the form of an anomalously hot, partially molten crust beneath the Altiplano-Puna plateau and Western Cordillera as indicated from wide angle seismic, gravity and electromagnetic data (Schmitz et al. 1997; Chmielowski et al. 1999), and high surface heat flow (Springer and Förster 1998; Springer 1999) (Fig. 3.3). This planar feature is (15–30 km thick and is referred to as the Andean low velocity zone (ALVZ), possibly the largest body of partially molten crustal rock known. The top of the ALVZ occurs at a depth of ~15–25 km and may represent the upper limit of compaction-porous vertical migration of fluids and granite melt that marks the present brittle–ductile transition (Schilling et al. 1997; Schmitz et al. 1997). The bottom of the zone occurs at ~38–46 km depth and may represent the upper boundary of dry, refractory lower crust (Yuan et al. 2000). The average percentage of granitic melt within the ALVZ is estimated to be ~20 vol % (Schilling and Partzsch 2001) giving a minimum volume of melt of some 18,000 km^3.

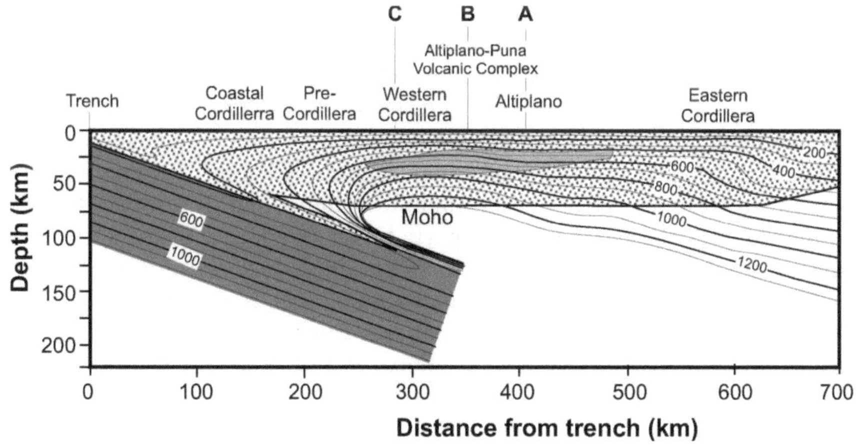

Figure 3.3. Section showing thermal structure across the central Andes of South America (after Springer 1999 and Fig. 9 of Brasse et al. 2002). Superimposed is the Andean low-velocity zone (ALVZ) (after Fig. 2c of Yuan et al. 2000).

3.1.2. *P–T* Conditions of Granite, Migmatite and Granulite Formation

The close association of amphibolite–granulite facies rocks, migmatite and granite that are commonly exposed in ancient orogenic areas strongly suggests that they are related through the process of in situ melting and granite magma retention. Thermobarometry of migmatites and granites worldwide indicates final equilibration conditions mainly within the sillimanite stability field of between 620 and 860°C and 5–28 km. Application of geobarometry to a number of granites and their aureole rocks indicates depths of crystallisation of between 4 and 20 km (Anderson 1996). The *P–T* conditions for migmatites and granites shown in Fig. 3.4A are bounded by 25–100°C/km linear geotherms.

Nair and Chacko (2002) and Pattison et al. (2003) have drawn attention to the probability that *T* estimates of granulite formation and the amphibolite–granulite transition defined by the appearance of orthopyroxene need to be revised upward by over 100°C on the basis of garnet-orthopyroxene thermobarometry that takes into account Al-solubility in Opx, corrected for late Fe–Mg exchange. However, the revised estimates for aluminous, intermediate and mafic granulite compositions given by Pattison et al. (2003), shifts their stability conditions out of the granite–migmatite field shown in Fig. 3.4B. This contrasts with earlier *P–T* fields of granulite equilibration given by Bohlen (1987) and Harley (1989) derived from conventional thermobarometry and the widely used amphibolite–granulite transition that occurs within the higher *T* part of the granite–migmatite field that should be expected if there is a genetic relationship between them. This discrepancy is clearly important and requires issues of reliability of thermodynamic data and its application (e.g. Al-solubility in orthopyroxene-based thermobarometry), whether fluid infiltration was involved, and details of mineral zonation and analyses (e.g. positioning of EPMA analysing points on coexisting minerals used in *P–T* thermobarometry computations) etc., to be addressed. For our purposes, we adopt the earlier estimates of granulite formation because they overlap the *P–T* fields of granite and migmatite genesis and because temperatures of 650–820°C are predicted at depths between ~20–30 km in thermal modelling caused by thickening and associated internal heating in collisional orogens (Chapter 2). The slightly higher pressure estimates from these granulite stability fields is compatible with a restite origin in the basal parts of partial melt layers that evolved to form granite magma layers that may be up to 15 km thick. If temperatures of granulite formation were higher, i.e. ~850 and above, much higher heat flow, melt generation and internal radioactive heat-producing rocks are required and available estimates of temperatures of migmatite and granite formation associated with granulites may need to be re-examined.

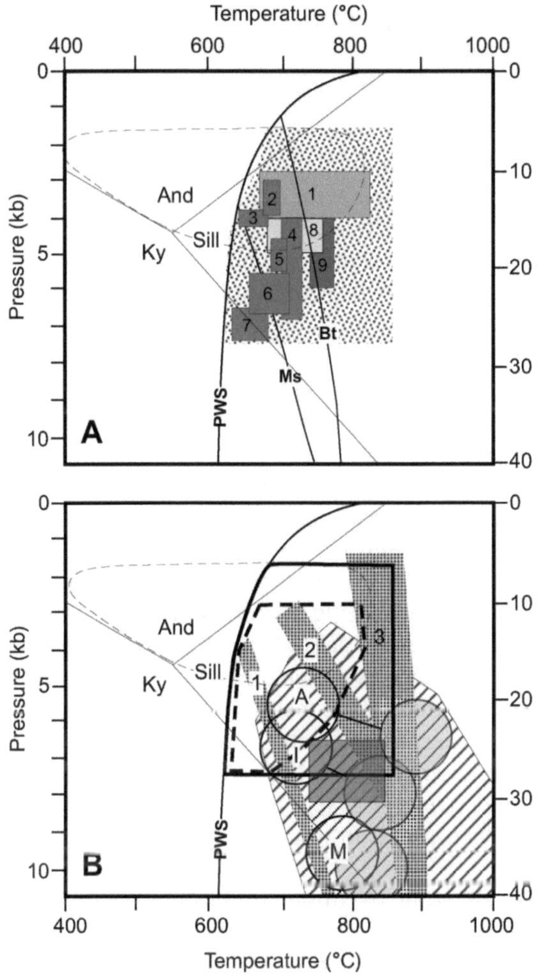

Figure 3.4. *T–P*–depth plots showing stability fields of granites, migmatites and granulites. In **A**, *P–T* estimates of various labelled and numbered migmatite terranes with, **1** = Mt.Stafford (Greenfield et al. 1996, 1998); **2** = Pyrenees (Wickham 1987); **3** = Maine (De Yoreo et al. (1989a,b); **4** = St. Malo, Brittany (Brown 1979); **5** = South Brittany (Jones and Brown 1989); **6** = West-central Massachusetts (Tracy 1978); **7** = Seiver hinterland (Patiño-Douce et al. 1990); **8**. Broken Hill (White et al. 2005); **9**. Calabria (Graessner and Schenk 2001). Fine stippled area = *P–T* field of granites. PWS = pelite wet solidus; Ms and Bt = muscovite and biotite dehydration melting curves, respectively. Al-silicate stability fields after Pattison (1992). High-*T*/low-*P* metamorphic field enclosed by dashed line. In **B**, stability fields of granite and migmatite enclosed by bold solid line and bold dashed line, respectively, compared with *P–T* estimates of granulites; diagonally lined area = granulite field (Harley 1989); grey shaded area = granulite field (Bohlen 1987); Open (uncorrected Fe Mg estimate) and shaded (corrected Fe Mg Al estimate) circles with tie lines = aluminous (**A**), intermediate (**I**) and mafic (**M**) granulites (Pattison et al. 2003). *Shaded strips* = granulite facies limiting melt-producing reactions; **1**. Ms + Qz ± Pl = AS ± Ksp + L; **2**. Bt + Sil + Qz ± Pl = Gt + Cd ± Ksp + L; **3**. Bt + Qz ± Pl = Opx ± Gt ± Cd ± Ksp + L and Hb + Qz ± Gt = Opx + Cpx ± Pl + L. Other symbols as in A. See text.

3.2. CRUSTAL MELTING I: INITIAL MELTING AND PARTIAL MELT LAYER

3.2.1. Formation of a Partial Melt Layer

As stated in Chapter 2, conditions for the production of optimum amounts of granitic melt are controlled by the ambient temperature, protolith fertility, i.e. how close the rock composition is to the granite minimum melt composition and the amount of free water available. The volume of melt produced is mainly determined by the duration and magnitude of the heat supply and by the length of time the rocks remain above their solidus.

Figure 3.5 shows the initial conditions of a partial melting system, in which the thickness of the upper crust is designated as 25 km with a thermal gradient of 30–33°C/km. Rock melting typically begins along grain boundaries of quartz and feldspars. Experimental work by Mehnert et al. (1973) clearly demonstrates rapid development of melt rims up to ~100 μm in thickness around contacting quartz and feldspars within a few days under hydrous conditions (Fig. 3.6). This initial stage of melting involves free water and occurs within the temperature interval between the protolith solidus and dehydration melting of muscovite, or biotite if muscovite is not present, or hornblende if both micas are absent. Under static conditions, the melt will largely remain at the site of melting and movement (flow) is predicted to be very low, as the creep rate of the unmelted matrix is large compared with the flow rate of the melt (Ribe 1987). However, deviatoric stresses would facilitate 'squeezing' the melt from its matrix towards regions of lower stress that are either orientated at a high angle to σ_1 or parallel to σ_3 towards areas of extensional stretching, e.g. as evidenced by cm- to dc-scale leucocratic segregations (leucosomes) in low-melt volume metatexites of migmatite terranes.

During this initial stage of crustal melting, the dominant mechanism of melt-solid separation is probably compaction. A possible corollary of this process is given from liquid phase sintering where it has been observed that after the establishment of a continuous, self-supporting skeleton of solid grains, a cylindrical sample of 1–2 cm in length will, over a period of hours to days at constant temperature, lead to an accumulation of its melt fraction at the top (Niemi and Courtney 1983). For this to occur, the solid skeleton must undergo compaction by a process of surface energy minimization whereby large grains tend to grow at the expense of smaller ones. As coarsening proceeds in the interconnected solid network of a partially molten material, some grains separate and settle a short distance through the intervening liquid. Over time this process leads to densification at the bottom and segregation of liquid in the top of the sample. In addition, coarsening of larger grains at the expense of smaller grains may allow melt to migrate upwards and crystals to settle (Miller et al. 1988).

When temperatures rise to intersect a hydrous mineral dehydration curve, a significant increase in the melt fraction occurs. Experimental melting of mica-bearing

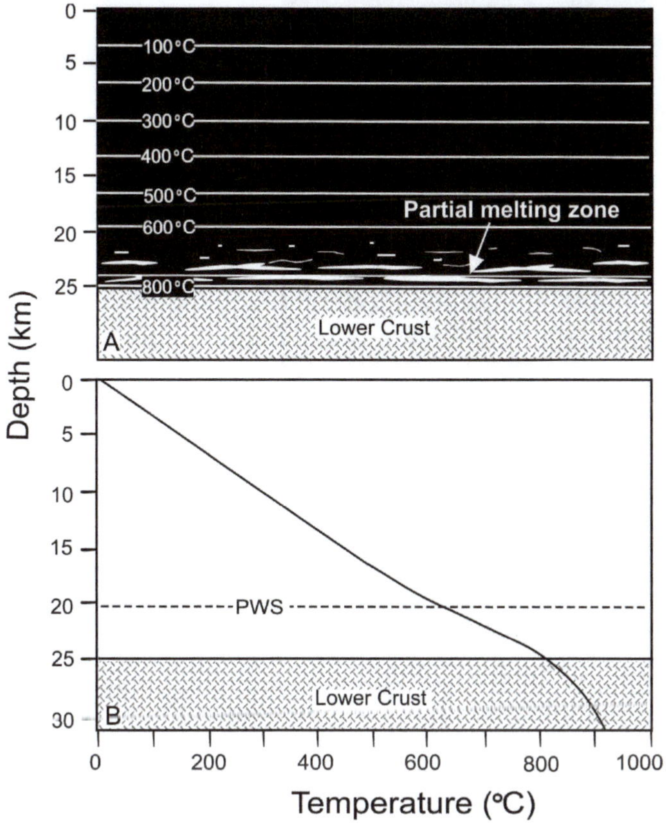

Figure 3.5. Initial stage of partial melting in the lower part of a 25-km-thick upper crust with a geotherm of 29–33°C/km. The lower crust is assumed to have a refractory mafic granulite composition and undergoes little or no melting due to steepening of the geotherm as shown. PWS = Pelite wet solidus. Mica-bearing metasedimentary wet solidi would be intersected at depths of ~21 km/620°C, with muscovite and biotite dehydration melting beginning at ~–22 km/670°C and ~23.5 km/760°C, respectively. Significant melting (>40%) takes place in rocks at the base of the partial melt layer at a temperature of ~800°C.

rocks indicates that there is a much larger volume change associated with dehydration melting of muscovite compared to biotite (Holyoke and Rushmer 2002), so that low volume granitic melt is able to migrate from reaction sites by way of an interconnected network of cataclastic zones and melt-induced fractures. In contrast, the biotite dehydration melting reaction is not associated with a large dilational strain and pore pressure increases more slowly with the melt migrating along grain boundaries causing grain disaggregation. The biotite–dehydration reaction also produces granite melt at a slower rate because of the effect of solid solution (it melts over a greater temperature interval; see Chapter 2) in comparison

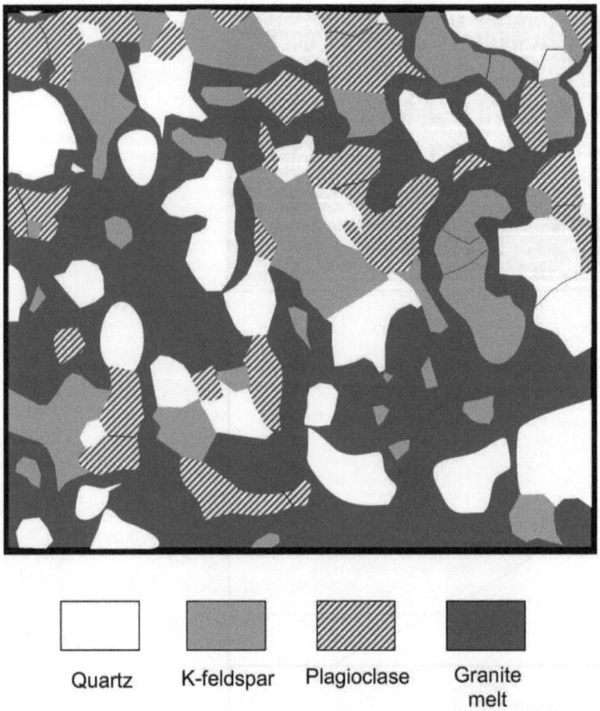

Figure 3.6. Drawing from a photomicrograph showing wet melting of quartz and feldspars in a felsic granulite at 760°C/2 kb/3 days (after Fig. 6 of Mehnert et al. 1973). The amount of melting is advancing into the rock by consuming H_2O from below (the edge of the sample).

to muscovite dehydration melting where a significant amount of melt can be produced rapidly just above the solidus as there is no (or only a very small) solid solution effect.

In a situation where isotherms have stabilised, a partial melt layer with an overall melt volume stratigraphy, such as that shown in Fig. 3.7 will be established. Because of the temperature difference between the top and base of a partial melt layer, a greater melt volume, i.e. >40%, of granite–adamellite–granodiorite composition (from biotite melting) will be generated in the lower part compared to the upper part with perhaps ~10% of leucogranite composition melt produced from muscovite melting, although this will depend on the modal amount of muscovite present.

In Fig. 3.8, *T*–depth sections constructed from Fig. 3.3 through the inferred partial melt layer beneath the Altiplano (A), the Altiplano-Puna volcanic complex (B) where the melt layer is thickest, and the Western Cordillera (C), Andes, are shown in relation to geotherms, a wet mica-bearing quartzofeldspathic solidi, muscovite and biotite dehydration melting curves. In A and B, the wet

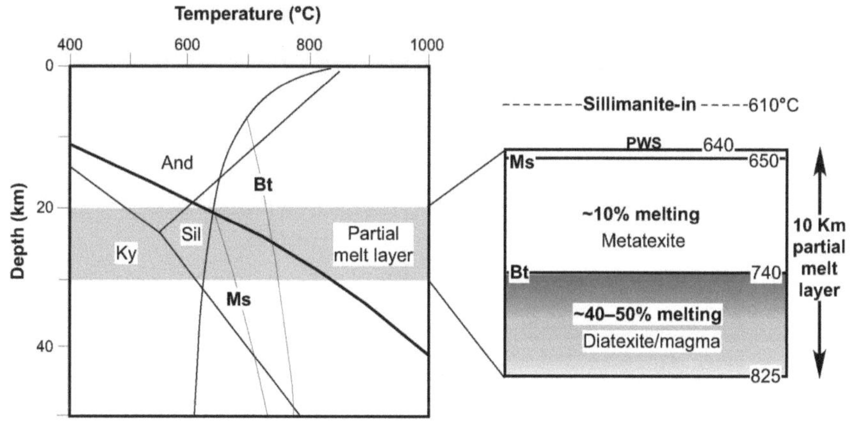

Figure 3.7. Ideal 'stratigraphy' of a partial melt layer developed in a 15-km-thick layer of mid-upper crustal mica-bearing metasedimentary rocks prior to the development of a magma layer, as determined from the relationship between *T*–depth, pelitic wet solidus (PWS), muscovite (Ms), biotite (Bt) dehydration melting and a crustal geotherm of ~28–31°C/km. Al-silicate stability fields after Pattison (1992). See text.

solidi do not correspond with top of the low-velocity zone implying that the upper part may have cooled with exsolution of fluids and the solidus is now at depths of approximately 32 and 25 km, respectively. In B, directly below the Altiplano-Puna volcanic complex, cooling may have started after cessation of dacitic ignimbrite eruptions. Only in C does the top of the partial melt layer correspond to the temperature of the wet solidi. We suggest that this is the youngest part of the partial melt layer that has propagated westwards across the Andean collision zone in response to shifting isotherms related to trenchward migration of mantle melting by asthenospheric upwelling due to steepening of the subduction angle.

3.2.2. Development of a Partial Melt Layer in Heterogeneous Crust

The upper crust is an anisotropic system caused by composition variation and structures of the rocks. It is clear from the experimental evidence presented in Chapter 2 that interlayered rocks of different compositions within areas undergoing partial melting are capable of generating variable amounts of granite melt at the same *T*–*P* (Figs. 2.14 and 2.16) with local rheologies exerting a strong control on the effectiveness of melts to separate and accumulate. In Fig. 3.9A, B, combined *P*–*T*–depth–distance–structure sections are used to illustrate melting relationships in a sequence of deformed rocks of different compositions with a linear thermal gradient of 40°C/km. In the depth–distance–temperature cross

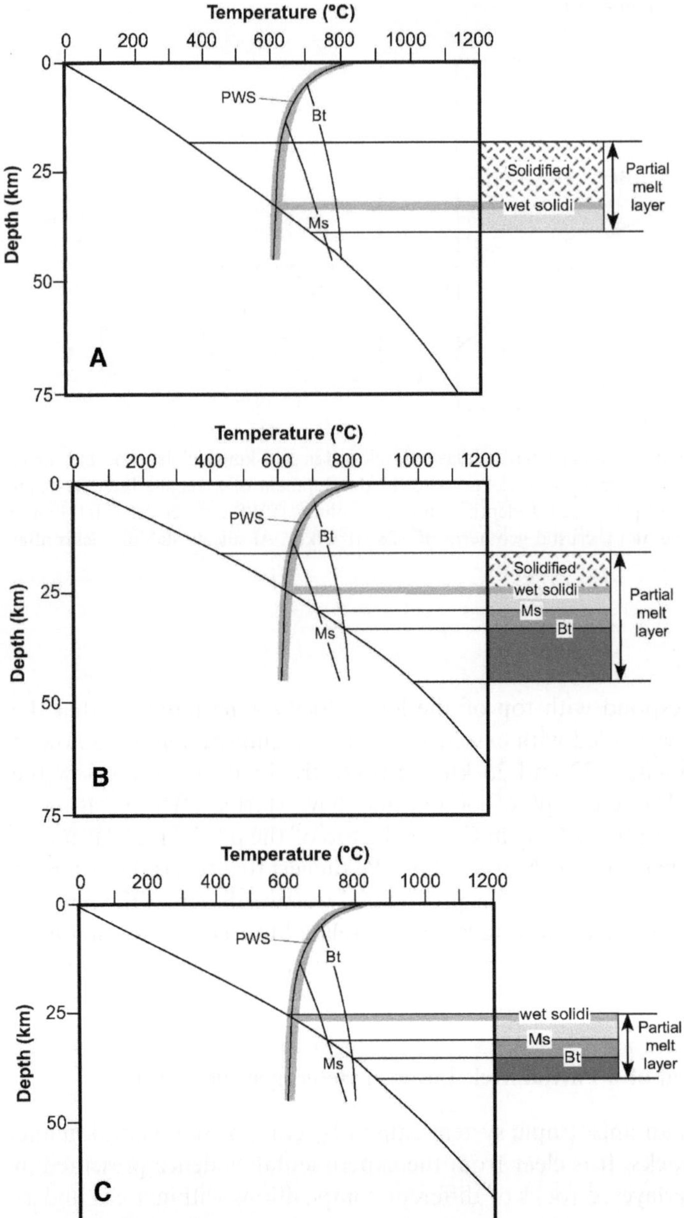

Figure 3.8. Interpretations of the partial melt layer 'stratigraphy' of the Andean low-velocity zone (ALVZ) based on geotherm structure below positions A, B, C in Fig. 3.3, and position of wet solidi of mica-bearing metasedimentary rocks (thick grey line with pelite wet solidus [PWS] curve shown as black line), muscovite (Ms) and Bt (Bt) dehydration melting curves. Different thicknesses and depth of the inferred partial melt layer are those of the ALVZ shown in Fig. 3.3. See text.

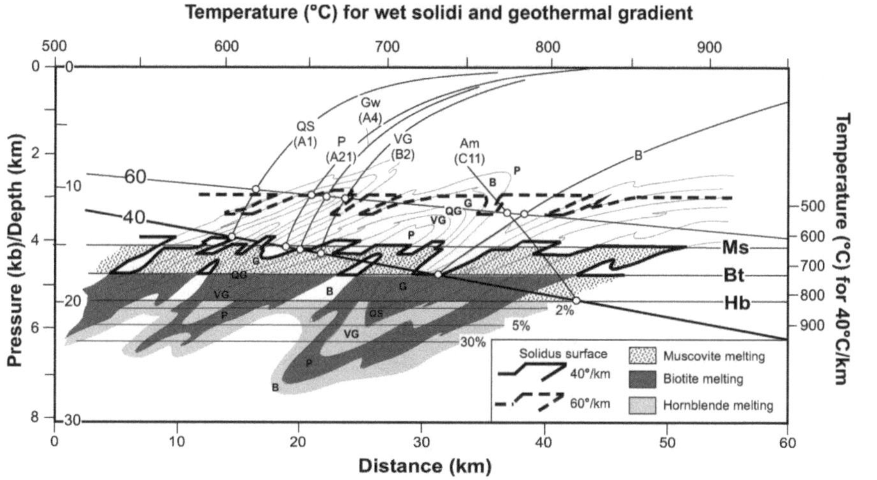

Figure 3.9. A. Combined plot of *P–T*–depth and *T*(40°C/km)–depth–distance (cross section) showing wet solidus curves for folded inter-layered lithologies, their intersection positions with respect to 60 and 40°C/km linear thermal gradients and cross-section form combined solidus surfaces for conditions of 60°C/km (dashed line) and 40°C/km (solid line). Solidus curve numbers refer to experimental rock compositions listed in Appendix 2A–C. Ms, Bt, Hb = positions of muscovite, biotite, hornblende dehydration melting temperature isotherms, respectively, for 40°C/km thermal gradient. Shaded areas denote regions where muscovite, biotite and hornblende melting occur in relation to rock composition and structure. 2, 5 and 30% lines indicate melt volumes in basaltic amphibolite. Lithologies of the folded metamorphic sequence are: QS = Quartz–plagioclase-rich layer of amphibolite grade schist; P = Pelitic muscovite-rich layer of amphibolite grade schist; Gw = greywacke metasandstone; VG = Volcanogenic greywacke; Am = Amphibolite; B = Basalt (metagabbro) (Lambert & Wyllie 1972). Note that with the linear thermal gradient of 40°C/km used in the construction rocks, at ~27 km depth the temperature would be 1,000°C, which is unlikely. See text.

section, isotherms are assumed to be horizontal and distance and depth scales are the same. In Fig. 3.9A, the protoliths are represented by five wet solidi and one H_2O-absent melting curve, and in Fig. 3.9B by seven H_2O-absent melting curves. Intersection of the solidus/reaction curves and the 40°C/km geotherm determines the depth–temperature position of the respective rock composition melting. The solidus surface marks the top of an extensive partial melt layer and is highly irregular with maximum amplitudes up to 3.5 km in A and up to 6 km in B with a skewed 'wave pattern' that reflects the inclined fold axes of the rocks. For comparison, the form of the combined solidi is also shown for a higher thermal gradient of 60°C/km in A. The base of the partial melt layer is not represented but can be assumed to be the depth limit of the folded sequence of 'fertile' lithologies shown.

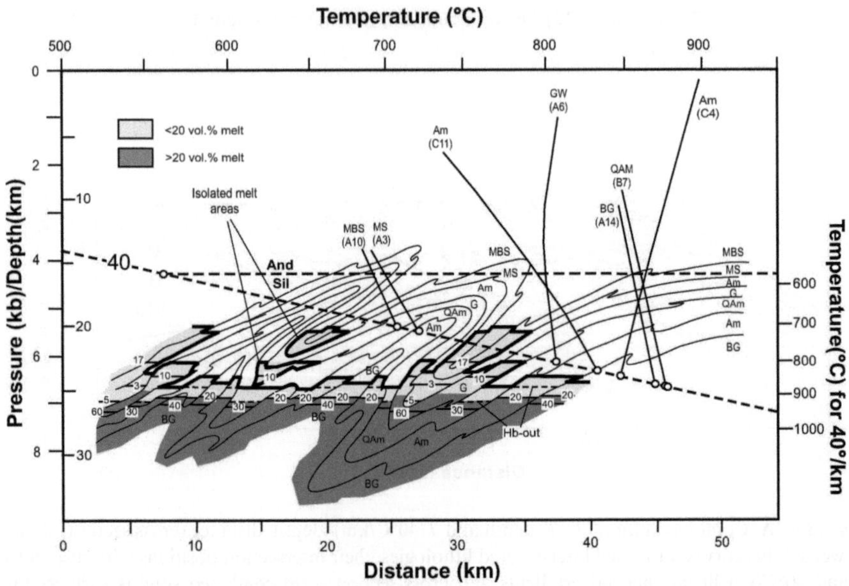

Figure 3.9. (*Continued*) **B**. Combined plot of *P–T*–depth and *T*(40°C/km)–depth–distance (cross section) showing dry solidus curves for folded inter-layered lithologies, their intersection positions with respect to a 40°C/km linear thermal gradient and cross-section form combined solidus surfaces (solid line). Numbered horizontal lines = vol. % melt in different lithologies. Horizontal dashed line = *P–T*–depth position of andalusite–sillimanite transition. In descending order the lithologies of the folded metamorphic sequence are: MBS = muscovite–biotite schist; MS = muscovite schist; Am = amphibolite; G = greywacke schist; QAm = quartz amphibolite; Am = amphibolite; BG = biotite gneiss. Sample numbers refer to bulk compositions described in Appendix 2A and C. Note that with the linear thermal gradient of 40°C/km used in the construction, rocks at ~27-km depth would be at 1,000°C, which is considered unlikely. See text.

Several important features illustrated by the diagrams are described below:

1. In contrast to the irregular form of the solidus, isotherm surfaces for dehydration melting of muscovite, biotite and hornblende (= dry amphibolite solidus) located at depths of ~15, 17.5 and 19.5 km, respectively, are horizontal in A, but not in B where these dehydration melting reactions represent the dry solidus of the various rock compositions.
2. Rocks in the *T*–depth interval between the muscovite and biotite dehydration isotherms in A can be regarded as metatexites and those below the biotite dehydration isotherm as diatexites. Maximum melt production (40–60%) in biotite-bearing lithologies would be expected to occur between ~850 and 950°C (Fig. 2.14) at ~20.5–23 km depth. Hornblende dehydration melting begins at ~825°C with 2, 5 and 30% melting at the 850, 900 and ~960°C isotherm depths, respectively.
3. Two 'rootless' areas of rock (greywacke and muscovite-biotite + muscovite schist) with only a small amount of melt (<5%) occur above the solidus surface in B and are an artefact of the fold pattern.

4. Variable mineral mode, grain size, texture, melt volume and H_2O content will result in significant rheological differences between adjoining rock units.
5. Melts in adjacent rock layers will be compositionally different and if they remain in situ will evolve differently due to variation in mineral mode and composition, H_2O-content and fO_2 at the same *P* and *T*. Compositions of experimental melts corresponding to the solidi and melting reactions used in the construction of Fig. 3.9B are plotted in Fig. 3.10 to illustrate the range of

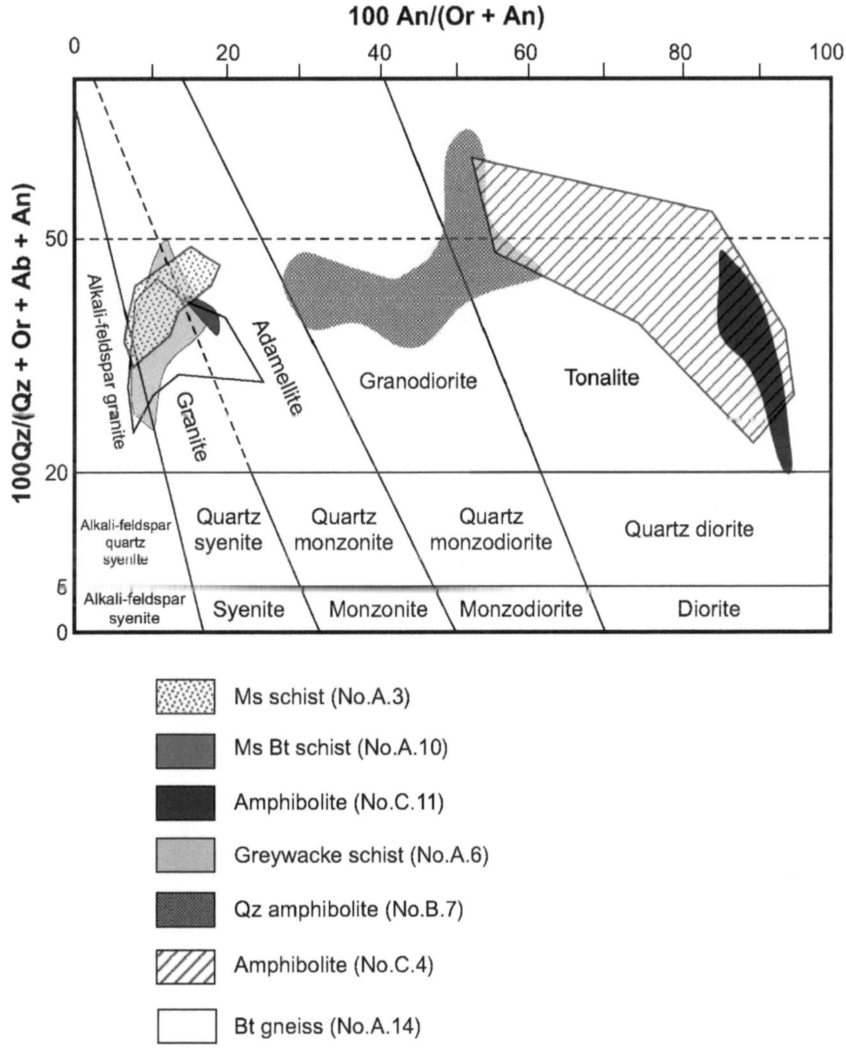

Figure 3.10. Normative Qz, Or, Ab An ratio plot of experimental melt compositions produced from partially melted lithologies in Fig. 3.9B. Sample numbers refer to rock compositions described in Appendix 2A and C.

melt compositions that could be expected in nature from partial melting of such a heterogeneous protolith. Clearly, this has important implications for inferences of magma origin based on chemical composition and isotopic characterisation. In this respect, it is pertinent to note that experimental results of Skjerlie et al. (1993) and Skjerlie and Patiño-Douce (1995) indicate the possibility of substantial elemental exchange between layered rocks undergoing dehydration melting that influences melt fraction and melt composition.

6. The layers of different composition and melt volumes will be subjected to different stress, with more competent layers supporting higher differential stress, if subjected to ductile deformation. This will create strain rate gradients, and as a consequence, pressure gradients that will drive (suck) the melt into low pressure dilatant regions such as boudin and shear zones and pressure shadows that develop in the deforming matrix of partially melted rock layers. In this way, it would be possible for melt from one rock layer to move into, and mix with, melt in an adjacent, less competent layer during this early stage of partial melting.
7. Given appropriate conditions for melt accumulation, an initially heterogeneous distribution of magma bodies of variable composition would result. With time, these separate magmas would likely combine and mix to form a more homogeneous, volumetrically larger body that would effectively become a magma layer.

3.3. CRUSTAL MELTING II: CONVECTION AND FORMATION OF MAGMA LAYER

3.3.1. Gravitational Separation and Formation of Magma Layer

During melting, the solid or restite framework fraction of a protolith will preserve an interconnected matrix until the volume of melt attains some critical value (the critical melt fraction or CMF) at which point the solid matrix becomes disaggregated (Arzi 1978; van der Molen and Paterson 1979) as evidenced in diatexites from migmatite terranes (Fig. 3.11). On disaggregation with further melting, restites will remain as isolated units within a melt-supported system and gravitational separation within the partial melt layer will occur because of the density difference between restite and melt. Once this begins, a restite-depleted magma layer can be expected to form as shown in Fig. 3.12.

3.3.2. Convection and Development of Magma Layer

Given the condition that the separated melt has a high melt-fraction relative to that of suspended solids (dilute suspension) and low strain rate, it will approximate an ideal Newtonian fluid (constant apparent viscosity independent of strain rate). In a uniform layer of incompressible homogeneous Newtonian fluid, the tendency for thermal convection to occur depends on the non-dimensional Rayleigh number (Ra) as given by

$$\mathrm{Ra} = (L^4 \, \alpha T^g \, \beta)/\eta K \tag{3.1}$$

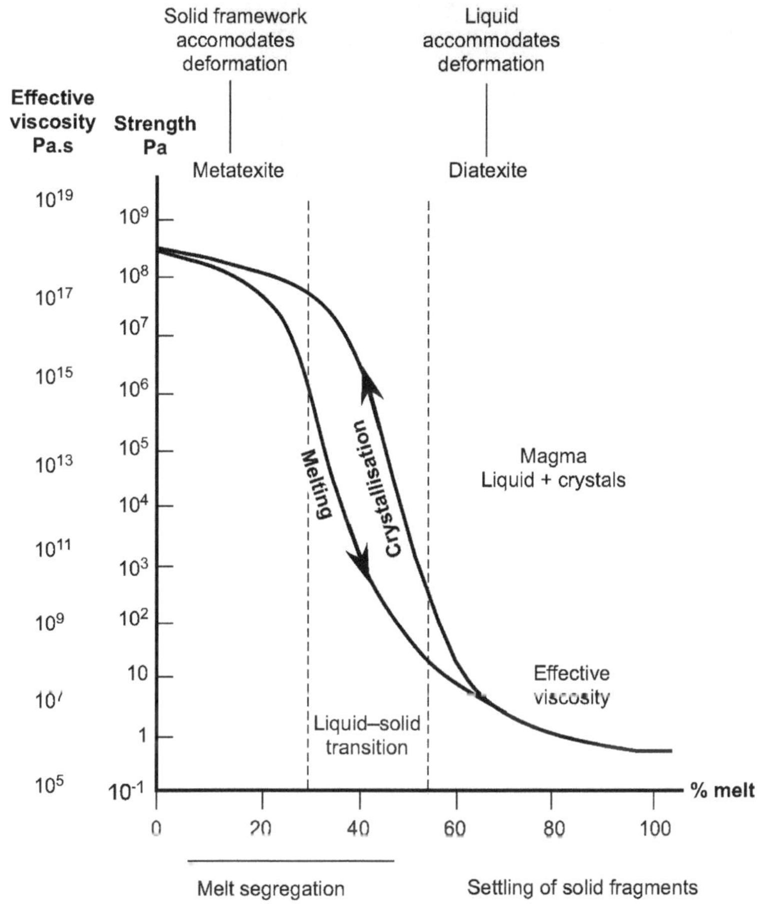

Figure 3.11. Diagram showing the behaviour of partially molten rock as a function of its liquid–solid proportions (after Fig. 1 of Vanderhaeghe 2001). Values on horizontal axis indicate geometrical thresholds of melt connectivity (controls the ability of the melt to separate from the solid framework) and continuity of the solid framework of the rock (controls melt/magma mobility and settling of solids in magma). Metatexites are solid dominated; diatexites are melt dominated.

where the driving force and destabilizing factors (buoyancy forces) are L = height of the layer in cm, αT = coefficient of thermal expansion, and the retarding factors (viscous resistance) are, g = constant of gravitational acceleration (980 cm/s), β = vertical temperature gradient in K/cm, η = kinematic viscosity (η/ρ), K = thermal conductivity of the magma in cal/gm K, and ρ = fluid density. Convection occurs when the Rayleigh number exceeds a critical value (Ra_c), which depends on the boundary conditions. If the top and bottom boundaries (isothermal

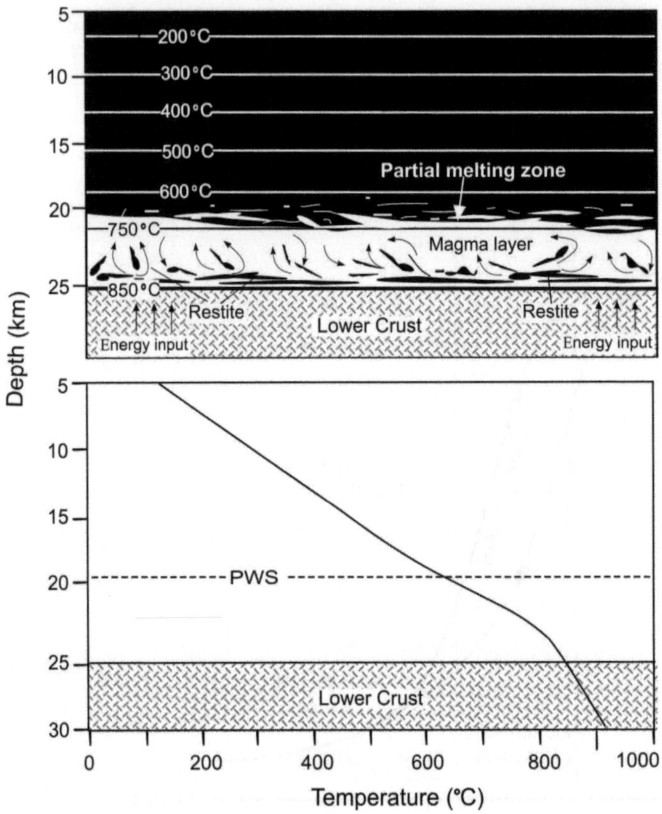

Figure 3.12. Diagram showing early-stage establishment of a granite magma layer and *T*-profile, with gravitational sinking of fragments of restitic roof rock in the upper part of the partial melting zone. At the same time, upward displacement of buoyant granite magma occurs that eventually results in the development of a compositionally stratified mafic (lower) to felsic (upper) magma layer within the partial melt zone.

planes) of the melt layer are free (movement can occur along them) but stationary, convection starts when the Ra > 657. If one boundary is free and the other is fixed (no movement along it), the Ra must exceed 1,700 for convection to occur. If both boundaries are fixed, $Ra_c = 1,708$. For Ra < 1,000, transfer of heat is predominantly by conduction. Steady convective heat transfer starts at ~Ra > 10,000, and strong eddying motion occurs when Ra = 100,000. Bodies with thickness > 10 m are likely to convect if the heat flux is significantly high and their heat loss is 10^{-5} to 10^{-3} cal/cm^2 s. Obviously, the thicker the magma layer and the lower its viscosity, the more likely that the convection will occur and for efficient heat dissipation to occur.

Gravitational differentiation of restites within a partial melting region not only results in formation of a magma layer (Fig. 3.12), but could also help to

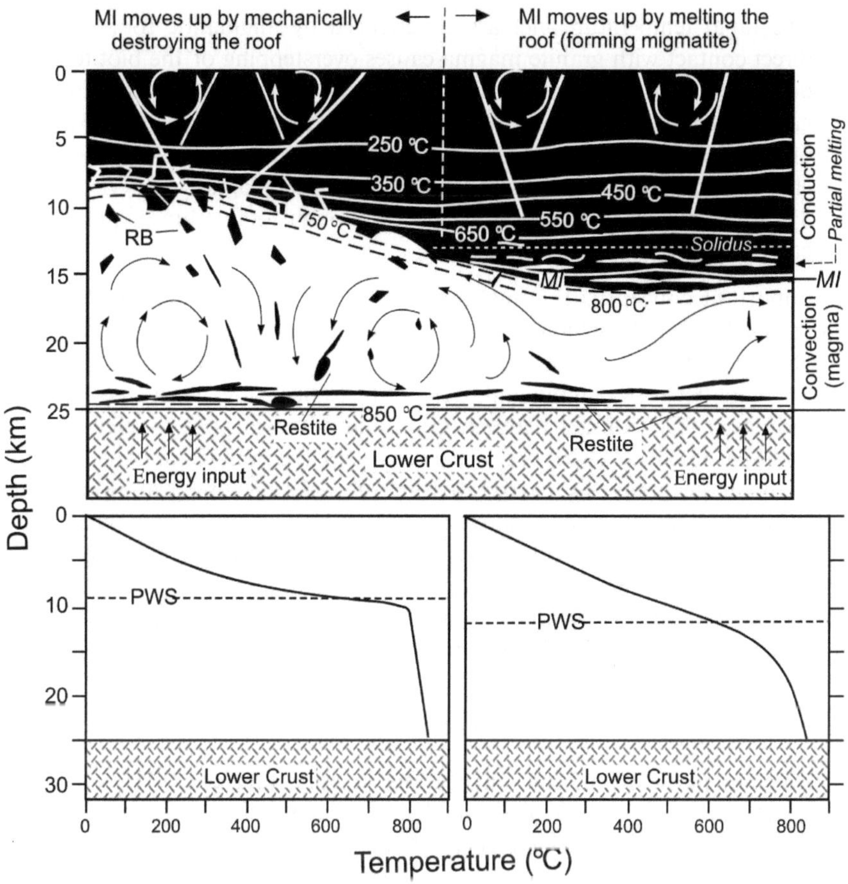

Figure 3.13. The intra-crustal convecting granite magma layer and corresponding *T*–depth profiles. Convection advects heat upwards within the magma layer and temperature variation of the convection system tends to become more homogenous (left). The MI defines the upper limit of the convecting magma layer and divides the convection (below MI) and conduction (above MI) heat flow modes. New migmatite is formed in the heat conduction zone (partial melting zone) over the MI (right). On the left, mechanical destruction of roof rock by the magma results in the MI reaching shallower levels than the original solidus of the roof rocks and thus coming into contact with low grade metamorphic rocks. Arrows in the crust overlying the MI indicate circulation of meteoric water.

initiate convection within the layer that immediately changes the thermal structure of the system (Fig. 3.13):

1. Temperature variation within the magma layer decreases because of convection.
2. The temperature gradient between the magma layer and its partially melted roof (the upper metatexite part of the original partial melting zone in which there is no convection) steepens markedly because of upward advection of heat by the convecting magma (Fig. 3.13);

(i) The rapid temperature increase experienced by migmatitic roof rocks in direct contact with granite magma causes overstepping of the biotite dehydration melting temperature and initiates extensive melting, with disaggregation allowing upward movement of the magma layer (Fig. 3.13). This process may be analogous to that outlined by Ahern et al. (1981) who modelled the upward migration of a 5 km thick granite magma layer (viscosity = 3×10^6 poise) that melts its way upward though some 5 km of pre-heated crust in only 45,000 years.

An important concept is the 'melting interface' (Chen and Grapes 2003) or 'magma interface' (MI), which refers to the upper surface of a crustal magma layer, physically defined as the upper limit of a crustal convection layer, marking the contact between magma and rock, and two heat flow modes, convection and conduction. The contact between granite and its wall rock is referred to a 'paleo-magma interface' (PMI, see Chapter 4). In the convection zone (below the MI), melts generated by partial melting of rocks become granite magma, whereas in the conduction zone (above the MI), melts generated by partial melting of rocks are preserved within the rocks and constitute the leucosomes of migmatite. In other words, thermal convection within a crustal partial melting region is essential for formation of granite magma. Without convection, partial melting generates migmatite, but not magma that forms granite batholiths.

3.3.3. Upward Thickening of Magma Layer

With constant input of thermal energy, the magma interface (MI) will move upward accompanied by increasing thickness of the convecting magma layer. Two modes for upward movement of the MI can be expected:

1. By melting of roof rock. This may occur where fractures in roof rocks are rare, in rocks below the brittle–ductile transition zone with melting dependent on rising isotherms. A relatively slow upward rate of movement of the MI leads to a longer period of roof rock heating, with partial melting taking place within rocks located between the MI and the solidus to form migmatite as shown in Fig. 3.13 (right). Most granite–migmatite terranes, such as those described in Chapter 4, are examples where there is an inferred upward/outward change from granite through diatexite, metatexite to unmelted amphibolite and greenschist facies grade metamorphosed rocks reflecting a gradual temperature decrease and macrostructural change overlying the MI (Fig. 3.7).
2. By mechanical destruction of roof rock (Fig. 3.13 [left]). Magma in the convection layer can intrude fissures/cracks/joints or along planer fabrics of the roof rocks as indicated by networks of granite-filled veins (Fig. 3.14A), which destroy the continuity of the roof and results in the formation of rock blocks (xenoliths) (Fig. 3.14B). With dislodgement of blocks of roof rock into underlying magma, the former positions of the blocks are immediately filled by

magma. In this way, the MI moves upwards and in many cases, it may reach a shallower depth than the solidus that defined the upper surface of the initial partial melt layer (Fig. 3.13). This could explain why many granite bodies are in contact with low grade (greenschist facies) wall/roof rocks (see Chapter 4).

Upward movement of the MI by mechanical destruction of roof rock is similar to magma 'stoping' (Daly 1914), i.e. thermal shattering induced by a hot magma coming into contact with relatively cold roof/wall rocks with invasion of fractures by the magma and consequent sinking of the shattered blocks. However, such stoping is limited by the thermal energy and 'room problem' of ascending magma into relatively 'cold' crust. In the case of in situ melting, the magma does not 'intrude' in the same sense, thus neither the energy consumption of the system nor the volume change of the magma resulting from addition of blocks of roof rock need to be taken into account. The upward motion of the MI is caused entirely by rising regional isotherms and thermal convection within the magma layer, which increase its thickness upwards. Intrusion of magma into fractures within unmelted hot roof rock is associated only with the uppermost part (boundary zone) of the magma layer and the mechanical destruction of roof rocks by the magma is only one aspect of the upward moving MI, which is the result, but not the cause of increasing crustal temperature.

The depth level that the upward moving MI can reach depends on energy input/output of the system. With constant input of thermal energy, the MI will keep moving up until it intersects the deepest extent of meteoric water circulation that will rapidly remove heat from the system and halt further upward movement of the MI.

3.4. COMPOSITIONAL VARIATION WITHIN MAGMA LAYER

When rock blocks (i.e. xenoliths in the crystallised magma) separate from the roof into the magma layer, they will be rapidly heated and partial melting will occur inwards from their surfaces (Fig. 3.14C). Because of their greater density, rock blocks and their restites sink gravitationally within the granite magma. Some of them may be almost completely melted during sinking because of the downward increasing temperature of the magma system, and others may reach the bottom of magma layer because of their more refractory (e.g. mafic) composition. Both cases result in the magma layer becoming compositionally stratified with a more mafic (granodiorite–tonalite–diorite) lower part and a more felsic (granitic) upper part due to upward displacement of buoyant felsic magma (see Chapter 5).

3.5. MAGMA LAYER, GRANITE LAYER AND GRANITE BODIES

A convecting magma layer that develops from the original mid-upper crustal partial melt zone may extend over an area of ~10^3–10^5 km^2 or even more as suggested from the intra-crustal low-velocity layers below the Andes and Tibet

Figure 3.14. Geological observations related to mechanical disruption of roof rocks of magma layer. **A**. Network of granite dykes before the fragmentation of roof rock. Tunnel floor exposure in the Bieshan granite, Guangzhou, Guangdong Province, SE China; **B**. Tunnel wall exposure of xenoliths in the Baimang granite, SE China, showing the destruction of roof rock by granite magma; **C**. A rock block that has undergone extensive partial melting to become a mafic enclave and with a dark reaction rim in contact with the granite, Baimang granite, SE China (see Chapter 4).

plateaus. The floor of this magma layer is determined by the lower limit of fertile rocks involved in anatexis and is assumed to be dry refractory mafic granulite rocks of the lower crust as shown in Fig. 3.12. Crystallisation of the magma produces an extensive layer of granite. Actually, as early as the 1980s, some geoscientists drew attention to the possible correspondence between granite belts and ancient intra-crustal low-velocity zones (Meyerhoff et al. 1996) and over the last 20 years geophysical investigations have shown that an increasing number of granite batholiths have a tabular form with thicknesses of typically less than 10 km (Vigneresse 1990; Petford and Clemens 2000; Petford et al. 2000).

The MI represents the upper limit of a convecting magma layer. In the unlikely case of a compositionally homogeneous upper crust, the MI would represent a particular isotherm. However, in layered rocks of variable composition, structure and rheology, an uneven rate of upward thickening of the magma layer would produce an undulating MI, similar in form to the undulating solidi surface shown in Fig. 3.9A, B. Thus, exposures of individual granite bodies may simply reflect the relationship of the erosion surface intersecting a series of dome-like structures (protuberances) developed in the roof of an extensive subsurface granite layer. This would explain the near-circular map plan shapes of many smaller granite bodies worldwide. Distortion of such domed structures will occur if their development was accompanied by regional deformation as shown in Fig. 3.15, of which there are abundant examples (see Chapter 4).

3.6. MI FLUCTUATION (REMELTING) AND GRANITE SEQUENCE

The position of the MI within the crust depends on the ratio of thermal energy input/output of the system. When the ratio >1, the MI will potentially move upwards (melting) (Fig. 3.16A). Where the ratio = 1 the MI will remain at a certain level, and where the ratio is <1, the MI will descend with crystallisation of magma layer from the top downward (Fig. 3.16B). During downward motion of the MI, if thermal energy input of the system increases (ratio of energy input/output >1), the MI will again move upwards by re-melting previously crystallised granite (Fig. 3.16C). If this younger MI rises to a higher level than that of the older one (PMI in Fig. 3.16C), the earlier granite will disappear and only evidence of the position of the younger MI is preserved. The variation of MI is therefore analogous to the fluctuating interface between water and ice cover of a lake with varying temperature, and the result could be a sequence of downward-younging granites (Fig. 3.16D) such as apparent in individual composite granite bodies (see Chapter 4) and the regional Mesozoic granite belts in SE China (see Chapter 7).

3.7. CONCLUSION

Extensive partial melting of rocks, as indicated by intracrustal low velocity layers in many young and old orogenic belts, may occur at mid-upper crust depths. Convection initiated from gravitational sinking of disaggregated restites within the partial melting region results in formation of a granite magma layer

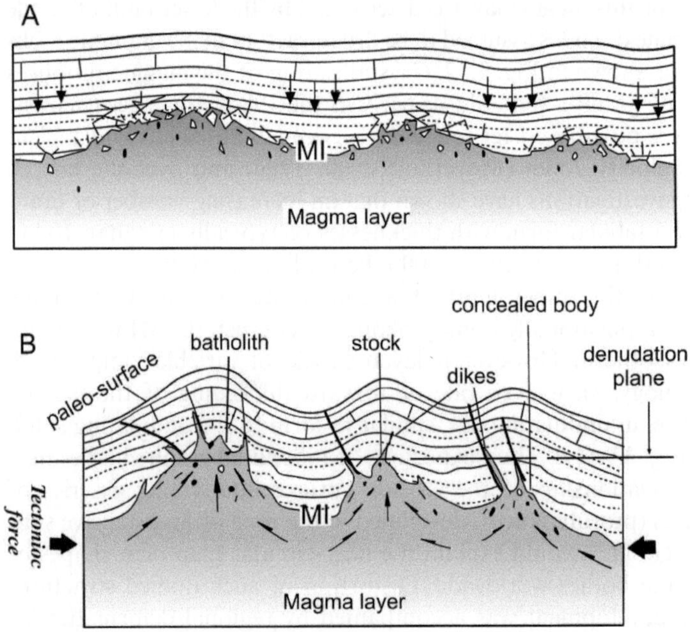

Figure 3.15. Diagrams showing the development of highs or domes in the upper surface of a granite magma layer. **A**. Assumed original shape of the MI, lateral variations in lithostatic load cause vertical deformation (vertical arrows) of rock overlying the magma layer; **B**. Deformation of the MI by lateral compression with magma (arrows) migrating to areas of lower pressure, causing updoming of overlying rocks and injection of magma along faults/fractures developed in roof rocks to form dykes. Note that isolated exposures of granite can be explained by intersection of the undulatory MI of the granite layer and the horizontal denudation plane.

that advects heat upwards by convection causing further melting of roof rocks. The MI determines the upper limit of a crustal convection layer, marking the contact between magma and rock, and two heat flow modes, convection and conduction. Granite is formed in the convection zone below the MI and migmatite is formed in the conduction zone above the MI. Without convection, partial melting generates migmatite but not magma that forms granite batholiths. Upward movement of the MI (upward thickening of the magma layer) occurs either by melting roof rocks that may produce new migmatite in the conduction zone, or by mechanical disruption of the roof that may result in the MI reaching shallower depths than the original solidus of the roof rocks as evidenced by granite in contact with relatively low grade, e.g. greenschist facies rocks.

Crystallisation of a convecting magma layer produces a granite layer, which suggests that individual granite bodies represent the exposures of an irregular upper surface (PMI) of the extensive granite layer, and their shape and size

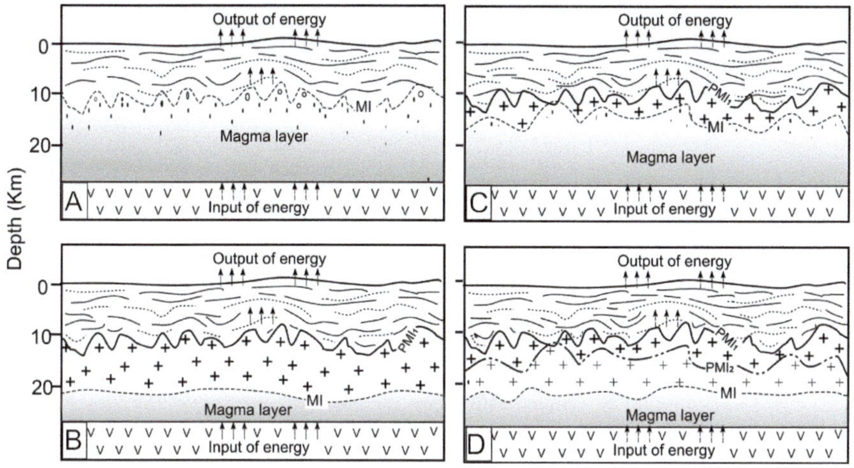

Figure 3.16. Diagrams to explain the formation of a downward younging granite sequence (after Chen et al. 2003). **A**. Energy input > output: isotherms and the MI rise; **B**. Energy input < output: isotherms and MI move downwards and upper part of magma layer crystallises. The paleo-magma interface (PMI) represents the original position of the MI; **C**. Renewed heating again results in energy input > output: the MI moves upwards, causing re-melting of older solidified granite. Only where the final level reached by the MI is lower than that of the PMI will evidence of the older granite be preserved; **D**. With cooling, downward movement of the MI results in formation of a younger granite layer below the older granite.

simply reflect the geometric relationship between the PMI and erosion surface (see Chapter 4). Variation in the position of the MI is related to the ratio of energy input/output of the convecting magma layer. Fluctuation of position of the MI involving remelting of the earlier-formed granite results in a downward-younging granite sequence.

CHAPTER 4

GEOLOGICAL EVIDENCE FOR IN SITU MELTING ORIGIN OF GRANITE LAYERS

4.1. MIGMATITE TO GRANITE

The magma interface (MI) divides regions of heat convection and conduction within the crust. Migmatites form the initial partial melt layer shown in Fig. 3.7 before establishment of a magma layer, and can also be produced in the conduction zone above the MI during subsequent upward thickening of the magma layer (Fig. 3.13). The commonly observed diatexite transition between lower granite and overlying metatexite migmatite is a zone of high melt-content that may preserve evidence of flow, possibly by sluggish convection, and can therefore be regarded as lying below the MI. Migmatite to granite sequences have been mapped in many areas and examples are described below that illustrate the complexities and variations of crustal melting.

4.1.1. Thor–Odin Dome, Canada

The Thor–Odin dome area is part of the Shuswap metamorphic complex, British Columbia, Canada, and exposes migmatite–granite of the Cordilleran orogen (Vanerhaeghe et al. 1999; Norlander et al. 2002) (Fig. 4.1). Pelitic (Sill + Ksp) migmatite (metatexite) with leucosomes and discontinuous layers of pelitic schist, calc-silicate, marble, amphibolite, quartzite are permeated by networks of pegmatitic and aplitic granite sills and dykes. Leucogranite and leucosomes in the migmatite indicate ~15–20% and locally >40% melting. The metatexite grades downwards over a few hundred metres into diatexite–heterogeneous granite that we interpret as the upper part of a magma layer. The occurrence of gedrite–cordierite rocks in boudinaged layers within metatexite and diatexite–granite (Fig. 4.1) suggests an upward movement of isotherms and MI causing melting of metatexite and remelting of leucogranite.

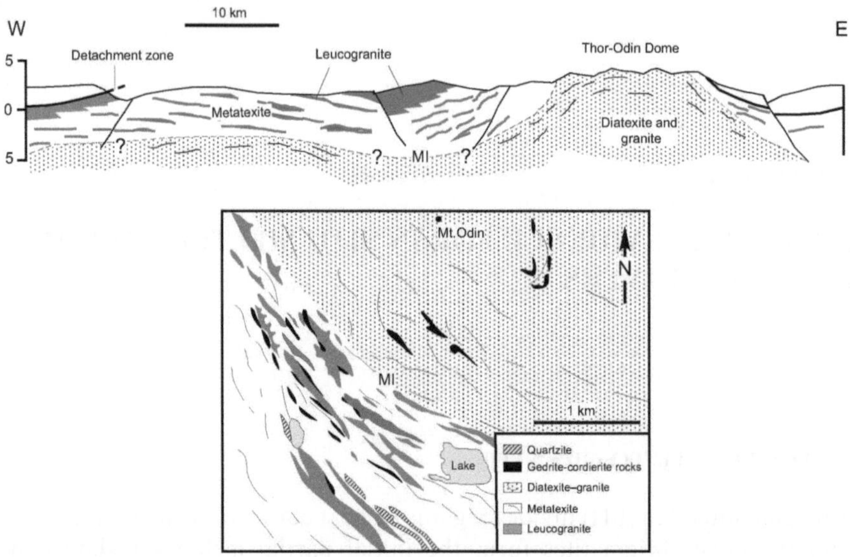

Figure 4.1. Above. Cross section across the Thor–Odin dome, Shuswap metamorphic core complex, Canadian Cordillera (see Fig. 2 of Norlander et al. 2002 for location), showing relationship between lower diatexite–granite and overlying metatexite with a network of leucogranite sills, dykes and larger bodies (laccoliths) beneath a low angle detachment fault (modified after Fig. 2c of Norlander et al. 2002). MI = inferred magma interface. *Below.* Geological map of part of the Thor–Odin dome centred on the area around Mt. Odin showing relationships between diatexite–granite, metatexite + leucogranite and distribution of boudins of gedrite–cordierite and garnet amphibolite (modified after Fig. 3 of Norlander et al. 2002). See text.

4.1.2. Broken Hill, Australia

The transition from metatexite to diatexite in a single rock type over a scale of tens to hundreds of metres is exposed at Broken Hill, New South Wales, Australia (White et al. 2005). All rocks contain the assemblage, Qz, Pl, Bt, Gt, Ilm, ± Ksp, ± Ms, but the mineral proportions and texture differ for each migmatite type. Metatexites contain cm-scale segregation of melt into leucosomes to form stromatic migmatite. The metatexite grades through schlieren diatexite characterised by disaggregation of the rocks to form schlieren migmatite, to massive diatexite/granite (interpreted as a proto-magma layer) that lacks schlieren (Fig. 4.2) and contains phenocrysts of plagioclase and K-feldspar. This transition records progressive disaggregation of the rock and is thought to reflect an influx of hydrous fluid at temperatures just above the solidus of the diatexite. The fluid is inferred to have been derived from nearby metapelitic rocks with higher temperature solidi undergoing sub-solidus dehydration reactions involving muscovite, chlorite and possibly staurolite and epidote. Continued heating during and after the influx of fluid resulted in melt contents of up to 60 mol % in the massive diatexite. The Broken Hill example

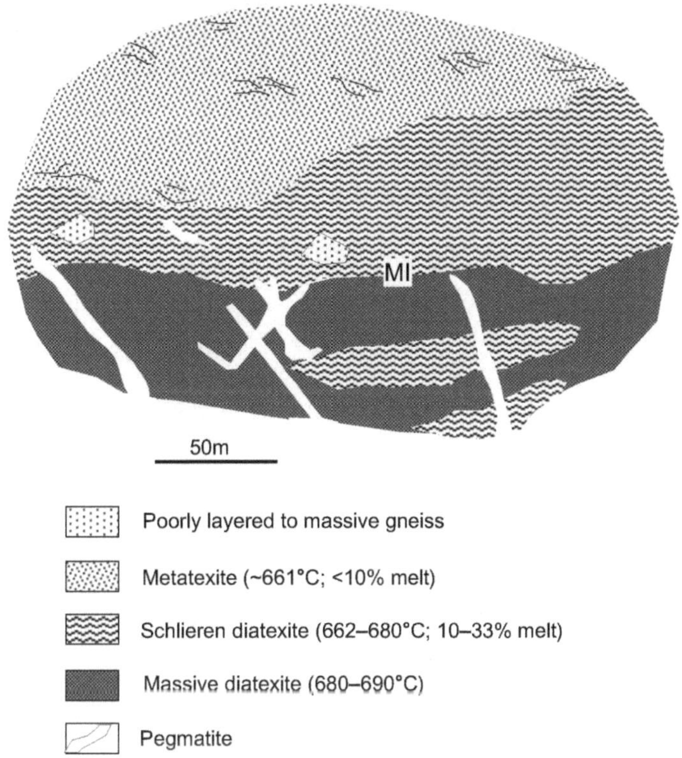

Figure 4.2. Simplified geological map showing an example of a metatexite–diatexite–granite transition at Broken Hill, New South Wales, Australia (redrawn from Fig. 2a of White et al. 2005). MI = inferred magma interface. See text.

shows that an influx of fluid at temperatures just above the wet solidus of a partially melted rock is possible because different rock compositions have different solidus temperatures and inter-layered supra- and sub-solidus rocks can coexist over a narrow temperature range. Thus, there should always be some free-H_2O available to enhance melting.

4.1.3. Mt. Stafford, Australia

Mt. Stafford is located in the Proterozoic Arunta block, Central Australia (Fig. 4.3A). The exposed sequence of low-*P*/high-*T* gneissic rocks can be characterised with increasing grade and partial melting in dominantly metapelitic–psammitic rocks as: Ms Qz schist; And Cd Ksp granofels with small melt segregations; Sp Sill Cd Ksp migmatite; Gt Opx Cd migmatite + minor diatexite; and Bt Cd Pl diatexite transitional to granite (Greenfield et al. 1996; Greenfield et al. 1998; White et al. 2003). *T–P* conditions are estimated at

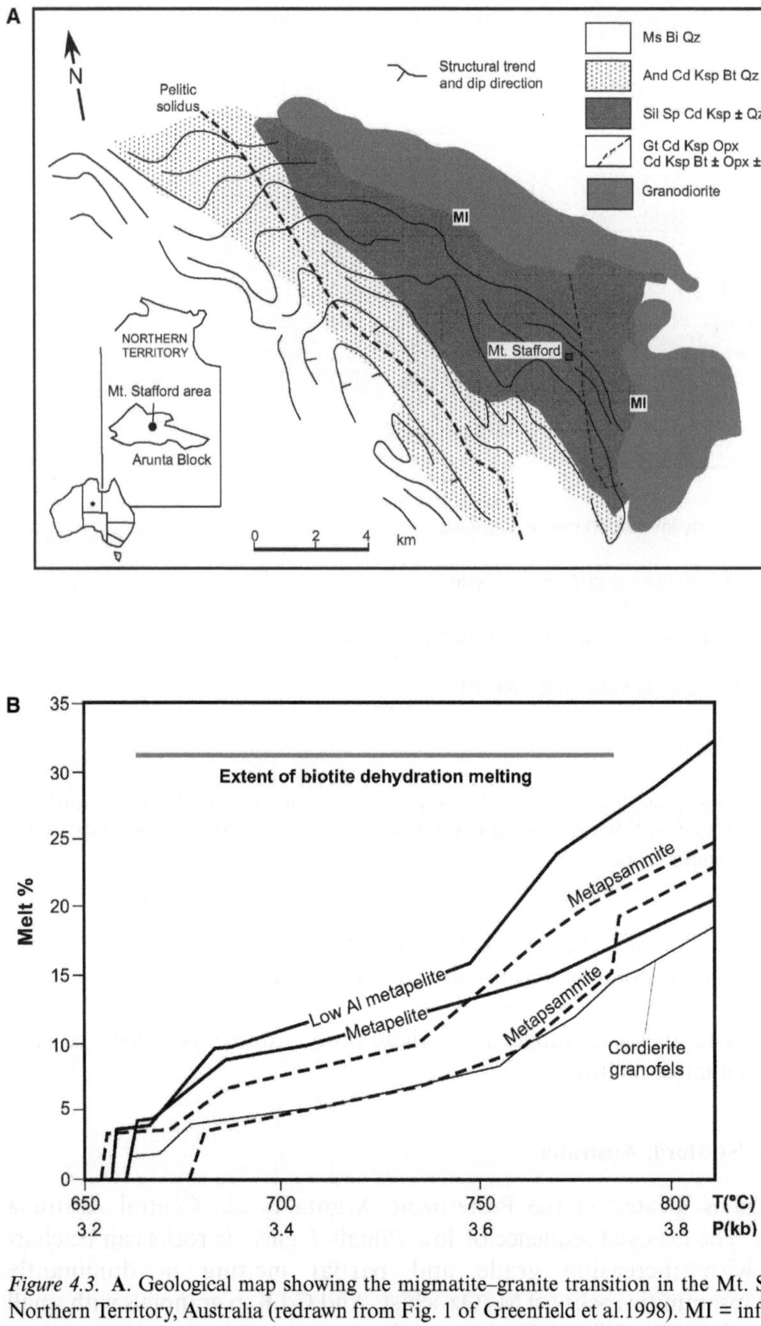

Figure 4.3. **A**. Geological map showing the migmatite–granite transition in the Mt. Stafford area, Northern Territory, Australia (redrawn from Fig. 1 of Greenfield et al.1998). MI = inferred magma interface. **B**. Plot of melt production as a function of *T–P* for bulk migmatite compositions, Mt. Stafford, Central Australia (after Fig. 15 of White et al. 2003).

640–810°C and 2.7–4 kb. An example of the amounts of melt produced in interlayered pelitic–psammitic lithologies over a *T*/*P* range of 150°C/0.5 kb is shown in Fig. 4.3B. Highest melt producers are low aluminous pelitic rocks and lowest melt production occurs in cordierite granofels (siltstone protoliths) because newly formed porphyroblastic cordierite consumes H_2O that would otherwise have been used to produce more melt.

4.1.4. Trois Seigneurs Massif, Pyrenees

The Hercynian low-*P*/high-*T* metamorphic terrane of the Pyrenees comprises Palaeozoic and older amphibolite–granulite facies metasediments, migmatite and granite exposed as elongate ESE-trending structural domes. The Trois Seigneurs massif exposes a continuous metamorphic sequence from chlorite-grade phyllites, through andalusite and sillimanite micaceous schist, to migmatite with leucosome/leucogranite segregations. Biotite granite–quartz diorite (solidified magma layer) is located at the deepest tectonic level (Wickham 1987) (Fig. 4.4). The ~4-km thick sequence implies very high thermal gradients of 80–100°C/km at shallow depths of between ~8 and 11 km although thermobarometric re-evaluation of pelitic assemblages elsewhere in the Pyrenees by Gibson and Bickle (1994) suggest lower gradients of 45–50°C with temperatures reaching 725 ± 25°C at 16 ± 2 km depth (Fig. 4.5).

Wickham (1987) considers that the formation of granite in the Trois Seigneurs massif involved in situ water-saturated melting of pelitic and more quartz-rich metasediment and probable mafic rocks. Stable isotope evidence supports infiltration of groundwater into the zone of melting allowing melts to be continuously saturated as they were generated (Wickham and Taylor 1985). Leucogranite melts were the first formed at 670–700°C, 3–4 kb under H_2O-saturated conditions and segregated into lenses and extensional fractures. They form a continuous gradation in size from cm-scale to m-scale within the migmatite and also occur within the overlying metasedimentary sequence. Biotite–sillimanite-rich restite resulting from this partial melting is now found as the dominant xenolith type in the underlying biotite granite together with other refractory lithologies such as quartzite, carbonate and minor amphibolite that comprise parts of the migmatitised amphibolite grade metasedimentary rocks. These observations suggest upward thickening (rising MI) of a convecting granite magma layer (biotite granite) involving melting and stoping of the overlying lower temperature migmatitic rocks.

4.1.5. Velay Dome, France

The 100-km diameter Velay dome is one of the largest granite–migmatite occurrences of the Variscan belt of Europe (Fig. 4.6), in which a granite core is enveloped by a zone of migmatite and country rock gneiss. Structural, petrological and geochronological data summarised by Ledru et al. (2001) indicates

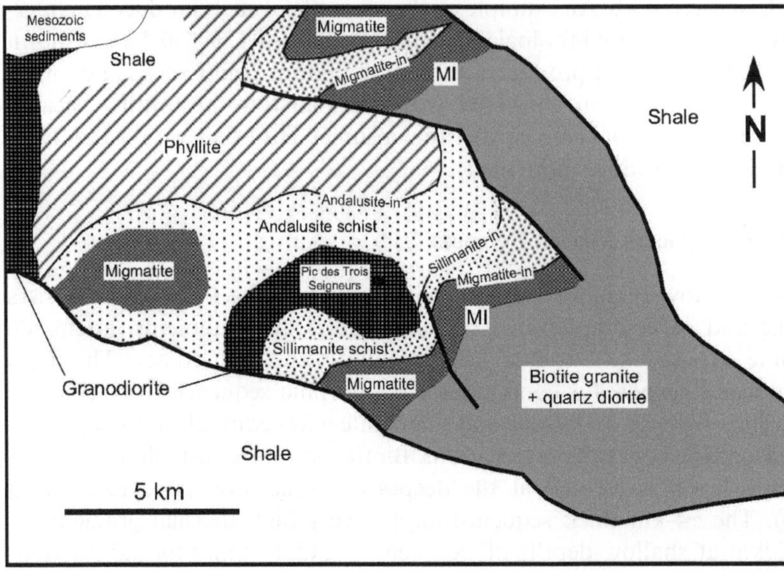

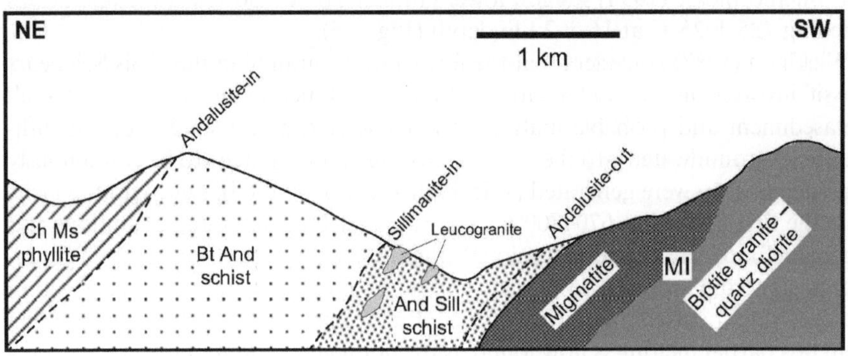

Figure 4.4. Geological map and cross section of the Trois Seigneurs Massif, Pyrenees, France (redrawn and modified after Figs. 2 and 8 of Wickham 1987, respectively). Heavy solid lines denote major faults. Small bodies of leucogranite occur within the high grade schist and migmatite and are not marked on the map. MI = magma interface.

that partial melting of dominantly metasedimentary rocks began ~340 Ma during thrusting in the hinterland of the Variscan belt, and ended ~40 Ma later when collapse of the orogenic crust occurred resulting in decompression, exhumation and deformation of the migmatite–granite. It is believed that thermal relaxation and increased radioactive heat production following crustal thickening are likely the main causes for an increase in temperature causing partial melting up to a maximum T of 850°C at depths of 15–25 km to form a granite magma layer.

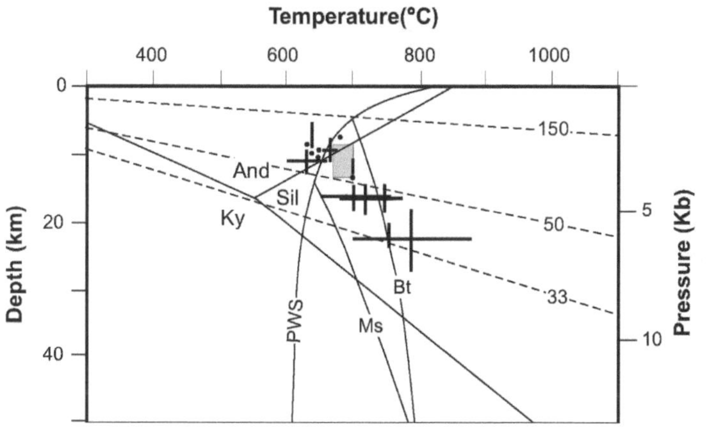

Figure 4.5. *P–T*–depth estimates of various Hercynian granite–migmatite localities in the Pyrenees (data from Table 1 of Gibson & Bickle 1994). Shaded box = Trois Seigneurs Massif (data from Wickham 1987). PWS = pelitic wet solidus; Ms, Bt = muscovite and biotite dehydration melting curves, respectively. Al-silicate stability fields after Pattison (1992). Linear geothermal gradients labelled 150, 50 and 33°C/km are given for reference.

4.1.6. Coastal Migmatite–Granite Zone, SE China

The Mesozoic metamorphic belt in Fujian Province, SE China, provides an example of a metamorphic sequence formed during upward migration of the MI. Massive granite–diatexite is located within the core of a major NE-trending anticlinal axis, and pass gradationally outwards through metatexites to greenschist facies rocks of Triassic–Jurassic age (Fig. 4.7) (GSF 1982). The $^{40}Ar/^{39}Ar$ cooling ages of metamorphic rocks on both limbs of the anticline are 98–99 Ma (Fu et al. 1989), the same as 96–99 Ma K/Ar cooling ages of the granite (GSF 1982; Fu et al. 1989) within the axial part of the anticline.

The granite–diatexite is considered to represent an Early Cretaceous magma layer, and the transition through lower temperature migmatites to unmelted greenschist facies rocks reflects an example of decreasing temperature in the conduction zone above the MI. The 4th episode of the Yanshanian deformation phase (~100 Ma) (Chen 1987) has deformed both the MI and the overlying metamorphic zones that formed synchronously with the development of the antiform in which the trend of leucosome/melanosome layering produced by partial melting is parallel to the anticlinal axis (Fig. 4.8).

4.1.7. Cooma and Murrumbidgee, Australia

The Cooma complex occurs within the Lachlan Fold belt of SE Australia, and comprises an approximately 10-km wide sequence of low-*P*/high-*T* metamorphic rocks surrounding a small body of granodiorite. Outwards from the granite

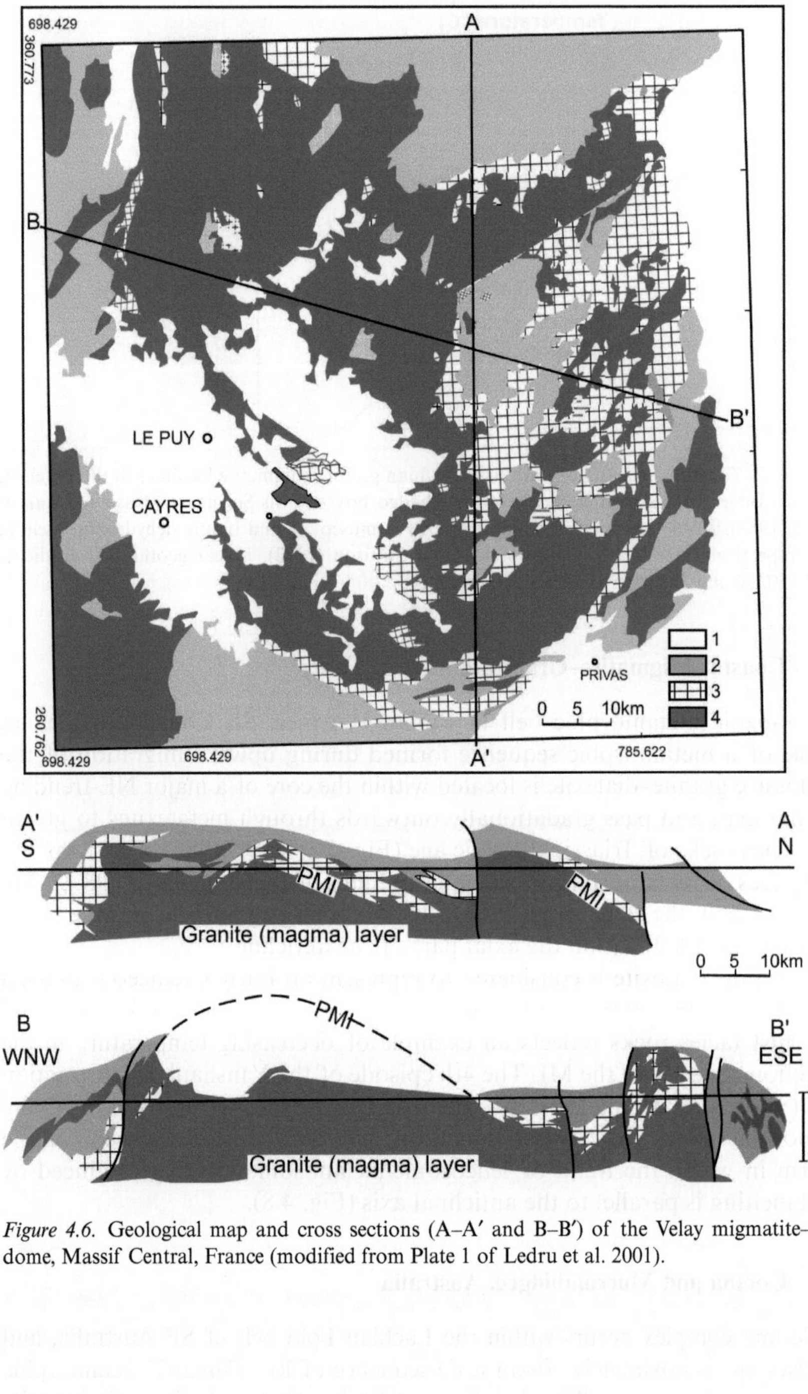

Figure 4.6. Geological map and cross sections (A–A′ and B–B′) of the Velay migmatite–granite dome, Massif Central, France (modified from Plate 1 of Ledru et al. 2001).

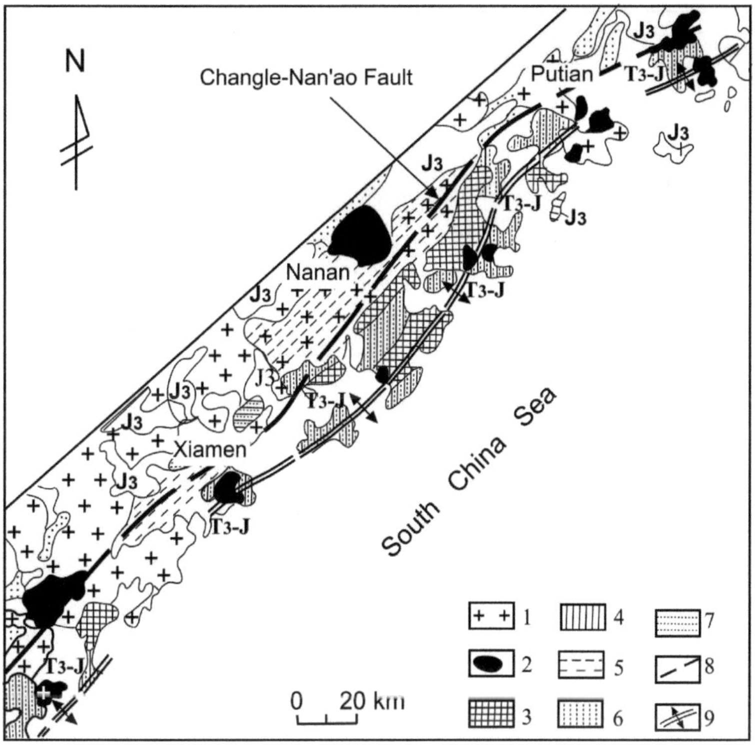

Figure 4.7. Geological map showing distribution of Mesozoic metamorphic rocks in SE Fujian Province, SE China. **1**. Pre-4th Yanshanian deformation phase granite, **2**. 4th Yanshanian deformation phase granite, **3**. Migmatitic granite, **4**. Migmatite, **5**. Hybrid granite, **6**. Amphibolite facies, **7**. Greenschist facies, **8**. Fault, **9**. Anticlinal axis. J_3 = Upper Jurassic; T_3–J = Upper Triassic–Jurassic

contact, the metamorphic rocks consist of Sil migmatite, And-Cd-Ksp gneiss, And-Cd, Cd, Bt and Chl schists (Fig. 4.9). The granodiorite is considered to have been derived from in situ partial melting of the same host rocks (Chappell and White 1974; Flood and Vernon 1978). The first indication of partial melting is the development of patchy leucosomes in metapelites. Some leucosome is localized along metapelite–metapsammite contacts. In the migmatite, prominent lenses and veins of leucosome are oblique to the main foliation, indicating that they are probably formed by fracturing, together with incipient boudinage of strong metapelitic layers (in contrast with more ductile metapsammite layers), providing low-pressure sites for leucosome segregation. As the amount of partial melting increases, the migmatites become coarser-grained and more stromatic, although they remain essentially bedded migmatites, and melting is largely confined to metapelitic layers. Locally, patches of granite occur in the highest-grade migmatites, and typically contain scattered metasedimentary xenoliths and migmatite remnants. The granite patches show only the latest foliation, as does the Cooma granodiorite, which

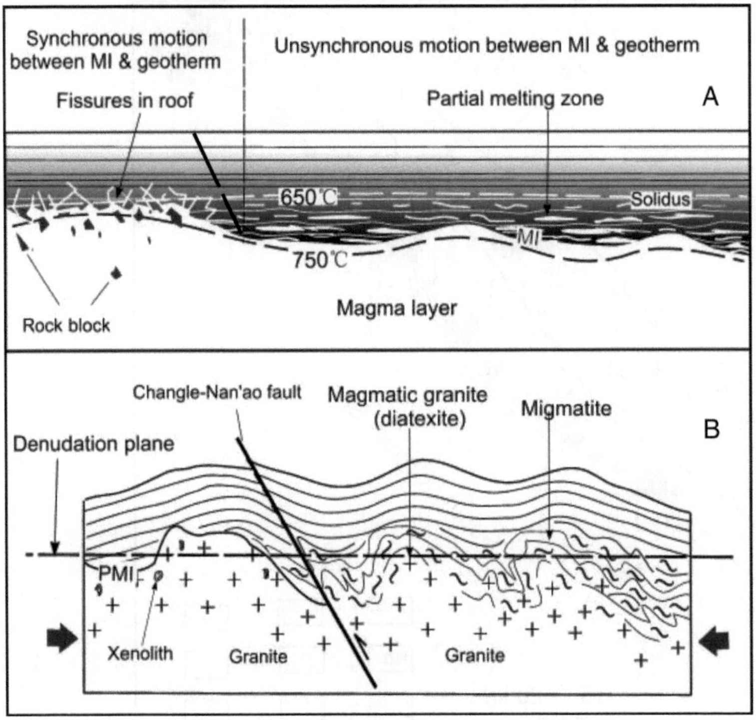

Figure 4.8. Schematic diagrams to illustrate the formation of the Mesozoic granite–migmatite belt of Fujian Province. **A**. Before the 4th Yanshanian deformation phase, showing the different situations on both sides of the Changle-Nan'ao Fault. *Left*. Synchronous upward movement of the MI and isotherms. A narrow contact metamorphic aureole is formed. *Right*. Isotherms move upwards faster than MI. Partial melting occurs and migmatite is formed; **B**. After deformation and erosion. During lateral compression, MI and metamorphic are deformed concordantly, so that contact aureole isograds and migmatite layering/foliation parallel the granite contact.

suggests that it formed in a similar situation, involving more extensive coalescence of leucosome and migration of portions of the resulting magma, with local preservation of gradational contacts into migmatite.

Most of the leucosomes consist only of quartz and feldspar, although porphyroblasts of cordierite occur locally. As muscovite is not present in non-migmatitic rocks of the Cd-Ksp and And-Ksp zones, muscovite dehydration melting did not contribute to leucosome formation. Alternative sources of granite melt are therefore from H_2O-saturated melting or biotite breakdown. The presence of high-grade hydrous cordierite at Cooma is consistent with the possibility of introduction of external water promoting the melting and reduced melting temperatures could have been facilitated by boron in the fluids (Johannes and Holz 1996), which is supported by locally abundant of

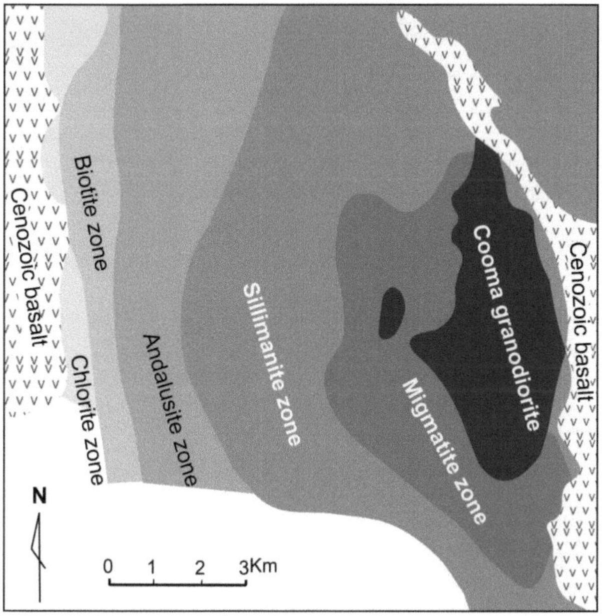

Figure 4.9. Map showing metamorphic zones surrounding the Cooma granodiorite, Lachlan Fold belt, SE Australia (redrawn from Chappell and White 1976).

tourmaline in leucosomes and pegmatitic patches, and by the presence of andalusite.

The metamorphic zones of Cooma can be traced continuously to the Murrumbidgee batholith to the north and along the unfaulted contact of the Murumbidgee batholith where they become progressively narrower and eventually disappear (Fig. 4.10) suggesting the possibility that they have been 'engulfed' by the granite (cf Trois Seigneurs massif, Pyrenees, described above) (see Fig. 3.5). Subsequent development and upward thickening of a magma layer (the Murrumbidgee batholith) may have resulted in melting/stoping of most of the migmatite cover towards the north and also andalusite and cordierite-bearing gneiss and schist some 50 km north of Cooma, where the granite is in direct contact with lowest-grade biotite and chlorite schist (Fig. 4.10). The formation of the Cooma granodiorite and its metamorphic zones has been explained as diapiric (Dixon 1975). If so, the diapiric magma could not have migrated for any significant distance, and may simply reflect local deformation of an underlying larger granite body continuous with the Murrumbidgee batholith. The field relationships shown in Fig. 4.10 suggest that the Cooma granodiorite and its relatively wide metamorphic 'aureole' represents the edge or lateral termination of an extensive layer-like granite body that is exposed by eastward tilting as the Murrumbidgee batholith.

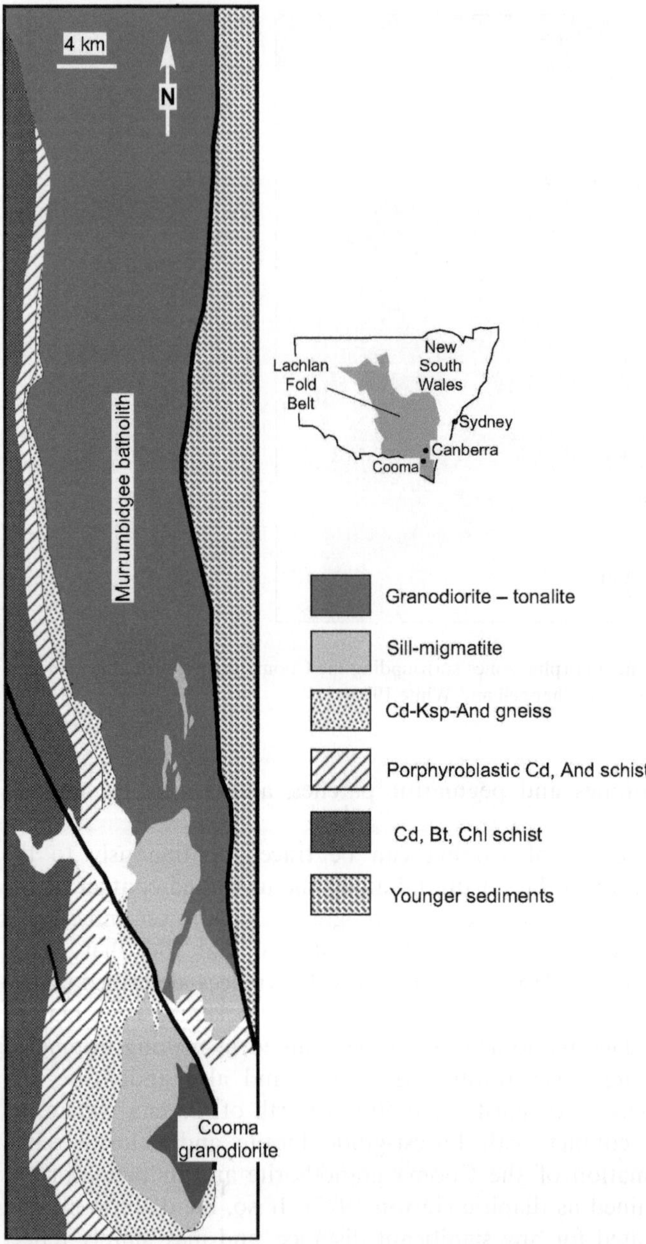

Figure 4.10. Geological map showing relationship between granite and contact metamorphic rocks of the Cooma granodiorite and eastern lobe of the Murrumbidgee batholith, Lachlan Fold belt, New South Wales, SE Australia (simplified after Fig. 1. of Richards and Collins 2002). See text.

4.1.8. Optica Grey Gneiss, Canada

Generation of granitic magma from migmatised amphibolite facies diorite, tonalite, granodioritic gneiss in the Archean Opatica sub-province of the Canadian Shield Superior Province is described by Sawyer (1998). The transition from paleosome grey gneiss to granite was an in situ melting process that generated an average of <30% melt. Initial stages of migmatisation involved formation of diffuse coarse-grained neosome patches in leucocratic parts of the grey gneiss that consist essentially of a fertile (minimum melt) Qz Pl Ksp assemblage. Other, more mafic bulk compositions show no evidence of partial melting. With increasing temperature, the neosome patches become larger and coalesce so that the grey gneiss appears as remnants. The earliest evidence of melt segregation is the presence of medium-grained pink leucosomes within small shear zones. With advancing migmatisation, bulk flow occurred to produce dark- and light-coloured schlieren enriched in biotite and plagioclase, respectively. These schlieren define flow banded diatexite units that may be up to 10 m, and rarely hundreds of metres in thickness confined between layers of unmigmatised gneiss. The more evolved diatexites are pink with abundant Ksp, euhedral Pl and have igneous-like textures and thus may have been ‘proto-granite magma’. The transition from metatexite with neosome patches to diatexite occurs both on a scale of hundreds of metres representing progressive regional-scale anatexis, to a few metres related to variation in paleosome fertility.

The actual transition from diatexite to magmatic granite within the migmatite zone is gradational over several tens of metres. The granite bodies may be several kilometres long and are apparently ponded beneath thick units of unmelted grey gneiss or occur within gently dipping backthrust zones. Above the migmatised rocks amphibolite to greenschist facies grade rocks are cut by dykes and sills typically of fine grained to pegmatitic leucogranite suggesting fracture-controlled intrusion into the cooler metamorphic envelope rocks. Similar sharp-bordered dykes within the migmatite indicate a late stage origin, presumably during the waning stage of anatexis. Other larger sheet-like bodies of medium-grained granodiorite–granite of between 1 and 20 km across represent areas of melt coalesce that we interpret to indicate early stage development of a granite magma layer. Presumably, continued growth of the magma layer was curtailed by regional cooling.

The large volume of granite in the Opatica terrane, and the very low biotite content of the grey gneiss protolith suggests H_2O-saturated melting of Qz Ksp Pl. Experimental evidence indicates that Ksp is the first mineral to disappear above the temperature of the wet solidi of granodiorite and tonolite, which means that modal Ksp will determine the amount of granitic melt produced in the absence of high enough temperatures to cause biotite dehydration melting. The Opatica grey gneiss paleosome contains between 3 and 13 vol % Ksp that would yield 10–48% granitic melt (at 5–7 kb; Winkler 1976). Because the partially melted layers were bounded by infertile, more competent grey gneiss layers, movement of the melt fraction was confined to the partially melted layers where much of it segregated and crystallised. Nevertheless, fracturing of the competent, unmelted wall/roof rock was able to occur allowing late-stage upward movement of some magma to crystallise as dykes and sills both within the cooling migmatised zone and in overlying non-migmatised rocks.

4.2. CONTACT METAMORPHISM

As shown in Fig. 3.13, if the MI moves upwards by mechanically destroying the roof, it will likely overstep the original solidus position of the roof rocks of the partial melt layer and come into contact with low grade, typically greenschist facies country rocks (Fig. 4.11). There are numerous studies of contact aureoles surrounding granite bodies and good summaries can be found in Kerrick (1991). Such granites are interpreted here as domed structures in the roof of an extensive subsurface granite. With deformation, both the MI and overlying metamorphic rocks will be deformed (Fig. 3.15), thus in map view, granite bodies are enveloped by metamorphic aureoles (Fig. 4.12) and some examples from SE China are listed in Table 4.1.

4.3. XENOLITHS TO MAFIC ENCLAVES

Microgranular mafic enclaves in granites are commonly considered to originate from a deep source or magma mixing (Mass et al. 1997; Vernon 1984; Vernon et al. 1988; Dorais et al. 1990; Sieger et al. 1993; Schöldlbauer et al. 1997; Weibe et al. 1997; Silva et al. 2000; Collins et al. 2000; Barbarin 2005), but are here regarded as the recrystallised mafic restites of extensively melted country rock fragments (Chen and Grapes 2003; Peng 2006). As shown in Figs. 3.14B and 4.13, xenoliths in granite are rock fragments derived from the country

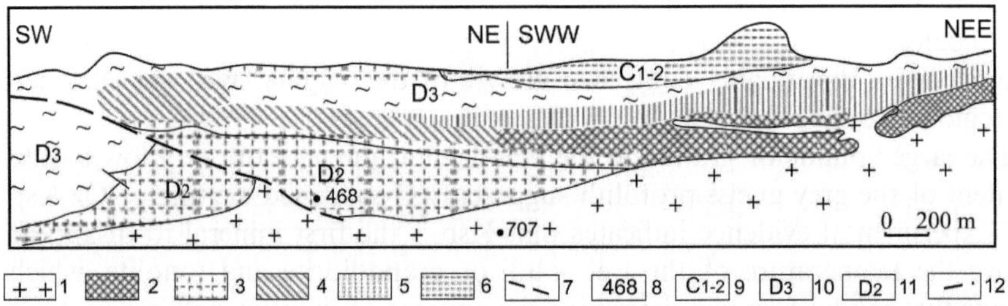

Figure 4.11. Metamorphic aureole rocks overlying the Longxianggai granite at Dachang, Guangxi Province, SE China (after BGMRGx 1985). **1**. Biotite granite; **2**. Skarn; **3**. Hornfels; **4**. Marble; **5**. Sillimanite hornfels; **6**. Silicified and sericitised rocks; **7**. Carbonatisation zone; **8**. Temperature (°C) of rock formation; **9**. Lower and Middle Carboniferous; **10**. Upper Devonian; **11**. Middle Devonian; **12**. Fault.

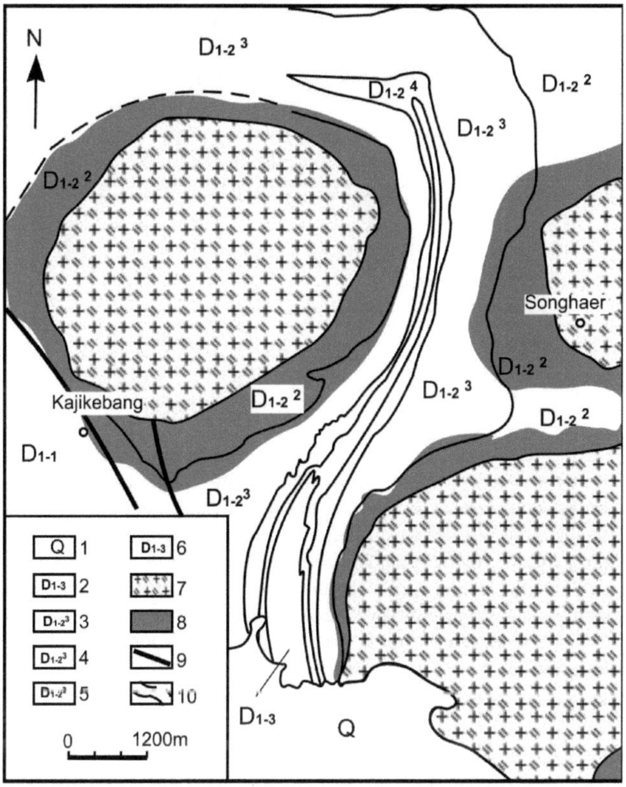

Figure 4.12. Contact metamorphic aureoles surrounding granite domes in XingJiang Province, NW China. **1**. Quaternary; **2–6**. Lower Devonian formations; **7**. Granite; **8**. Contact metamorphic aureole; **9**. Fault; **10**. Stratal contact.

Table 4.1. Contact metamorphic aureole zones and their map widths (m) surrounding granite outcrops in SE China **(after 2nd Party of Metallurgical Geological Exploration Corporation of Jiangxi Province 1977)**

Granite Subzone	Xihuashan	Zhang-dongkeng	Piaotang	Muziyuan	Zhong-shukeng	Dalong-shan	Huangsha	Xialong
Bt hornfels	250	200–300	100–200	80–210	200	240	0–50	120
Ms hornfels	150	100–200	150–300	125–315	350	300	50	110
Spotted slate	200	160–230	160–230	60–100	120	150	10–30	

rock roof. With gravitational sinking in the magma, partial melting will begin from the surfaces of the rock blocks inwards and with progressive melting, mafic enclaves may be formed as shown in Fig. 4.14A, B.

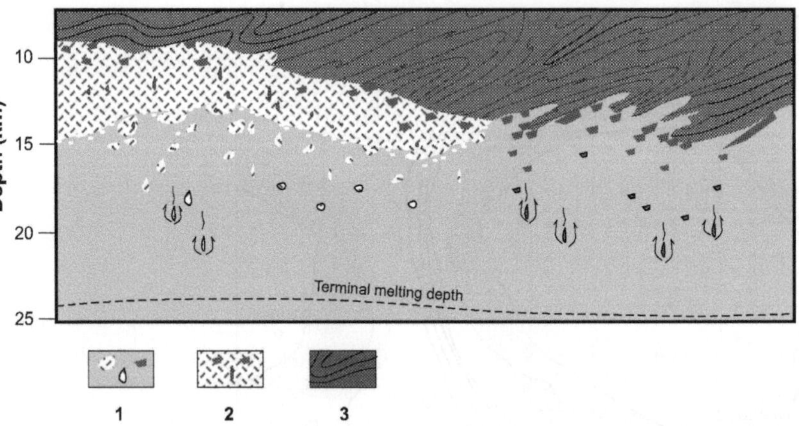

Figure 4.13. Schematic diagram to illustrate the incorporation of xenoliths (not scaled) derived from stoping and melting of roof rocks composed of metasediments and granite by an upward-moving MI of a convecting granite magma layer. The xenoliths undergo melting and disaggregation during downward movement to form rounded, reconstituted mafic enclaves as described in text. Thick black lines surrounding deeper enclaves represent biotite-rich selvages. No enclaves/xenoliths occur below the terminal melting depth within the granite magma. Small arrowed lines indicate displacement direction of granite magma around descending xenoliths/enclaves. **1**. Granite magma with xenoliths of older granite, metasediment and their enclave equivalents; **2**. Older granite country rock that contains xenoliths and enclaves; **3**. Folded metasedimentary county rock.

Roof rocks of crustal-scale magma layers presumably consist of various sedimentary, metamorphic and igneous rocks with a wide spectrum of compositions. Thus, various types of enclave are expected to occur within a granite body such as shown in Fig. 4.14C where more than ten kinds of enclaves (differences based on visual appearance of colour, texture, main mineral composition and grain size) coexist in an outcrop area less than 2 m^2. The largest has a diameter is about 15 cm and the smaller ones are only 1–2 cm in diameter.

Investigation of numerous granite bodies in SE China (Chen et al. 1994; Peng 2006) indicates that their mafic enclaves can be divided into two categories: homogenous and inhomogenous, with the latter more common and typically mantled by a dark (biotite-rich) rim (Fig. 4.15A, B). The compositionally zoned nature of many mafic enclaves indicates that speculation of their source or origin based with a single chemical/isotopic analyses representing the composition of an entire enclave is questionable (among many others, Eberz et al. 1990; Elburg 1996; Silva et al. 2000; Tod et al. 2001; Campos et al. 2002; Yang et al. 2004; Gomes and Neiva 2005). Also, if inhomogenous enclaves are derived by crystallisation of a hybrid mafic–silicic magma as commonly concluded, enclaves within in a granite body should have similar compositions, grain size and texture. However, this is not generally the case.

An example from the Luogang granite, Gunagdong Province, is a complete somewhat flattened spindle-shaped enclave with an undulating central edge

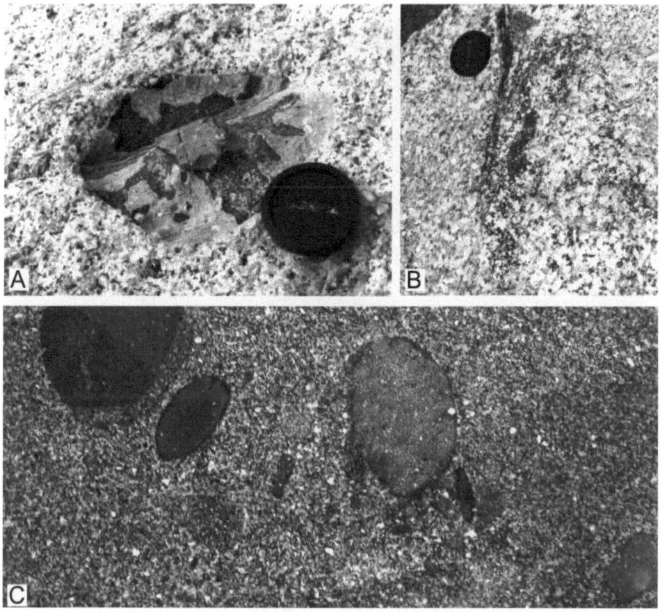

Figure 4.14. **A**. A zoned mafic enclave in the Wangmu granite exposed along the coast of Dapeng Bay, SE China, with a hornfels core representing unmelted protolith, and the dark envelope representing the outer part of the enclave that has undergone extensive melting and diffusive exchange with the surrounding granodiorite magma. The white specks on the fractured surface are salt from evaporation of seawater; **B**. An indistinct, extensively melted and disaggregated enclave with a darker restite core; **C**. Enclaves of different compositions within a small outcrop area of the Fengkai granite, Guangdong Province, SE China.

(Fig. 4.15C). A cross-section (Fig. 4.15D) indicates that grain size and modal feldspar decrease from the core outwards and that there is a distinct dark, 2-cm-thick biotite-rich rim. Four samples from the enclave (Nos A, B, C, D in Fig. 4.15D) have been analysed (Peng at al. 2005) and indicate decreasing Fe_2O_3 (as total iron), MgO, TiO_2, CaO, Na_2O, and increasing SiO_2, Al_2O_3, K_2O outwards with a marked increase in MgO, Fe_2O_3, K_2O and decrease in SiO_2 in the rim reflecting the high biotite-content (Fig. 4.16). Oxide variation in the main part of the xenolith reflects relatively small changes of decreasing modal hornblende, plagioclase (CaO, Na_2O decreasing) and increasing K-feldspar and quartz (increasing SiO_2, Al_2O_3, K_2O). An extensive study of inhomogenous enclaves from numerous granite bodies of SE China indicates that they have similar core-rim compositional variation and, as in the above-described enclave, consist of two generations of both major and accessory minerals (relic cores and overgrowths/matrix) (Peng 2006). Later-formed minerals (Hb, twinned Pl, Qz, Ksp, Bt, Tn) have a hypidiomorphic granular texture implying crystallisation from a melt and enclose Pl, Qz, Hb, Bt, Ilm, Mz, Zr relics/restite of xenolith melting. In particular, relic plagioclase cores typically exhibit the characteristic

Figure 4.15. **A, B**. Typical compositionally zoned mafic enclaves in granites of SE China; **C**. A spindle-shaped enclave from the Luogang granite, Guangdong Province. The horizontal white line through the middle of the enclave indicates the cross section shown in **D** where portions labelled A, B, C and D are the analysed bulk samples plotted in Fig. 4.16. Scale length = 30 cm.

frittered or fingerprint texturé of melting. The later mineral assemblage has crystallised from a melt of granodioritic composition indicating advanced (higher *T*) melting of possible intermediate plutonic or volcaniclastic metasedimentary xenoliths, e.g. experimental melting of samples B.2, 3 in Appendix 2 and Fig. D2a. Two-way element diffusion involving inward movement of Si, K and outward movement of Mg, Fe, Al, Ti occurred in response to both compositional and temperature gradients between the granite magma host and partially melted rock. Late-stage formation of biotite-rich rims (highest K, Fe, Mg, Ti; lowest Si, Ca) occurred during cooling when element diffusion profiles were effectively 'frozen' and crystallisation occurred.

Enclaves tend to be more commonly distributed in the upper part of granite bodies. They are rare or absent where erosion has exposed deeper parts of the granite, such as the Dadongshan and Jiufeng batholiths in Guangdong Province (Chen et al. 1994; Peng 2006), where the terminal melting/disaggregation depth of xenoliths/enclaves has presumably been exceeded (Figs. 4.13 and 4.17A). Evidence for the downward movement of enclaves is commonly observed as shown in Fig. 4.17B where biotite-rich 'flow' lines extending outwards from one side of the xenolith may preserve a sinking trail within the granite magma.

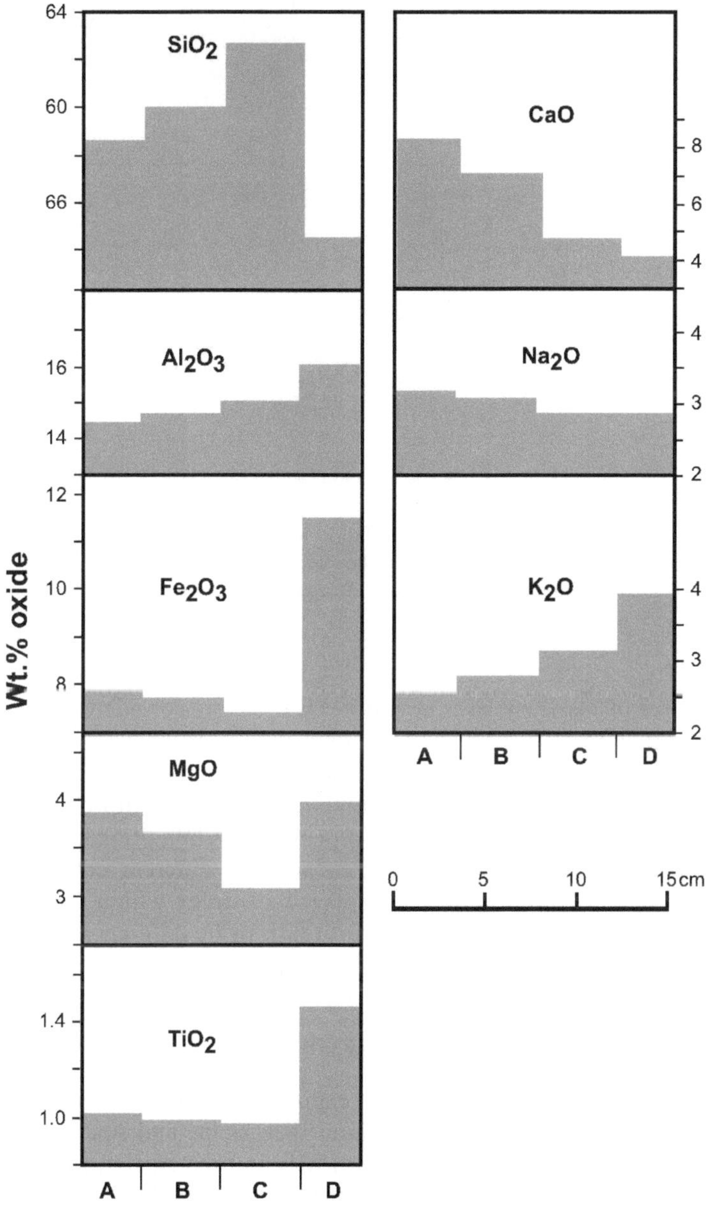

Figure 4.16. Major element variation (wt. %) with respect to core and margin of the enclave from the Luogang granite, Guangdong Province, shown in Fig. 4.15D (after Peng 2006).

Figure 4.17. **A**. Various types of enclaves and their evolution in granite (redrawn and modified after Durr 1960; reproduced in Didier 1973); **B**. Mafic enclave with biotite trail defining an inferred 'flow' path that suggests sinking within the granite magma, Luogang granite, Guangdong Province.

The primary source of mafic enclaves in granites is therefore considered to be rock blocks (xenoliths) derived from the roof of a granite magma layer that have undergone a high degree of melting. The melt largely remains within the confines of the enclave (globules of melt + unmelted minerals), with element diffusion occurring with the granite magma host prior to crystallisation of their granodioritic mineral assemblage. Obviously, enclaves of different compositions, sizes, shapes and textures would be expected to occur within granite formed by partial melting of a heterogeneous protolith or by different degrees of melting.

4.4. GRANITE LAYER AND GRANITE EXPOSURES

As mentioned in Chapter 3, individual granite exposures represent the domed parts of an extensive subsurface granite layer and their shape and size merely reflect the geometric relationship of the MI (or PMI) and the erosion surface. Figures 4.18, 4.19, and 4.20 show some examples of this. In Fig. 4.18, the exposed area of the Xihuashan granite in Jiangxi Province, SE China, is no more than 20 km^2 and there are some small granite outcrops in the Piaotang and Zongshukeng areas northeast of the Xihuashan granite. Geophysical investigations and drilling indicate that all the individual granite bodies in the area are domed structures developed in the upper surface (PMI) of an extensive granite layer (BGMRJx 1984). In the Pingying area, Hunan Province, SE China,

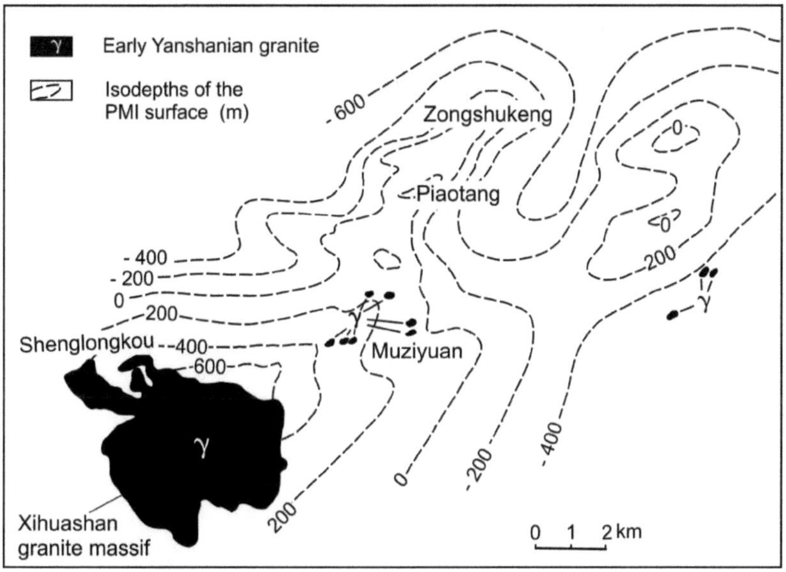

Figure 4.18. Map showing depth contours (m) delineating an irregular subsurface continuation of the Xihuashan granite, Xihuashan–Zongshukeng area, Jiangxi Province, SE China (simplified after BGMRJx 1984). g = small granite outcrops.

three granite bodies are surrounded by a number of smaller granite occurrences and subsurface granite is indicated by geophysical methods and confirmed by drilling (Fig. 4.19), showing that all the granites are connected, and that they also form domed parts of the upper surface of a large granite batholith. Similar relationships are common worldwide. For example, the numerous outcrops of Hercynian granite, the south Sardinia, Italy (Fig. 4.20), are associated with Paleozoic country rocks and are connected at depth (Biste 1982).

Granite 'plutons' in west-central Maine, USA consist of two-mica granite, biotite granite and granodiorite with crystallisation ages between 408 Ma and 389 Ma and they are surrounded by contact metamorphic aureoles (Fig. 4.21A) (Brown and Solar 1998b, 1999). The large granite bodies are considered to have been derived by partial melting of metasedimentary migmatite protoliths via muscovite dehydration melting at 700–720°C and ~25 km within a crustal-scale shear zone system (Central Maine belt) and to have ascended via fractures and shear planes in the migmatite to be emplaced at ~12–15 km depth by ductile flow and/or stoping, and inflation (ballooning) involving floor subsidence, to form tabular bodies in the vicinity of the brittle–ductile transition zone. Based on the idea that the granites represent exposed protruding parts of the surface of a continuous granite layer at depth and migmatites represent the remnant of the partial melting zone above the granite, cross-section interpretations are shown in Fig. 4.21B, C. The mylonite zone that borders the Central Maine belt

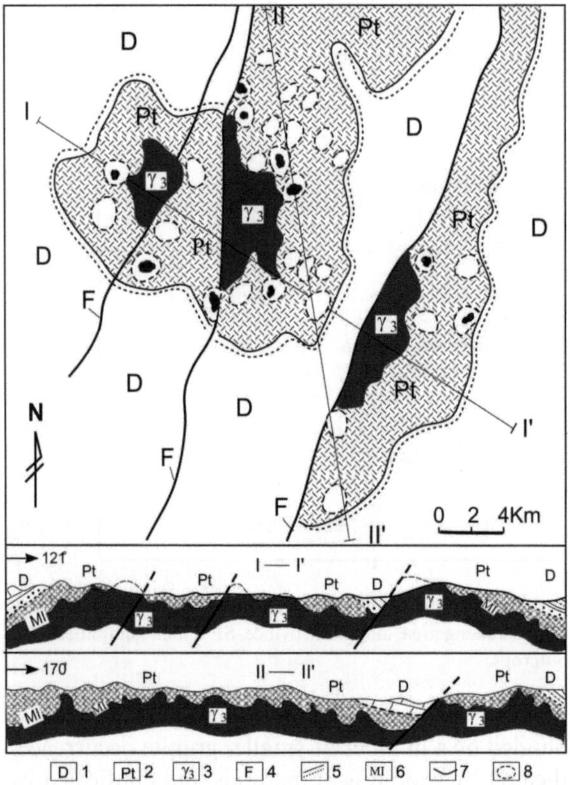

Figure 4.19. Geological map and cross sections of the Pingying area, SE China, showing exposures of granite and subsurface granite. **1**. Devonian; **2**. Paleozoic metamorphic rocks; **3**. Granite; **4**. Fault; **5**. Unconformity; **6**. Granite contact; **7**. Stratal boundary; **8**. Subsurface granite from drilling.

is considered to be a pre- or syn-magma structure and thus is limited to the roof rocks (Fig. 4.21C).

Differences of rock composition/structures, the presence and variation in fracture/joint density in the roof rocks overlying the granite magma layer may result in the formation of an irregular MI. In areas where rocks have lower initial melting temperatures or contain numerous fractures/faults/shear zones, the MI may move up more rapidly to produce a highly irregular MI as shown schematically in Fig. 4.22 and illustrated by an example from Shaaxi Province, NW China, where a series of granite ribbons are inter-layered with Devonian metasediments (Fig. 4.23). If faulting occurs in roof rocks on a comparatively large scale, the MI will likely move up faster along the fracture zone and 'stoping' may more readily occur within shattered rock of fault zones as shown in Fig. 4.24.

An undulating MI caused by variation in crustal thickness above the magma layer resulting in uneven distribution of lithostatic pressure, in addition to lateral

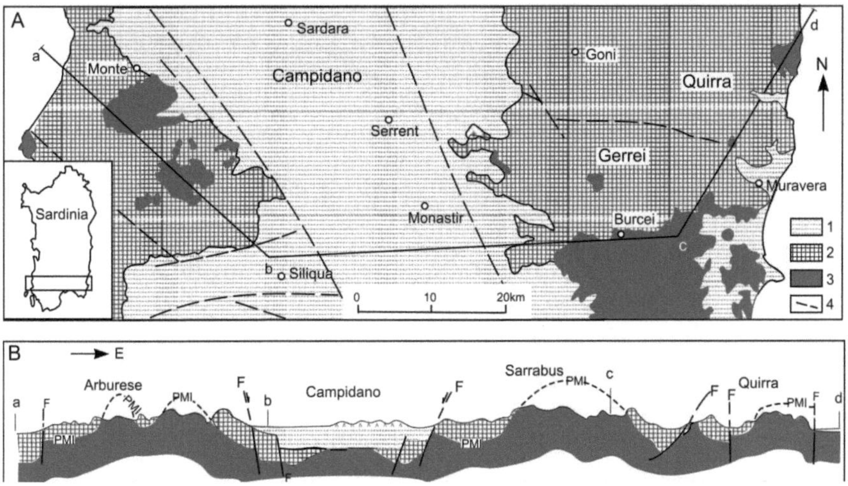

Figure 4.20. Geological map (**A**) and cross section (**B**) of south Sardinia, Italy, showing that isolated granite occurrences represent the protruding parts of an undulating surface of a granite layer (modified after Biste 1982). **1**. Mesozoic sediments; **2**. Paleozoic metasediments and volcanics; **3**. Hercynian granite; **4**. Fault.

compression (Fig. 3.15), will result in flow of magma towards anticlinal areas of the MI (Fig. 3.15B). An example of the relationship between anticlines and granite, in which nearly all the anticlines or domes have granite cores, is shown in Fig. 4.25. In the coastal region of SE China, although faulting has often affected the original shape of most granite bodies, and of anticlinal structures, there is a close spatial relationship between many granite outcrops and major fold structures e.g. Lianhuashan (Fig. 4.26A) and Hetian granites (Fig. 4.26B), Guangdong Province, SE China. A similar relationship is apparent between regional anticlines and the distribution of granites in Cornwall, SW England (Fig. 6.11).

4.5. FLUCTUATION OF MI AND DOWNWARD-YOUNGING GRANITE SEQUENCE

When the energy input/output ratio of the magma layer becomes <1, convection is expected to slow or cease, with resultant descent of isotherms and the MI because of top-downward crystallisation of the magma layer. Examples indicating the downward (inwards in map plan view) crystallisation of the domed parts of a granite layer includes the Jiufeng biotite granite, north Guangdong Province, SE China, which formed in the early Jurassic with K–Ar ages indicating a cooling interval of ~25 Ma (197–172 Ma) from margin to centre (Fig. 4.27).

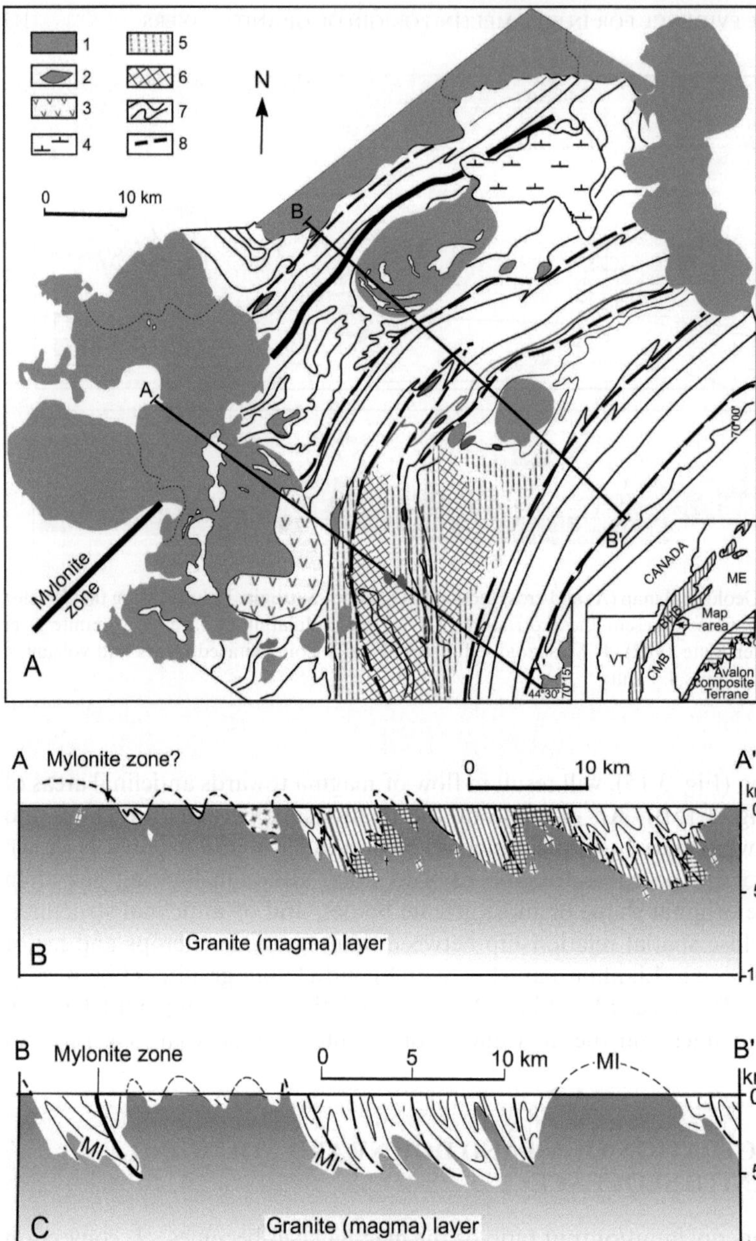

Figure 4.21. **A**. Geological map of a portion of western Maine and adjacent New Hampshire, USA (simplified after Brown and Solar 1999), showing the distribution of granite 'plutons' and migmatite. Thin solid lines indicate trends of steep to moderately dipping foliation in metasedimentary rocks. **1**. Granite; **2**. Smaller granite occurrences (leucogranite, pegmatite); **3**. Granodiorite; **4**. Gabbro; **5**. Stromatic migmatite; **6**. Inhomogeneous migmatite; **7**. Metasedimentary rocks; **8**. Boundaries between steeply dipping (regular thin solid lines = high strain zones) and moderately dipping (irregular, folded thin solid lines = low strain zones) parts of metasedimentary rocks. **B**, **C**. Cross sections (A–A′ and B–B′, respectively) are reinterpretations based on the in situ melting model of granite formation, in this case representing the deformed upper surface of a subsurface granite layer.

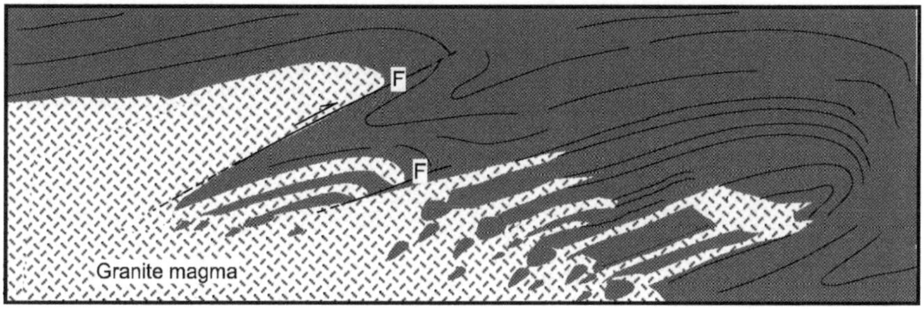

Figure 4.22. Schematic diagram to illustrate the complex form of the upper surface of a granite magma layer MI. F = Fault.

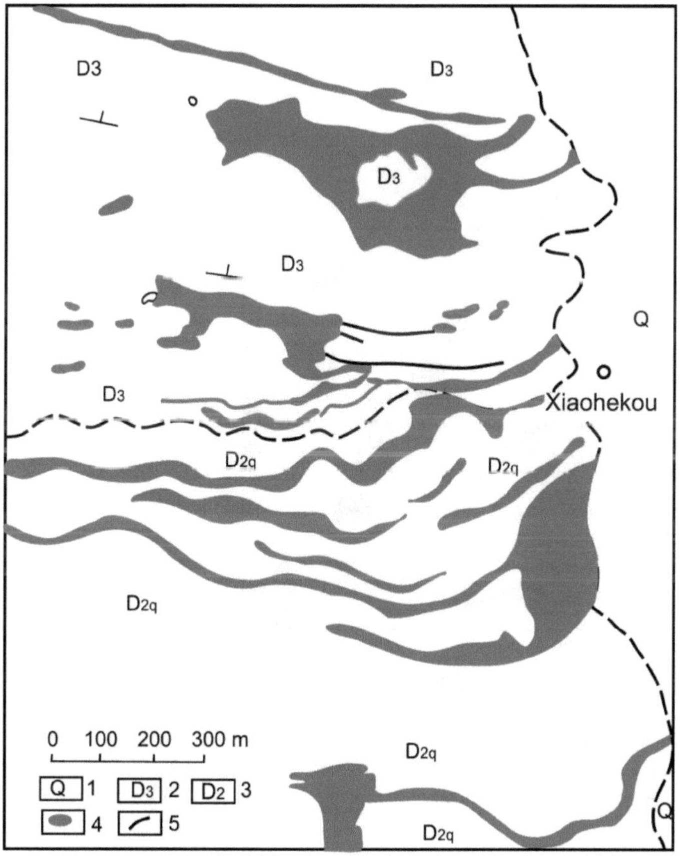

Figure 4.23. Swarm of granite layers parallel to the strike of Devonian metasediments in Xiaohekou, Shaanxi Province, NW China (modified after Yan 1985). **1**. Quaternary; **2**. Upper Devonian; **3**. Middle Devonian; **4**. Granite and granodiorite; **5**. Boundary of metamorphic aureole.

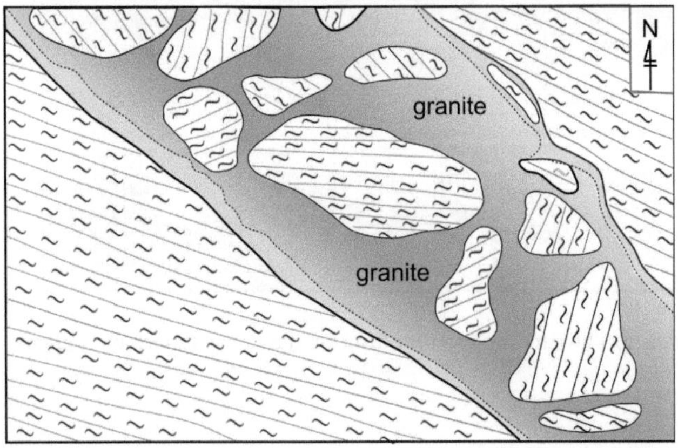

Figure 4.24. An example of stoping along a fault zone at Tongchang, Jiangxi Province, SE China (modified after Zhu et al. 1983). A scale is not given with the original map.

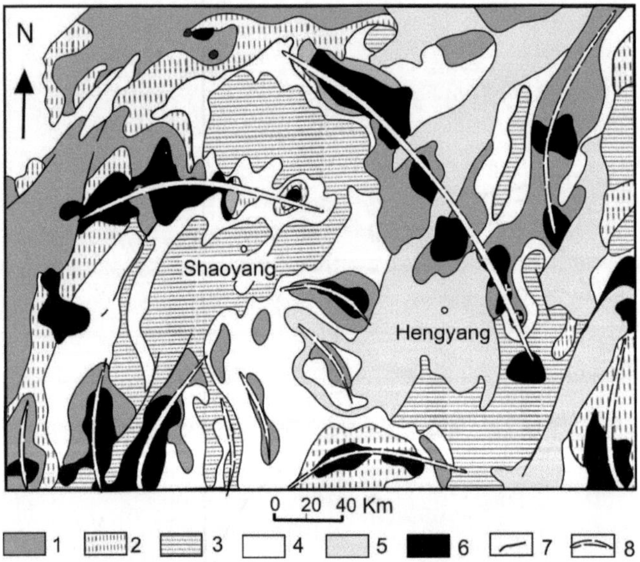

Figure 4.25. Anticlinal structures in the Shaoyang area, central-south Hunan Province, SE China, that coincide with outcrops of granite. **1**. Pre-Paleozoic; **2**. Lower Paleozoic; **3**. Upper Paleozoic–early Triassic; **4**. Middle Triassic–Jurassic; **5**. Cretaceous–Tertiary; **6**. Granite; **7**. Faults; **8**. Anticlinal axes.

As discussed in Chapter 3, remelting of granite results in the formation of a sequence of downward-younging granites (Fig. 3.16) that can reasonably explain the interrelationship of granite bodies observed in SE China. For example, a geological section from Lankou to Tangjiao, Zhijin County, Guangdong

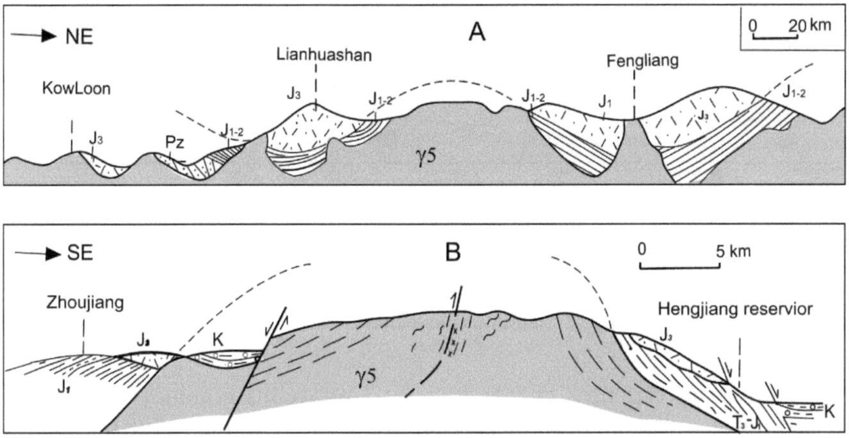

Figure 4.26. **A**. Section from Kowloon (Hong Kong) to Sanheba across the Lianhashan granite, Guangdong Province, SE China. Pz = Paleozoic; J_1 = Early Jurassic; $J_{2,3}$ = Middle–late Jurassic; γ_5 = Mesozoic granite. **B**. Cross section of the Hetian granite, east Guangdong Province, SE China, showing dome-like structure of the granite and anticlinal structure in upper Triassic (T_3)–lower Jurassic (J_1) metasedimentary country rocks. J_3 = Upper Jurassic volcanics; K = Cretaceous sediments. Horizontal = vertical scale.

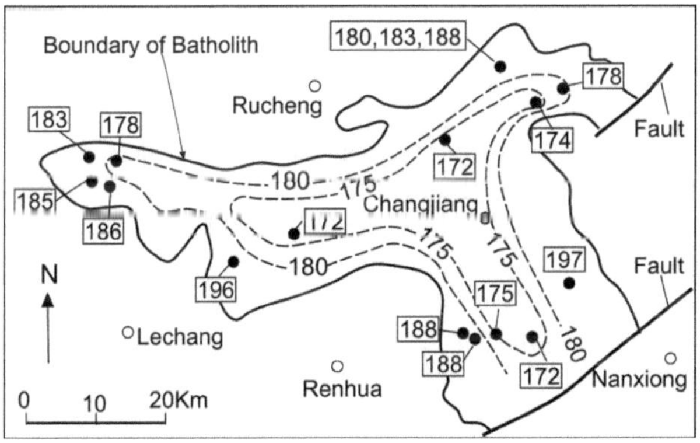

Figure 4.27. Simplified map of the Jiufeng biotite granite, north Guangdong Province, SE China, showing location and K–Ar ages (Ma value in boxes). Dashed lines depict possible cooling isochrons.

Province transects 15 separate granite bodies that are considered to be individual intrusions. A reinterpretation based on the idea of remelting of granite, indicates that there are three granite layers younging downwards from Triassic, through Jurassic to Cretaceous (Fig. 4.28). Another example of downward-younging granites is provided from a seismic survey carried out by the Institute of Geophysics, Chinese Academic Sciences (Li 1992; Yin et al. 1995) along a 400-km long north to south traverse in the southern part of Guangdong Province, that

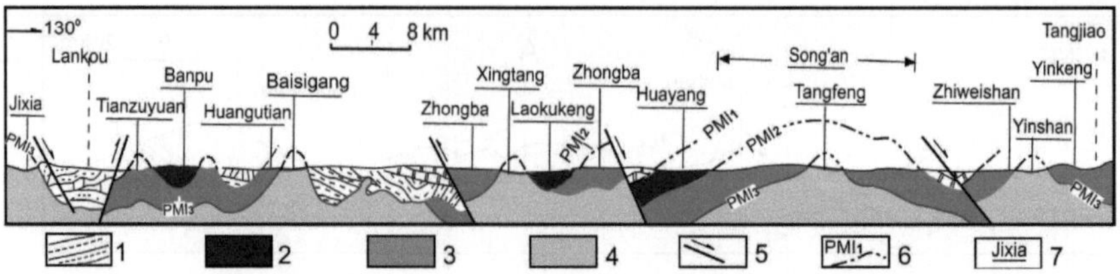

Figure 4.28. Section from Lankou to Tangjiao, Zhijin County, Guangdong Province, SE China, showing three granite layers younging downwards. **1**. Metasedimentary cover; **2**. Triassic granite; **3**. Jurassic granite; **4**. Cretaceous granite; **5**. Fault; **6**. Paleomelt interface (PMI), the number representing its relative formation age; **7**. Name of granite body.

crossed numerous outcrops of late Jurassic–Cretaceous granite (Fig. 4.29A). A 2D crustal structure model indicates five velocity layers (Fig. 4.29B); the mantle with P_n = 8.05 km/s, lower crust with P_n = 6.6–6.8 km/s; a mid–upper crustal lower velocity zone with P_n = 5.8 km/s, metasedimentary cover with P_n < 5.9 km/s, and between the lower velocity zone and metasedimentary cover, a 'basement layer' of P_n = 5.9–6.3 km/s. According to a geological interpretation (Ma 2006), the P_n = 5.9 km/s velocity interface equates with the upper surface of a Jurassic granite layer (PMI_1), and the velocity interfaces of P_m = 6.0 and 6.1 km/s are considered to represent the tops of early and late Cretaceous granite layers (PMI_2 and PMI_3), respectively. Fluctuation of the velocity interfaces in the shallow part of the seismic profile is considered to mirror the irregular shapes of the PMI's and the low-velocity zone below the late Cretaceous granite layer to represent the youngest cooling magma layer (Fig. 4.29B).

Because of engineering construction, several tunnels have been drilled into the interior of the Shangji granite located in the northeast part of Boluo County, Guangdong Province, close to the seismic traverse (Fig. 4.30A), revealing the contact between an upper late Jurassic granite (whole-rock K–Ar age: 156 Ma; Ma 2006) and a finer-grained lower late Cretaceous granite (85.5 Ma) (Ma 2006) (Fig. 4.30A). The contact between the two granites is marked by a zone of pegmatite segregations ranging in thickness between 20 and 50 cm (Fig. 4.30B), above which fractures in the late Jurassic granite contain zoned assemblages of quartz–fluorite–calcite (Fig. 4.30C). Contacts between the different-aged granites both in this example and elsewhere are typically well defined by colour, composition, texture and grain size differences suggesting that they are the product of remelting of earlier-formed granite and/or renewed fusion of a partially molten body caused by vertical fluctuation of the melting interface in response to changing isotherms (Fig. 3.16), possibly facilitated by accumulation of exsolved volatiles in the upper part of the new magma layer (Chen et al. 1996). Thus, composite granite bodies are not considered to be the result of multiple intrusion of magma from a deep source along the same conduit and geochemical/isotopic evidence supporting this is described in Chapter 5.

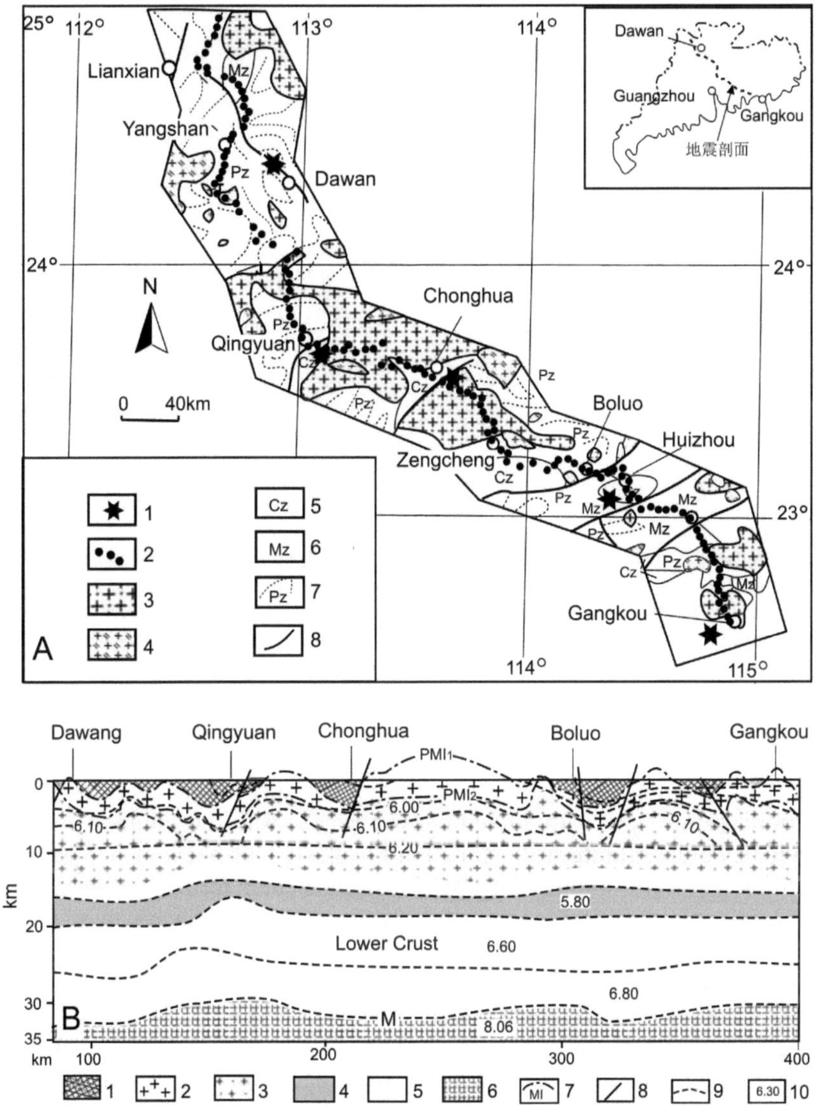

Figure 4.29. **A**. Geological map along the line of the Liangxian–Gangkou seismic profile (after Ma 2006). **1**. Detonation point; **2**. Recording station; **3**. Fault; **4**. Jurassic granite; **5**. Cretaceous granite; **6**. Cenozoic sediments; **7**. Mesozoic sedments; **8**. Paleozoic metasediments. Dotted lines = stratigraphic strike trends. **B**. Geological interpretation of the Lianxian–Gangkou seismic profile (after Ma 2006). **1**. Metasedimentary rocks; **2**. Jurassic granite; **3**. Cretaceous granite; **4**. Low velocity layer; **5**. Lower crust; **6**. Mantle; **7**. Paleo-magma interface; **8**. Fault; **9**. Velocity interface; **10.** P_n (km/s).

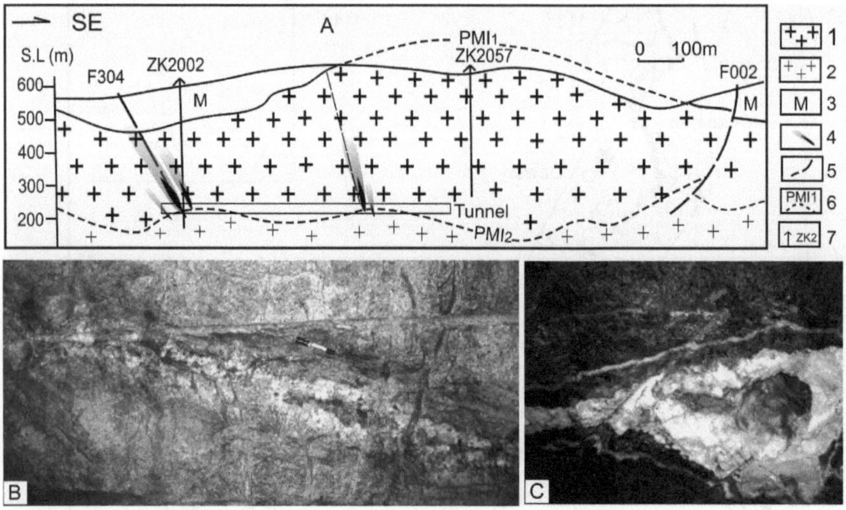

Figure 4.30. **A**. Section of the Shangji composite granite, Guandong Province, showing the contact between upper (late Jurassic) and lower (late Cretaceous) granite (redrawn from Ma 2006). **1**. Jurassic granite; **2**. Cretaceous granite; **3**. Migmatite; **4**. Quartz–calcite–fluorite vein; **5**. Fault; **6**. Paleo-magma interface (PMI); **7**. Drill hole. **B**. Pegmatite occurring along contact between the two granites exposed in a tunnel wall. **C**. Quartz–fluorite–calcite vein formed along the fracture zone in the upper older granite.

Most of the granites in SE China are composite bodies in that they consist of two or more granite types with different ages, typically with the youngest granite surrounded by older granite as illustrated in Fig. 4.31. This phenomenon can be explained as exposure of different PMI's at a particular locality as shown in Fig. 4.32. Petrologically, a composite granite (CG) usually evolves from biotite/hornblende granite/granodiorite to two-mica/muscovite granite with decreasing age (Mo et al. 1980; Zhang and Sun 1988; Meng 1993). Four well-studied CG examples in SE China are described below (Fig. 4.33).

Jiufeng. In addition to the age relations of the biotite granite described above, the Jiufeng granite is also a CG that hosts two other younger granites, namely, the Jietou and Huangdong granites. The main Jiufeng batholith (g^{5a}) (only part of which is shown in Fig. 4.31a) is a coarse-grained porphyritic biotite granite of early Jurassic (195–172 Ma) age (Chen et al. 2003b). The Jietou granite is a medium-grained two-mica granite (g^{5b}) of latest Jurrasic age (K–Ar age of 153 Ma, BGMRHn 1988) and the youngest Huangdong granite (g^{5c}) that occurs within the Jietou granite is a fine-grained muscovite granite with a K–Ar age of 138 Ma (BGMRHn 1988) (Fig. 4.33A).

Dengfuxian. The oldest member (g^{5a}) of this CG is a coarse-grained-porphyritic biotite granite (230 Ma/U-Pb, Deng et al. 1986), which contains a medium-grained two-mica granite (g^{5b}) (135±12 Ma/Rb-Sr, Deng et al. 1986)

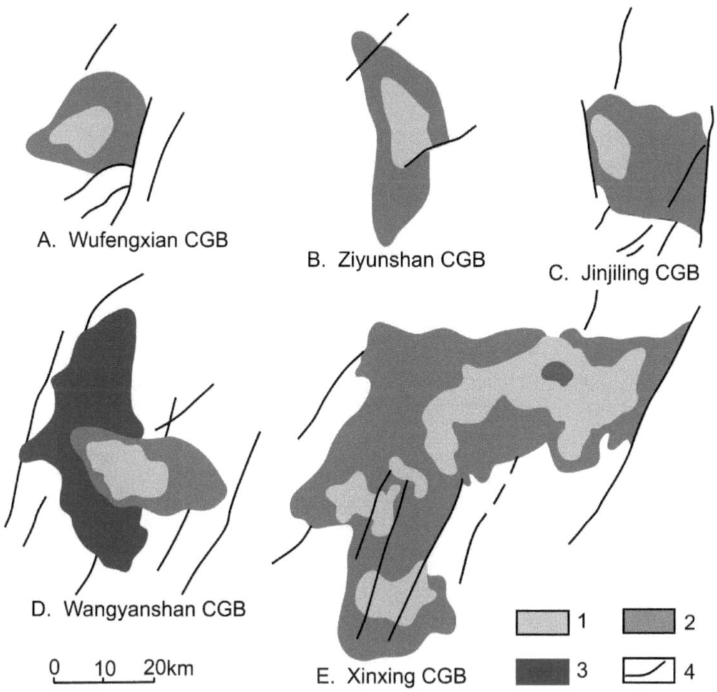

Figure 4.31. Sketch maps of typical composite granite bodies in SE China. **1**. Youngest granite; **2**. Older granite; **3**. Oldest granite; **4**. Fault.

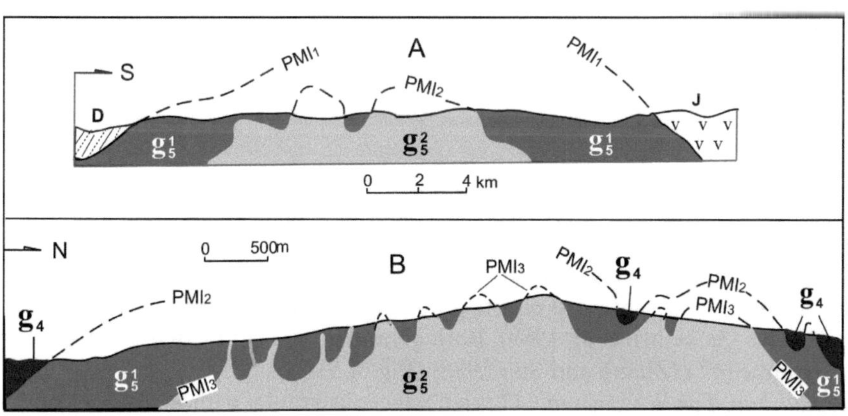

Figure 4.32. **A**. Cross sections of the Yangcaogang CG, eastern Guangdong Province and **B**. Tong'an CG, southern Jiangxi Province. g_4, g_5^1 and g_5^2 denote the oldest, younger and youngest granite members, respectively. D = Devonian (metashale–sandstone); J = Jurassic (volcanic rocks). Horizontal scale = vertical scale.

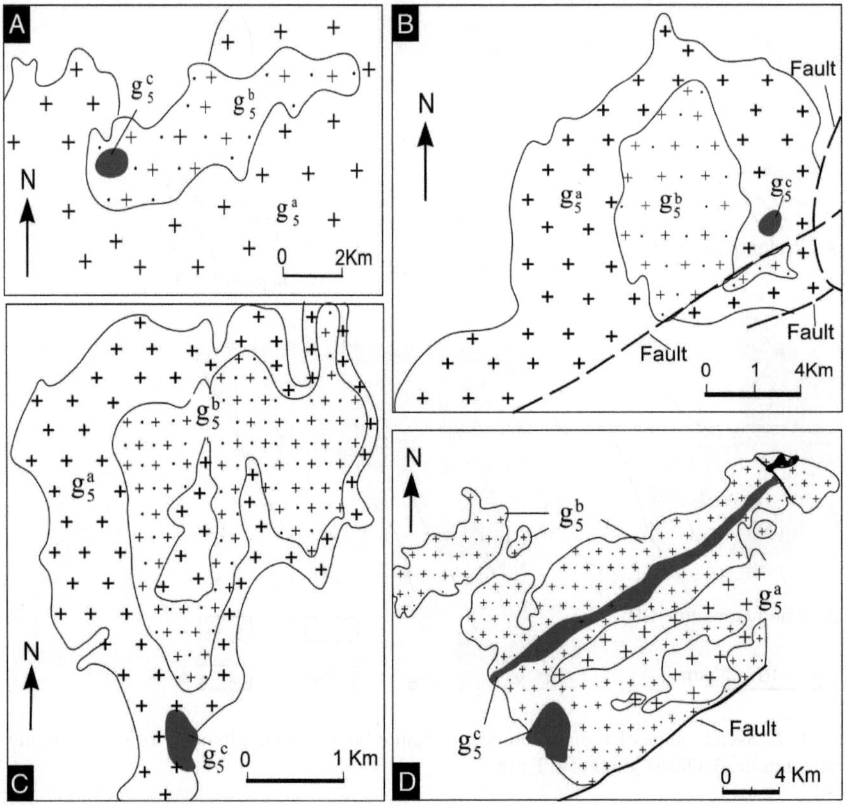

Figure 4.33. Simplified geological maps of four Mesozoic composite granite (CG) bodies in SE China. **A**. Part of the Jiufeng CG, northern Guangdong Province; **B**. Dengfuxian CG, eastern Hunan Province; **C**. Qianlishan CG, southeastern Hunan Province; **D**. Wuguishan CG, middle-south Guangdong Province; g_5^a = oldest, g_5^b = younger, and g_5^c = youngest granite member.

and a smaller body of fine-grained muscovite granite (g^{5c}) dated at 109 Ma (muscovite K-Ar age) (Fig. 4.33B).

Qianlishan. The oldest member of this CG is a porphyritic biotite granite (g^{5a}) of early Jurassic age (162 Ma/Rb-Sr; 172 Ma/K-Ar), that contains a second stage medium–coarse-grained biotite granite (g5b) dated at 139 Ma (K–Ar) and 141 Ma (Rb–Sr) (Chen et al. 1989). Both g^{5a} and g^{5b} are cut by fine-grained two-mica granite (g^{5c}) (Zhang and Sun 1988) (Fig. 4.33C).

Wuguishan. The Wuguishan CG is a Cretaceous granite with K–Ar ages ranging from 136–108 Ma (Cai and Zheng 1997). The oldest granite (g^{5a}) occurs within a younger granite (g^{5b}) that is cut by the youngest granite (g^{5c}) (Fig. 4.33D). All the granites are porphyritic biotite granites although their porphyritic crystal contents decrease from old to young (Cai and Zheng 1997).

CHAPTER 5

DIFFERENTIATION OF MAGMA LAYER: GEOCHEMICAL CONSIDERATIONS

5.1. INTRODUCTION

Over the last three decades, numerous studies have attempted to identify the sources of granitic magmas using bulk rock, mineral chemistry and isotope systematics (among many others, Chappell and White 1974; Faure 1977; Ishihara 1977; Taylor 1978; Ayres et al. 1997). Such studies are based on the prerequisite that granite magmas are sourced from deeper levels and that the magmas have ascended to a shallow depth of emplacement. Yet, as shown in Chapter 1, none of the key models of the two knowledge systems, e.g. continental geology and plate tectonics, explain the origin of granite (Fig. 1.3). This means that the current paradigm of the magma intrusion model to explain the origin of granite is uncertain, and if granites are not derived from exogenic sources then geochemical data relating to their origin needs to be re-evaluated.

Unlike the granite intrusion model, the in situ melting model for granite formation regards the mid-upper crust as a closed system that requires only the input of energy rather than material from deeper levels. On a crustal scale, as discussed in Chapters 3 and 4, exposures of individual granite bodies are explained as the protruding parts of the upper surface of an extensive layer-like body of granite. On this basis, we discuss aspects of the chemical and isotopic evolution within the granite magma layer.

5.2. COMPOSITIONAL VARIATION

Experimental data given in Chapter 2 and Appendix 2 clearly demonstrates that granitic melt compositions change with temperature, pressure, fO_2 and whether free H_2O is present or not, depending on the composition of the protolith undergoing melting. Thus, a wide variety of granite magma compositions can be expected as summarised in Fig. 2.8. As shown in Fig. 3.11, in a convecting magma layer mafic enclaves tend to move down and hot felsic liquids move upward which results in vertical compositional variation of the magma layer. This variation is reinforced during cooling with fractional crystallisation of near-liquidus phases such as calcic plagioclase, biotite, hornblende, pyroxenes, Fe–Ti-oxides, and accessories such as monazite, titanite, apatite, zircon etc., that move downward gravitationally (cognate cumulates) (Wall et al. 1987) relative to a buoyant, largely felsic liquid component. This means that chemical and isotopic compositions of granites more likely reflect in situ differentiation of the magma rather than different magma sources.

Chemical and isotope data from a deep borehole drilled into the 137 Ma Xiaotaoyuan biotite granite, North Guangdong Province, SE China, indicates increasing total iron + MgO, CaO, ΣSr, Ce, δEu and decreasing SiO_2, Na_2O and initial Sr-ratio (ISr) with depth, representing vertical compositional zonation of the granite (Fig. 5.1). Mineralogically, these changes essentially reflect downward increasing biotite and Ca-enrichment of plagioclase, and upward increasing quartz and more Na-rich plagioclase. There is abundant evidence of similar top-downward, and margin-inward, compositional variation in granite bodies in SE China and worldwide. Downward changes from muscovite (or two-mica) granite to biotite granite or granodiorite are commonly present in many SE China granite bodies, Zengjialong (Fig. 5.2A) , Niulingao (BGMRJx 1984), Huangsha (Xia and Liang 1991), Xianghualing and Laizhiling (Wang and Huang 2000).

The Weiya granite, located about 135 km southeast of Bomi city, Xinjiang, NW China, is K-Ar dated at between 200 and 259 Ma. The central part of the granite body consists of K-feldspar granite that grades outwards through less calcic to

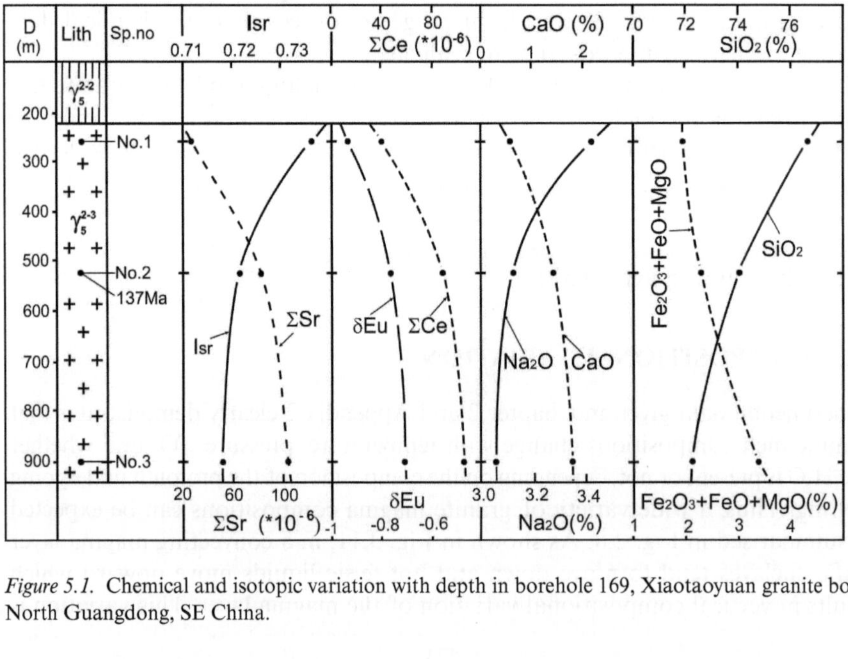

Figure 5.1. Chemical and isotopic variation with depth in borehole 169, Xiaotaoyuan granite body, North Guangdong, SE China.

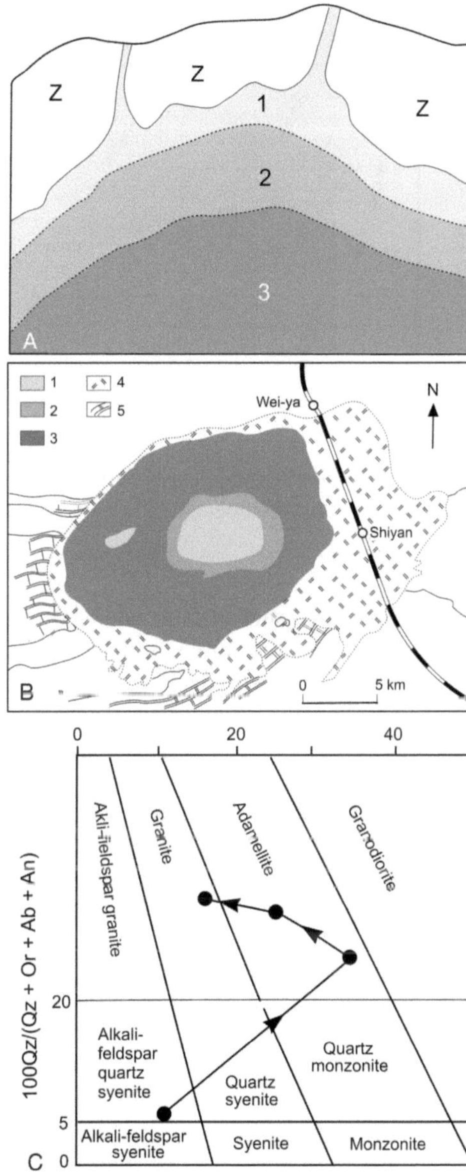

Figure 5.2. **A**. Cross section of the Zengjialong granite, Jiangxi Province, SE China, showing vertical compositional variation. Z. Sinian metasedimetary rocks; 1. Muscovite granite; 2. Two mica granite; 3. Biotite granite (redrawn after Zhu et al, 1981). **B**. Geological map of the Weiya zoned granite, Xinjiang, NW China (after BGMRxj 1993). **1**. K-feldspar granite; **2**. Adamellite; **3**. More calcic adamellite; **4**. K-feldspar quartz syenite, **5**. Carbonate wall rock in the SW and with older granodiorite, diorite, and Proterozoic metasedimentary country rock elsewhere. **C**. Normative ratio plot of granite compositions associated with Weiya composite granite in B, with arrows indicate direction of younging (between 1 and 2) and differentiation (between 2 and 4).

more calcic adamellite that has a sharp, fine-grained contact with an outer and older K-feldspar quartz syenite (Fig. 5.2B, C). The inner granite–adamellite part of this domed composite granite may be explained by in situ differentiation with upward increasing SiO_2, K_2O, and downward increasing Al_2O_3, FeO, Fe_2O_3, MgO and CaO implying cooling from the roof downward and is considered to be the result of remelting of the K-feldspar quartz syenite.

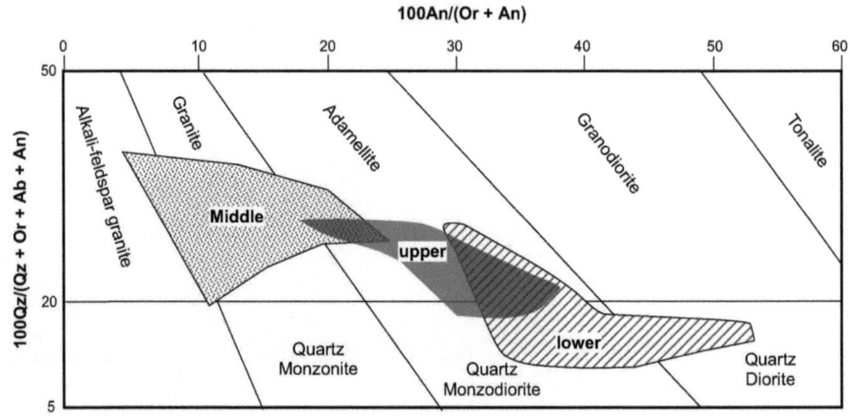

Figure 5.3. Normative ratio plot of granite compositions in the vertically zoned Searchlight pluton, Nevada, USA (data from Bachi et al. 2001). The quartz monzonite is considered to be the undifferentiated magma composition.

Examples of other granite bodies that show composition zoning include the Searchlight, Fichtelgebirge, and Kurobegawa granites. The 80 km^2 Searchlight Pluton, Nevada, USA (Bachi et al. 2001), provides a rare exposure of a vertically zoned ~10 km thick granite body with fine-grained quartz monzonite below the roof, ~2 km of medium-grained granite in the centre underlain by ~6 km of coarser, more mafic quartz monzonite. Field relations, elemental and isotope data suggest that the upper part formed as a descending solidification front from the roof and represents the primary granite magma composition. In the more slowly cooled interior, the granite and mafic quartz monzonite parts represent complimentary segregated melt and crystal-rich cumulate, respectively (Fig. 5.3).

The Fichtelgebirge granite located on the borders of Germany and the Czech Republic, consists of muscovite granite, two-mica granite and porphyritic biotite granite (Fig. 5.4). Hecht et al. (1997) consider that the different granites were derived from different sources. As there is no significant age difference between the granites (326 ± 2 Ma; Carl and Wendt 1993), it is difficult to understand why and how the source regions for the different granite magmas changed at the same time and why these magmas were 'emplaced' at the same location? An alternative explanation for the distribution of the granite types could be that it refects the erosion difference of various lithologic zones formed in the upper part of granite body as shown in Fig. 5.2, which can be reasonably explained as the result of in situ differentiation with Fe, Mg, Ca and Ti, increasing and Si, Na decreasing from muscovite granite through two-mica granite to biotite granite (Table 5.1).

The vertically-zoned Kurobegawa granite, Hida Mountains, central Japan (Wada et al. 2004), is exposed as a ~6 km-thick tilted section. The upper part consists of fine-grained porphyritic granite–adamellite that grades downwards to a medium grained equigranular–porphyritic adamellite–granodiorite (Fig. 5.5).

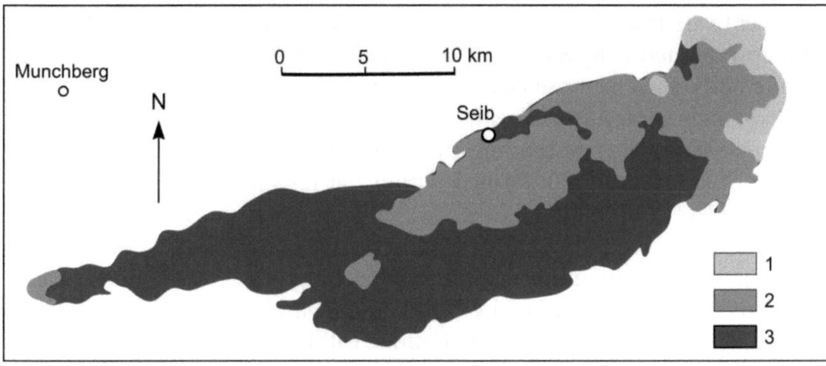

Figure 5.4. The Fichtelgebirge granite body, central Europe, showing distribution of: **1**. Muscovite granite; **2**. Two-mica granite; **3**. Porphyritic biotite granite, (redrawn from Hecht et al. 1997).

Table 5.1. Mean values (wt %) of whole-rock major elements of OIC, Fichtelgebirge granite

Lithology	S. N	SiO_2	TiO_2	Al_2O_3	Fe_2O_{3tot}	MgO	MnO	CaO	Na_2O	K_2O	P_2O_5
Biotite	21	68.29	0.63	15.11	3.71	1.09	0.059	2.04	3.24	4.59	0.31
granite	11	70.83	0.45	0.27	2.57	0.73	0.042	1.45	3.17	4.74	0.25
Two-mica	8	72.61	0.27	14.60	1.60	0.45	0.033	1.03	3.50	4.79	0.20
granite	14	73.38	0.13	14.50	1.07	0.24	0.021	0.53	3.44	4.75	0.32
Muscovite granite	6	73.86	0.07	14.57	0.56	0.07	0.024	0.44	3.83	4.27	0.36

(Data from Hecht et al. 1997)

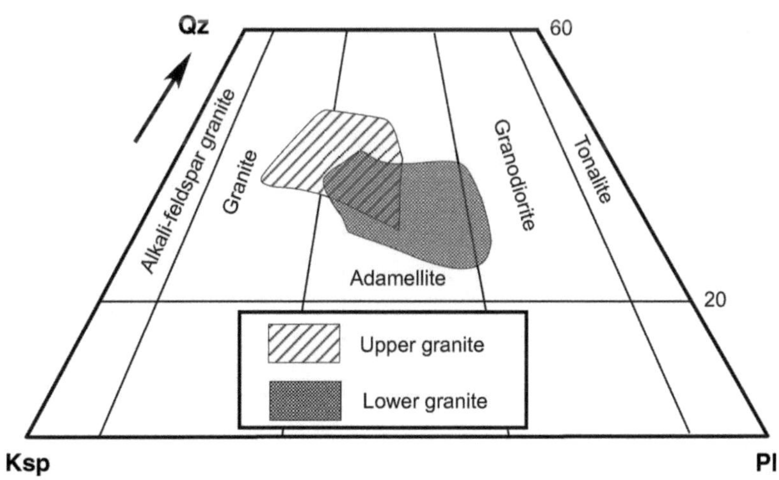

Figure 5.5. Modal compositions in terms of Qz-Ksp-Pl components for the upper and lower granites of the vertically zoned Kurobegawa granite, Hida Mountains, Japan (redrawn from Fig. 3 of Wada et al. 2004).

Major element variation indicates downward increasing Al_2O_3, Fe_2O_3, MgO, CaO, TiO_2 and upward increasing SiO_2, K_2O consistent with fractional crystallisation.

A granite body, such as those described above, represents the domed part of a PMI (paleo-magma interface). Because of differentiation of magma below the MI, i.e. mafic components descending and felsic components ascending, at a particular level of erosion granite formed within the region of melting should be more felsic, or less mafic than its host rock. Where country rock lithologies are distinctly different, there is often a correlation between granite-type and host protolith. A case in point are granite bodies of the Carmanville area, northeastern Newfoundland, Canada, described by Currie and Pajari (1981) in terms of in situ melting of different protoliths (Fig. 5.6). Granite within mafic rocks has a tonolite–granodiorite composition whereas those outcropping within adjacent psammitic–semipelitic rocks are biotite granite with porphyritic to megacrystic Ksp, and granite within more aluminous semipelitic–pelitic rocks is muscovite–biotite granite, muscovite granite occasionally with garnet. One granite body associated with metavolcanic and semipelitic–pelitic rocks is concentrically zoned with a core of granodiorite, grading outward through two-mica granite to muscovite– garnet leucogranite (Fig. 5.6). In north China, quartz monzonite and granodiorite

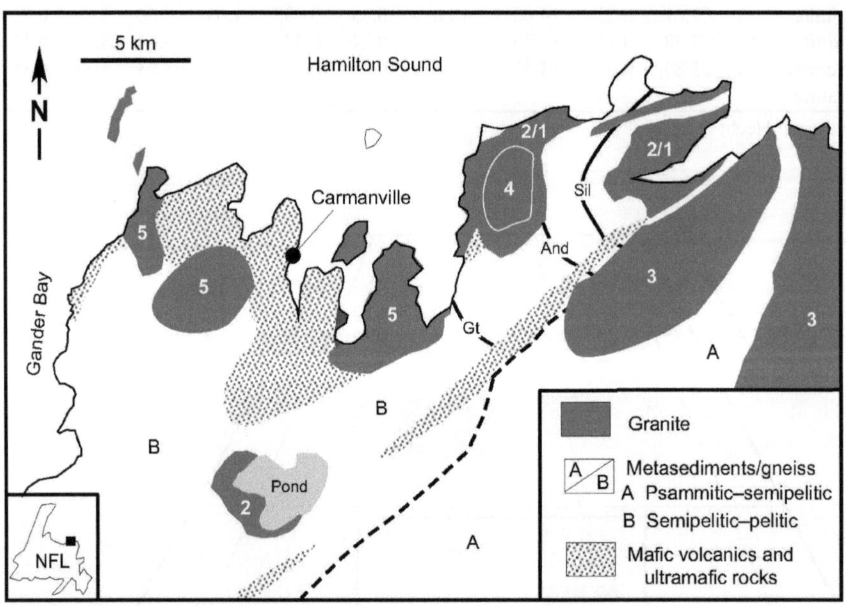

Figure 5.6. Geological sketch map showing distribution of granite, metasediments and metabasaltic rocks, Carmanville area, northeastern Newfoundland, Canada (after Fig. 1a of Currie and Pajari 1981). For the granites, **1.** = Muscovite–granite leucogranite; **2.** = Muscovite–biotite granite; **3.** = Biotite granite; **4.** = Granodiorite; **5.** = Tonalite + minor granodiorite. Isograds; Gt = Garnet, And = Andalusite, Sil = Sillimanite.

bodies are associated with metasedimentary sequences that contain calcareous lithologies such as limestone and volcanics (Table 5.2).

If the country rock protoliths are relatively homogeneous, e.g. granite, then geochemical evidence supporting melting to form new granite is stronger, despite limitations imposed from sampling. Experimental evidence and natural examples of melting of granites indicate that new melts are distinctly more Si and K rich than the parent compositions as shown in Fig. 5.7, the exception being where the

Table 5.2. Country rocks of quartz monzonite-granodiorite in North China

Granite body	Lithology	Country rock	Age (Ma)
Fangshan	Granodiorite	Limestone, sandstone-shale	Kb130–145
Heishanzhai	Quartz monzonite	Limestone, sandstone-shale	
Heijiachuan	Quartz monzonite	Limestone, intermediate volcanics	Kb134
Maoshan	Granodiorite	Metamorphic rock	
Yangshudixia	Granodiorite	Overlain by J_3 volcanics	Kb148
Shaoguodi	Granodiorite	Metamorphic rock	Kb157
Baijiadian	Granodiorite	Limestone, metamorphic rock	Kw174
Shouwangfen	Granodiorite	Limestone, volcanics	Sr129
Wanganzhen	Granodiorite	Limestone, volcanics	Kb138
Jianchang-Caobai	Granodiorite	Volcanics, metamorphic rock	
Daheishan	Quartz monzonite	Limestone, volcanics	Kw108–133
Shahewan	Quartz monzonite	Slate	
Penglai	Quartz monzonite	Metamorphic rock	
Tongguanshan	Quartz diorite	Limestone	

(After Zhang and Sun 1988); Kb = biotite K-Ar age; Kw = whole rock K-Ar age; Sr = Rb-Sr age

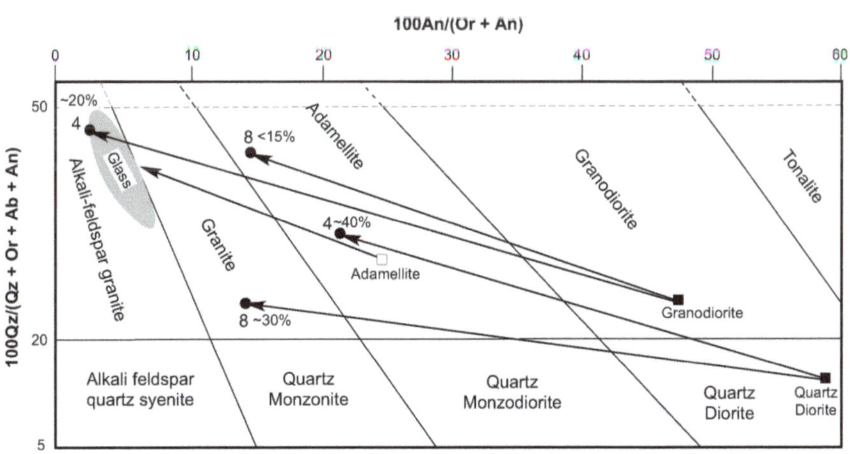

Figure 5.7. Normative ratio plots of melt compositions derived from an example of natural melting of an adamellite, Sierra Nevada (Davies and Tommasini 2000), and experimental melting of granodiorite and quartz diorite at 4 and 8 kb (labelled points) (data from Patiño-Douce 1997). Pressure and volume % melt produced in the experimental examples are indicated.

host granite is a K-feldspar granite or K-feldspar quartz syenite (Fig. 5.4). In the Hunan Province, several Mesozoic granite bodies (γ_5) occur within Paleozoic Caledonian-age granite (γ_3). Mean compositions of three younger granites and their granite protoliths are given in Table 5.3 and all analyses are plotted in Fig. 5.8A. As expected, the younger granites have higher SiO_2, K_2O and lower iron, MgO, CaO, TiO_2 that their inferred adamellite–granodiorite Caledonian granite protoliths. Examples where composite granites comprise three granites of

Table 5.3. Average compositions (wt %) of Mesozoic granite (γ_5) and respective Caledonian-age granite country rocks (γ_3) in Hunan Province, SE China

	Qijia granite		Liangfengjie granite		Douzhashan granite	
	γ_3	γ_5	γ_3	γ_5	γ_3	γ_5
SiO_2	70.77	72.74	68.94	72.93	70.09	73.77
TiO_2	0.46	0.32	0.52	0.20	0.31	0.10
Al_2O_3	13.39	14.56	14.97	14.38	14.30	13.49
Fe_2O_3	1.12	0.52	3.16	1.42	3.05	0.60
FeO	2.60	1.63	2.57	1.36	2.36	0.71
MnO	0.05	0.04	0.20	0.05	0.04	0.05
MgO	0.95	0.47	1.61	0.56	1.20	0.14
CaO	1.86	0.51	2.81	1.45	2.18	0.48
K_2O	4.61	4.96	3.59	4.36	4.13	4.43
Na_2O	2.68	2.86	2.80	2.77	2.36	3.10

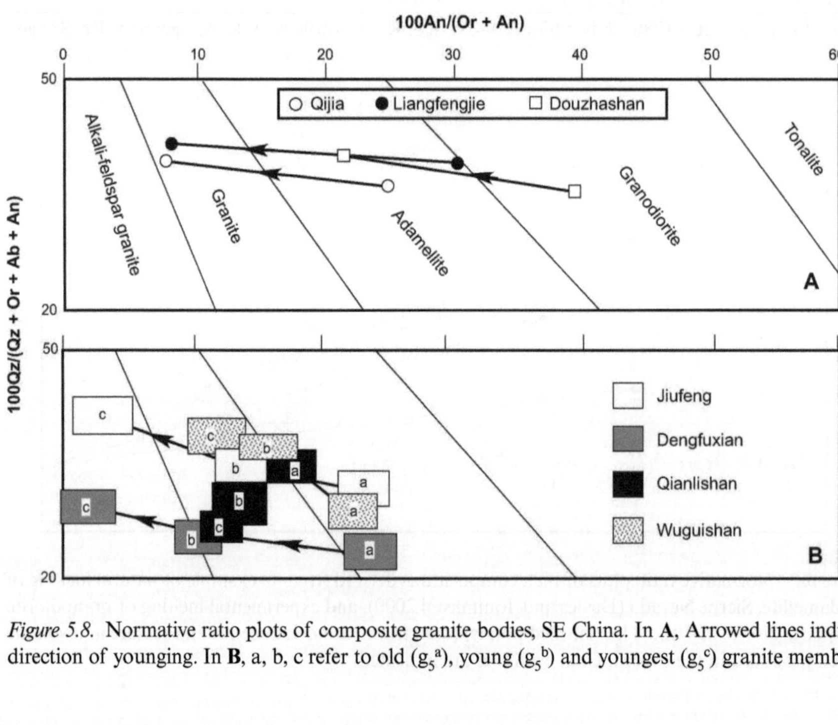

Figure 5.8. Normative ratio plots of composite granite bodies, SE China. In **A**, Arrowed lines indicate direction of younging. In **B**, a, b, c refer to old (g_5^a), young (g_5^b) and youngest (g_5^c) granite members.

different ages are the Jiufeng, Dengfuxian, Qianlishan and Wuguishan granites described in Chapter 4 and shown in Fig. 4.33. Average analyses of these composite granites are listed in Table 5.4 and all data are plotted in Fig. 5.8B. The younger granites are invariably more silica-rich, and they have higher D.I. and alkali/alumina (NK/A) ratios, generally higher Fe_2O_3/FeO, and lower total Fe, CaO (normative An), MgO, TiO_2 and P_2O_5, consistent with their less mafic, more felsic (and oxidized) compositions that would be expected at approximately the same level of erosion in the in situ melting model of granite formation. Compositions change from adamellite (oldest) to granite and K-feldspar granite for the younger granites. Most of the Mesozoic composite granite bodies in SE China show similar compositional variations as those listed in Table 5.4 (Mo et al. 1980; Chen 1987; Yu et al. 1987; Zhang and Sun 1989; Meng 1993), and in this respect it is interesting that Paleozoic (Caledonian), Early Mesozoic

Table 5.4. Average compositions (wt % and norms) of composite granite bodies, SE China

Granite	Jiufeng			Dengfuxian			Qianlishan			Wuguishan		
Age	a	b	c	a	b	c	a	b	c	a	b	c
SiO_2	71.07	73.16	75.04	65.04	73.16	74.88	74.20	75.05	74.69	71.70	75.60	76.08
TiO_2	0.34	0.21	0.06	0.63	0.14	0.03	0.19	0.06	0.04	0.38	0.15	0.19
Al_2O_3	13.95	12.26	13.05	16.00	14.16	14.05	12.68	12.71	13.33	13.43	12.28	12.19
Fe_2O_3	0.45	0.36	0.20	0.33	0.47	0.38	0.64	0.52	0.34	0.82	0.46	0.50
FeO	2.76	2.24	0.96	3.47	1.25	0.97	1.50	1.09	0.80	2.06	1.33	1.07
MnO	0.08	0.06	0.09	0.07	0.06	0.12	0.07	0.04	0.04	0.10	0.09	0.10
MgO	0.74	0.46	0.07	1.22	0.33	0.18	0.25	0.10	0.08	0.66	0.17	0.29
CaO	1.83	1.10	0.28	2.18	0.84	0.28	1.21	0.92	0.71	1.59	0.95	0.74
Na_2O	2.78	2.84	2.82	3.16	3.49	4.49	3.13	3.39	4.32	3.26	3.14	3.57
K_2O	4.96	5.16	5.26	5.19	4.66	2.99	5.03	5.00	4.75	4.90	4.80	4.51
P_2O_5	0.13	0.07	0.00	0.24	0.19	0.26	0.05	0.03	0.02	0.09	0.03	0.01
LOI				2.15	1.27	1.37				0.65	0.47	0.61
Total	**99.09**	**97.92**	**97.83**	**99.68**	**100.02**	**100.00**	**98.95**	**98.91**	**99.12**	**99.64**	**99.47**	**99.86**
Q	29.51	33.08	38.04	19.11	21.75	36.84	33.38	33.80	29.60	28.93	35.96	35.63
Or	29.61	31.1	31.73	31.90	27.95	17.91	30.02	33.44	28.31	29.25	28.60	26.89
Ab	23.78	24.54	24.37	27.42	29.95	38.50	26.74	29.02	36.89	27.84	26.82	30.46
An	8.33	5.10	0.98	9.48	2.98		5.73	4.41	2.98	7.40	4.57	3.66
C	0.98	0.19	2.42	1.85	2.33	3.49	0.01	0.11	0.40	0.09	0.35	0.10
Hy	6.21	4.84	1.88	8.43	2.64	2.10	2.69	1.81	1.21	4.36	2.67	2.33
Ilm	0.65	0.40	0.11	1.23	0.27	0.06	0.36	0.11	0.08	0.72	0.28	0.17
Mt	0.65	0.54	0.29	0.49	0.70	0.57	0.94	0.77	0.49	1.20	0.67	0.72
Ap	0.30	0.17	0.16	0.58	0.44	0.60	0.12	0.07	0.05	0.21	0.07	0.02
D.I.	91.2	93.9	95.1	87.5	93.6	93.3	95.9	97.1	97.8	93.4	96.0	96.6
X_{Mg}	32.4	24.2	9.7	36.6	25.7	19.4	17.5	10.1	11.3	29.7	21.5	25.3
X_{Ca}	26.7	17.6	5.3	27.6	11.7	3.3	17.6	13.0	8.4	21.3	14.3	10.3
NK/A	0.72	0.83	0.79	0.68	0.76	0.75	0.84	0.86	0.92	0.80	0.85	0.88
Fe_2O_3/FeO	0.16	0.16	0.21	0.10	0.38	0.39	0.43	0.48	0.43	0.40	0.35	0.47

D.I. = Differentiation Index (normative: Qz, Or, Ab, Lc, Ne)

X_{Mg} = Mg/(Mg + Total Fe); X_{Ca} = CaO/(Ca + Na); NK/A = molar $Na_2O + K_2O/Al_2O_3$

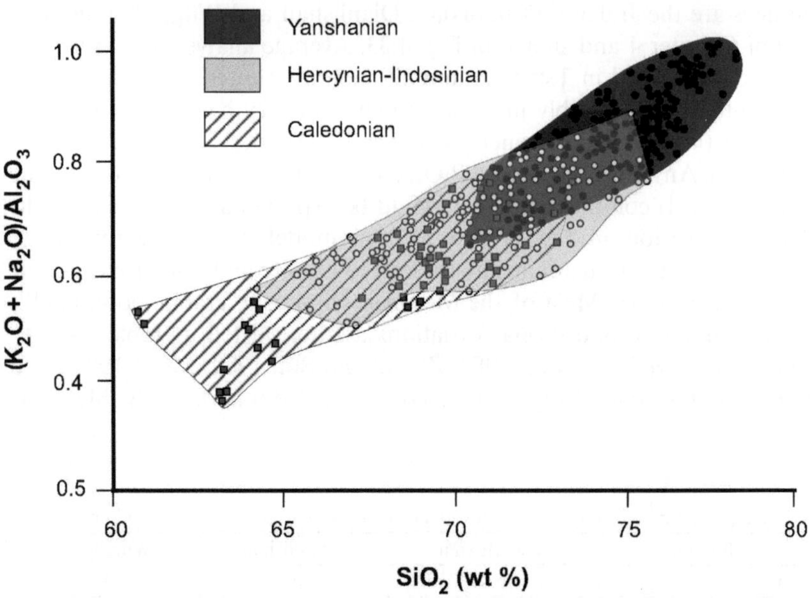

Figure 5.9. SiO_2 verses $(K_2O + Na_2O)/Al_2O_3$ showing composition fields for Caledonian, Hercynian–Indosinian, and Yanshanian granites from the Nanling area, SE China.

(Hercynian–Indosinian) and Late Mesozoic (Yanshanian) granites in SE China show a trend of increasing SiO_2 and alkali/alumina ratio with younging (Fig. 5.9), possibly indicating episodes of crustal-scale heating involving granite production and remelting. Similarly, compositional variation with younging of the Late Mesozoic granites involves increasing SiO_2, Fe_2O_3/FeO, decreasing Al_2O_3, FeO, MgO, TiO_2 and K_2O (Table 5.5).

The chemical variations described above underscore the premise that in situ crustal melting involves simultaneous melting and crystallisation in the source region. The rate of melting of roof rock determines the heat lost from the underlying magma and differentiated granite bodies such as those described above suggest low heat-conductivity of the roof rocks during their crystallisation (Figs. 5.10A1, B1). If the rate of heat loss is high enough, melt near the MI will likely crystallise rapidly to produce a relatively mafic and fine-medium grained marginal facies (Fig. 5.10A2) with the effect of creating a thermal barrier that insulates the main melt body and prolongs the rate of cooling/crystallisation (Fig. 5.10B2).

A cross section of the Shalang–Xuemuna granite body, Tibet (Fig. 5.11) illustrates the case depicted in Fig. 5.10B2. Biotite granite is exposed in the central part of the granite body and grades outwards through granodiorite to a quartz diorite border facies that contains abundant xenoliths and mafic enclaves (EGCAS 1981). An analogous example, the Criffell granite, southern Scotland,

Table 5.5. Mean compositions and CIPW norms of Mesozoic biotite granites, Nanling area, SE China

Granite	1	2	3
n	88	166	46
SiO_2	72.46	73.46	74.09
TiO_2	0.28	0.22	0.20
Al_2O_3	13.59	13.35	13.20
Fe_2O_3	0.94	0.72	1.02
FeO	1.99	1.88	1.55
MnO	0.07	0.07	0.06
MgO	0.48	0.40	0.39
CaO	1.16	1.23	1.06
Na_2O	3.13	4.10	3.39
K_2O	5.06	4.91	4.65
LOI	0.68	0.58	0.55
Total	99.84	100.92	100.16
Q	30.78	26.85	33.06
Or	30.14	28.90	34.03
Ab	26.74	34.61	28.77
An	5.80	3.49	5.26
C	0.86	2.23	0.67
Hy	3.77	2.46	2.77
Ilm	0.53	0.42	0.38
Mt	1.38	1.04	1.48
D.I.	93.5	93.9	94.7
***X*Mg**	23.0	22.1	21.9
***X*Ca**	17.0	14.3	14.7
NK/A	0.78	0.90	0.81
Fe_2O_3/FeO	0.25	0.38	0.76

1,2,3 refer to oldest, younger and youngest granites respectively.
D.I. = Differentiation Index (normative: Qz, Or, Ab, Lc, Ne)
*X*Mg = Mg/(Mg + Total Fe)
*X*Ca = Ca/(Ca + Na)
NK/A = molar $Na_2O + K_2O/Al_2O_3$

is illustrated in Fig. 5.12A (differentiation index contours) and B which shows a central biotite granite surrounded by marginal granodiorite. The contact between the two granites is delineated by the appearance of alkali feldspar megacrysts that characterise the biotite granite (Stephens and Halliday 1979). Differentiation Index contours are concentric to the map centre of the body and show a steep gradient between both rock types. This data could be interpreted as crystallisation of the magma inward and downward from the contact with earlier formation of the marginal granodiorite that slowed the rate of heat loss resulting in the formation of a central K-feldspar megacryst-bearing biotite granite with high Differentiation Index. A further example, the Shouwangfen massif in Hebai, North China, exhibits an increase in silica (modal quartz) towards the central part of the granite (Fig. 5.13), interpreted to represent downward and inward crystallisation beneath a domed MI.

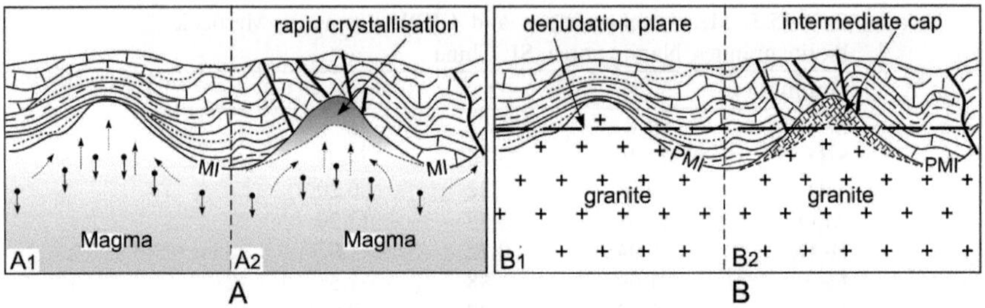

Figure 5.10. Model for chemical and petrographic zonation of a granite body; MI = magma interface. **A**. During crystallisation. **(A1)** Little heat is lost through country rocks above the domed MI, below which the magma differentiates. Arrowed lines terminated with black dots denote downward movement of roof rock xenolith material and early-crystallising Ca-plagioclase and mafic minerals; Arrowed dashed lines indicate direction of movement of felsic melt component; **(A2)** Below areas where there is more rapid heat loss, perhaps due to fracturing of the roof rocks and cycling of meteoric water in rocks above the MI, magma in the region of the domed MI crystallises to form a thermal barrier below which the magma can also differentiate. **B**. Following crystallisation. **(B1)** (corresponding to A1). Granite body with normal compositional zonation; **(B2)** (corresponding to A2). A cap of intermediate granitic rock (tonalite to granodiorite) forms in the domed region of the paleo-melt layer (PML). If the erosion surface intersects the PMI, the exposed granite (in plan view) will have a fine-medium grained marginal part of diorite–granodiorite composition.

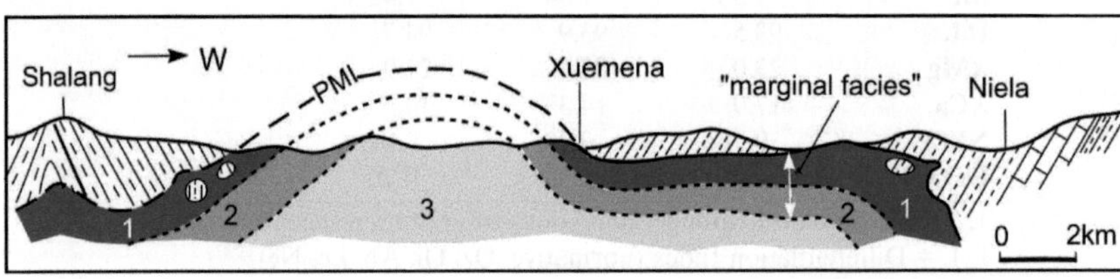

Figure 5.11. E-W section across the Shalang–Xuemuna granite, Tibet, showing marginal zoning (modified after EGCAS 1981). **1**. Quartz diorite; **2**. Granodiorite; **3**. Biotite adamellite. Horizontal scale = vertical scale.

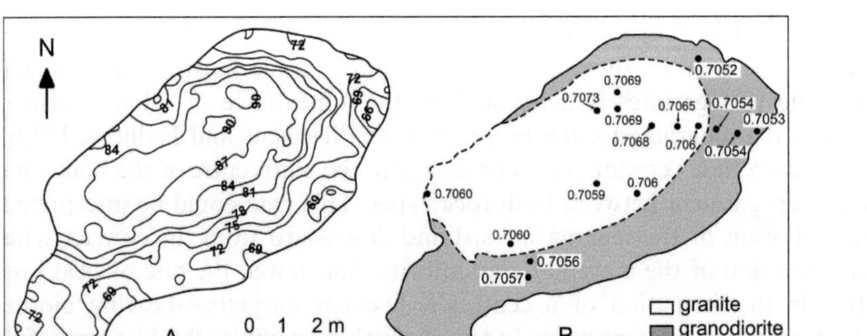

Figure 5.12. The Criffell granite, Scotland (modified from Stephens and Halliday 1979). **A**. Contours of Differentiation Index (Σnormative Qtz Or Ab); **B**. Distribution of granodiorite and biotite granite together with initial $^{87}Sr/^{86}Sr$ (ISr) values.

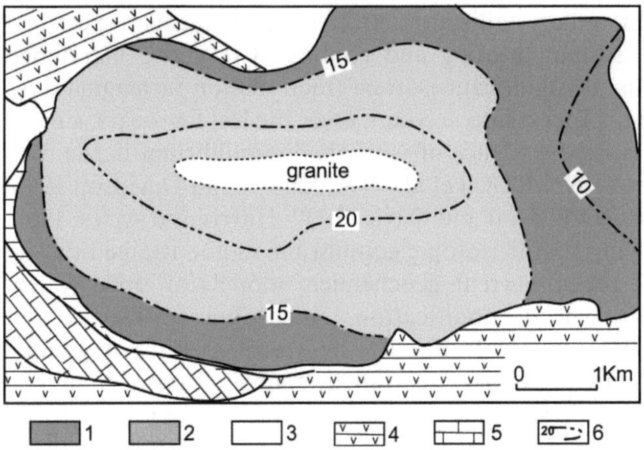

Figure 5.13. Geological map of the Shouwangfen granite, Hebei Province, North China (modified from Zhang and Sun, 1988). **1**. Quartz monzonite; **2**. Granodiorite; **3**. Granite; **4**. Tuffaceous metasediments; **5**. Marble; **6**. Modal % quartz contours.

It can be noted that if the border facies of the Shouwangfen, Criffell or Shalang–Xuemuna 'plutons' are significantly thick and the level of erosion has not intersected their central parts, only diorite/tonalite or granodiorite will be exposed. This means that such granites do not have to be derived from a deep (lower crustal) mafic source, and could indicate either a high degree of melting of a calcareous mica-bearing metasedimentary or a volcanoclastic protolith with sufficiently high CaO (see Chapter 2). In this respect, it is interesting to note that the country rocks and inferred protoliths of most quartz monzonite–granodiorite bodies in eastern China are calcareous metasedimentary sequences (Zhang et al. 1988) (Table 5.3).

5.3. STRONTIUM ISOTOPES

Characterisation of granite sources using initial $^{87}Sr/^{86}Sr$ ratios (ISr) is based on the following assumptions (Faure 1977);

1. The $^{87}Sr/^{86}Sr$ ratio of mantle rocks is lower than crustal rocks.
2. There is no significant isotope fractionation in magmas.

Many crustal rocks have present-day $^{87}Sr/^{86}Sr$ ratios (PSr) approximating that of the present-day mantle (0.704 ± 0.002; Faure 1977), e.g. Cenozoic granites of Japan (Shibata et al. 1978) and the Dexing granodiorite in SE China (Chen et al. 1996), which have a PSr of between ~0.704 and 0.706. Wherever such rocks were derived from, if they melt without isotopic exchange, their future ISr should be the same as their pre-melted PSr values. This ideal situation indicates that granites with lower ISr are not necessary derived from the mantle or a mafic lower crust.

Melting of a protolith and crystallization of the melt represents respectively different states of the system: heating and cooling. The above mentioned assumption that there is no significant isotope fractionation in magma does not take the rock-melting process into account. Over the last few years, a number of studies have presented evidence of isotopic disequilibrium in the rock melting process (Barbero 1995; Brown et al. 1996; Knesel and Davidson 1996; Harris et al. 1997, 2000; Tommasini and Davies 1997; Harris and Ayres, 1998; Sallet et al. 2000) implying that if isotopic equilibrium is not attained during the melting of crustal rocks, current geochemical models of open-system magmatic processes will need modification. According to Davies and Tommasini (2000), melts produced from a low degree of melting of granite, have significantly higher ISr values than their source granite. Similarly, feldspar rims have higher ISr values than (relic) cores. This data implies that initial melts in crustal rocks likely have higher $^{87}Sr/^{86}Sr$ ratios than the residual protolith. Gravitational differentiation of residual (restite) assemblages from derivative felsic magma will thus tend to segregate radiogenic Sr and in highly evolved granite bodies, ISr values should decrease downwards and inwards from the contact.

The isotopic abundance of ^{87}Sr is ~7.04% and Sr^{2+} usually replaces Ca in Ca-bearing minerals such as calcic plagioclase (in particular), calcic amphibole and pyroxene, epidote, apatite and titanite (Liu et al 1984) that concentrate as residual phases during melting and also form early crystallising phases in granodiorite–tonalite–quartz diorite magmas. This implies that variation of Sr should be inverse to that of ISr, and show a downward/inward increase in a particular granite body, e.g. the Xiaotaoyuan granite where ISr values decrease downward (Fig. 5.1) while Sr increases.

Another example is the Xingluokeng granite that is separated by a narrow screen of metasediment into a northern body of fine-grained muscovite granite (marginal facies) and a southern body of medium to coarse-grained biotite granite (central part) (Fig. 5.14). The marginal facies has been eroded along the southern contact. Whole rock compositions, ISr and isochron ages (BGMRFj 1985) of both granites (Fig. 5.14) indicate that the biotite granite is slightly younger with lower ISr and higher Sr consistent with higher CaO, and FeO + MgO reflecting a deeper erosion level. Similarly, in the Criffell granite (Fig. 5.12B), where the marginal facies is granodiorite, ISr values are lower (0.7052–0.7057) than in the central K-feldspar-bearing biotite granite (0.7060–0.7073).

As the compositions of crustal rocks that have undergone partial melting will vary from terrane to terrane (and also within a single in situ melting terrane), ISr values of derivative granite magmas should be regionally variable (Chen et al 1996). Such variation has been documented in a number of areas (Kistler and Peterman 1973; Shibata et al. 1978; Zhang and Sun 1988). The Mesozoic granites and related volcanic rocks of the western North American Cordillera, USA, for instance, have ISr values between 0.7031 and 0.7094, generally increasing towards the SE (Fig. 5.15). This is believed to reflect compositional differences in Late Precambrian to Paleozoic protoliths of the granites (Faure 1977) because rocks

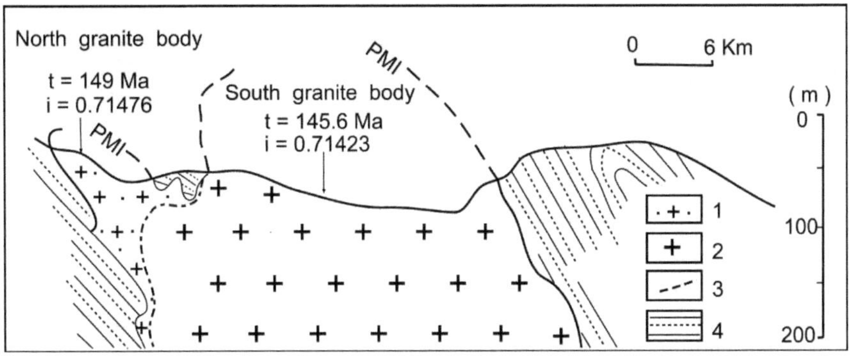

Figure 5.14. Schematic N-S section through the Xingluokeng granite, Fujian Province, SE China (redrawn after BGMRFj 1985); **1**. Muscovite granite; **2**. Biotite granite; **3**. Contact between the two granites; **4**. Cambrian metasedimentary rocks. PMI = Paleomagma interface.

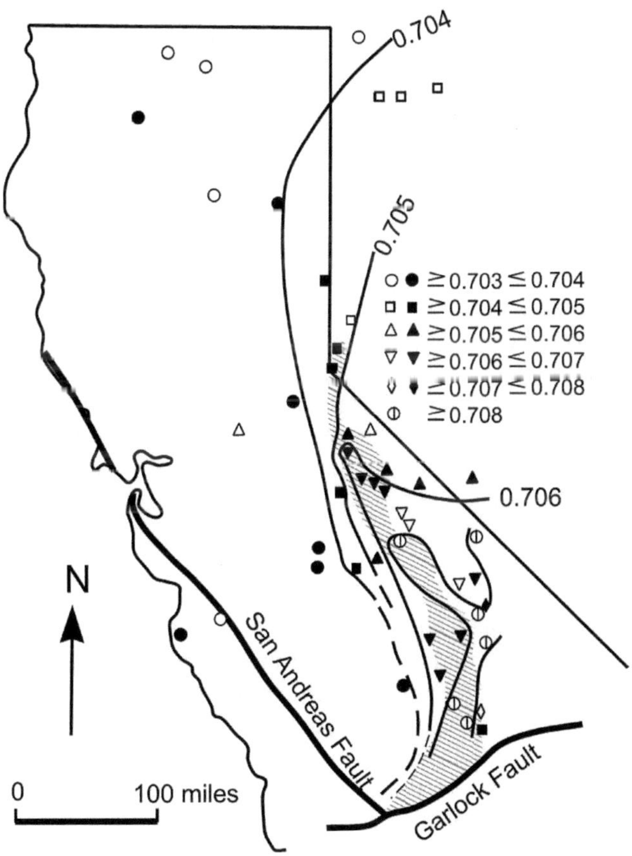

Figure 5.15. ISr contours related to North American Cordillera Mesozoic granite and volcanic rocks (after Kistler and Peterman 1973). Shaded area = undeformed platform sediments.

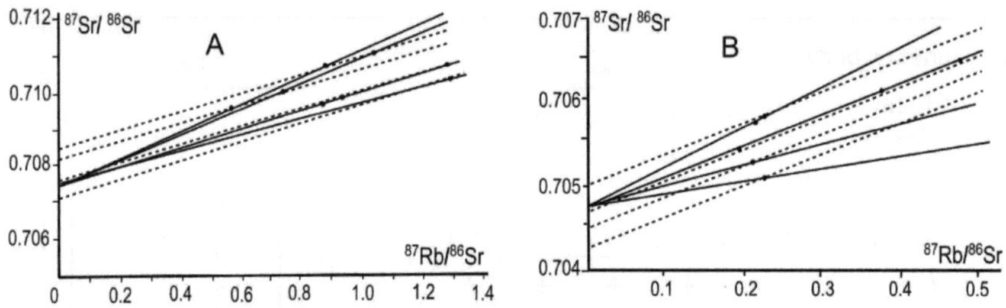

Figure 5.16. Possible Rb-Sr isochrons of the Xiaohekou granite (A) and Mangling granite (B) in Shanxi Province, NE China, showing bulk rock samples as having either different initial ISr values or different crystallisation ages (data from Yan 1985).

SE of the 0.706 ISr contour line in Fig. 5.15 are relatively undeformed sandstone-shale and carbonate, and rocks west of the line are deformed subduction zone greywacke, volcanics and chert.

Rb-Sr isochrons relating to single granite bodies are usually dispersive (Yan 1985). For a given ISr, samples will have different ages, forming a radial pattern on an $^{87}Sr/^{86}Sr$ versus $^{87}Rb/^{86}Rb$ diagram such as that shown in Fig. 5.16 in relation to the Xiaohekou and Mangling granites, NW China. However, for a given age, samples will have different ISr values, depicted as a series of parallel lines with the same slope but with different intercepts (Fig. 5.16). If the effect of late-stage alteration is excluded, granites from both bodies should have either different crystallisation ages or different ISr values. The latter seems more probable because in both cases the dated samples are from restricted localities where crystallisation ages are not expected to exceed the dating error.

It is therefore concluded that ISr values can not be used to constrain granite source because the Sr-isotope composition of a rock is related to that of the protolith or more likely a heterogeneous protolith involved in partial melting, the degree of melting and crystallisation, and the depth of erosion in the granite bodies from where the samples were collected.

5.4. OXYGEN ISOTOPES

Hydrogen in rocks is mainly combined with oxygen either in the form of bonded OH or unbonded H_2O in minerals, and as free H_2O in pores and along grain boundaries. When rock blocks are incorporated into the magma, water in them is subject to high temperatures, and isotopic hydrogen should preferentially migrate into the melt. This means that δD values of the rock block (xenolith) should be generally higher than that of the surrounding magma. Isotopic ^{16}O should also preferentially migrate into the melt. However, the total amount of oxygen combined with hydrogen to form water forms only a trivial part of a rock and most of the oxygen occurs in silicates or metal oxides requiring destruction of the crystal lattice to be released. Therefore, expulsion of ^{16}O during the dissociation of water should not significantly influence the bulk rock $\delta^{18}O$ value. Within a magma layer, the difference in mass between ^{18}O and ^{16}O would result in relative upward movement of ^{16}O and its possible depletion in the lower part of the magma layer. If so, the following phenomena should be evident: $\delta^{18}O$ values of country rock xenoliths in granite should be generally lower than their host granite and the opposite should be the case for δD values. A detailed hydrogen and oxygen isotope study of the Berridale batholith, Australia, by O'Neil and Chappell (1977), shows that the average δD values of xenoliths is higher and the $\delta^{18}O$ values generally lower than that of the host granite (Table 5.6). The same situation is found in the Hoedjiespunt granite described by Harris et al. (1997).

The tendency for upward movement of ^{16}O in a magma layer results in the '^{16}O-shift' effect and an increase in ^{18}O values from the top of a granite body downward or inward from the border. This is the case for the Xiaotaoyuan granite (Fig. 5.1). The $\delta^{18}O$ values of three samples (1, 2 and 3) from borehole 169 are 10.43‰, 10.57‰ and 11.44‰, respectively. Figure 5.17 shows the variation of $\delta^{18}O$ from the top downwards in the Suzhou granite, eastern China (Wang et al. 1998) that also demonstrates the same relationship. Two further examples from SE China, the Xianghualing and Dajishan granites, have $\delta^{18}O$ values of their marginal and central granite facies of 8.9‰, 16.95‰ (Chen et al. 1989), and 9.5‰, 10.3‰ (Sun et al. 1989), respectively.

Theoretically, with cooling, residual magma that accumulates in domed areas of the MI (Wang and Huang 2000; Chen and Grapes 2003) would exsolve a gas phase. Since ^{16}O migrates from magma to the gas phase faster than ^{18}O, the concentration of ^{16}O in the residual magma should decrease. Evidence for this is provided from a study of oxygen isotope variation within a one metre-long quartz crystal in a granitic pegmatite from Volyn, Central Kazakhstan (Dontsova et al. 1972). From the centre of the crystal outwards, $\delta^{18}O$ progressively increases with

Table 5.6. $\delta^{18}O$, δD and H_2O^a of granites and their xenoliths, Berridale Batholith, Australia

Rock body	Lithology	No. of samples	$\delta^{18}O$	δD	H_2O (wt %)
Buckleys	Granite	4	9.45	–80	0.38
Lake	xenolith	2	8.95	–77	0.85
Delegate	Granite	2	9.94	–88	0.43
	xenolith	2	8.45	–81	1.14
Dalgety	Granite	6	10.06	–62	1.04
	xenolith	3	8.74	–54	1.53
Tara	Granodiorite	3	8.12	–75	0.67
	xenolith	3	7.00	–58	1.43
Finister	Granodiorite	2	8.14	–59	1.49
	xenolith	1	8.69	–55	1.81
Cootralatra	Granodiorite	4	9.71	–62	1.21
	xenolith	5	9.08	–54	2.45

[a]Average values. After O'Neil and Chappell (1977)

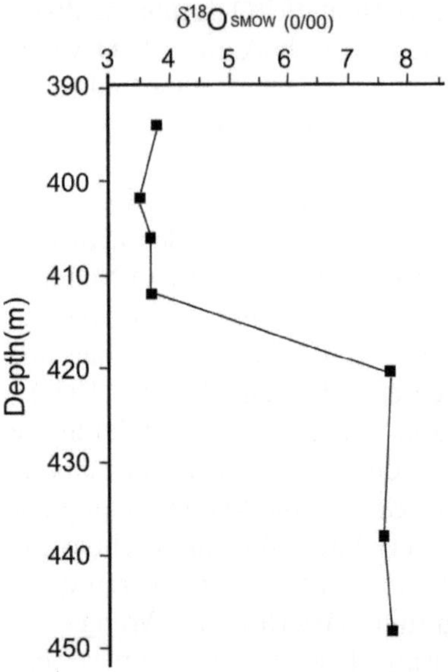

Figure 5.17. Vertical variation of $\delta^{18}O$ in the Suzhou granite, eastern China (after Wang et al. 1998), indicating an '^{16}O-shift' towards the top of the granite.

Table 5.7. Crystallisation temperature and $\delta^{18}O$ values across a large quartz crystal, Volyn, Central Kazakhstan, Russia

	Position measured		T°C	$\delta^{18}O$ (%)
	3	Outer zone	150	14.2
	2	Middle zone	400	12.5
	1	Core	600	11.6

Data from Dontsova et al. (1972)

decreasing crystallisation temperature (Table 5.7), indicating continuous depletion of ^{16}O in the melt phase during the growth of the crystal.

Exsolution of a gas phase from residual melt in the domed parts of the upper surface of a granite layer is expected to enhance the intrinsic pressure of the system (Chen et al 1996). Once this exceeds the confining pressure of the surrounding rocks, hot and ^{16}O-rich gas-fluid will migrate upward along fractures to form hydrothermal veins (Chen et al. 1996; Zhuang et al. 2006). For this reason, high-temperature hydrothermal veins should have lower $\delta^{18}O$ values than their related granite, which is the case in SE China (Zhang 1989) (Fig. 5.18) (see Chapter 6). When ^{16}O-rich hydrothermal fluids enter fractures in the rocks overlying a cooling granite, energy and isotope exchange is expected to occur between the hot fluid and wall rocks. The further from the fracture, the weaker the isotopic exchange, which is related to variation of both temperature (Taylor 1974) and ^{16}O concentration in the fluid. This may be the reason that $\delta^{18}O$ values of rocks hosting hydrothermal veins usually increase away from the veins (Zhang 1989) (Fig. 5.19).

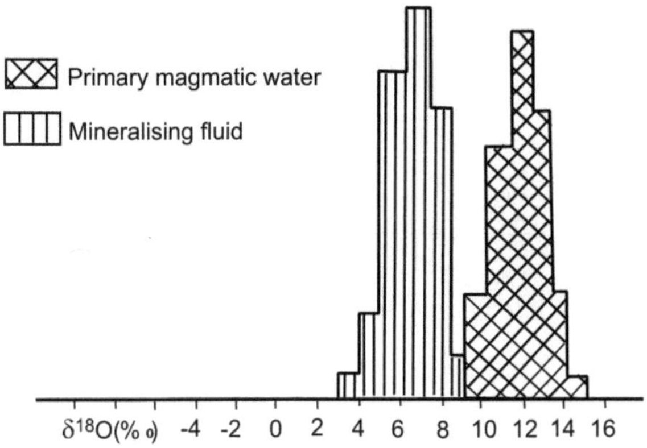

Figure 5.18. $\delta^{18}O$ values of mineralising fluid involved in the formation of vein-type tungsten deposits associated with the granites in SE China (redrawn after Zhang 1989).

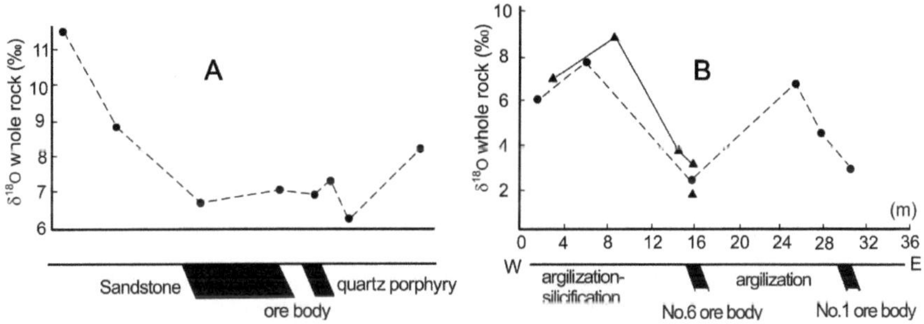

Figure 5.19. Whole rock $\delta^{18}O$ variation between hydrothermal ore bodies and host rocks. **A**. Lianhuashan tungsten mine, South China; **B**. Bainaibiao tungsten mine, Inner Mongolia (modified after Zhang 1989).

If the cover rocks remain unfractured, ^{16}O-rich hydrothermal fluid accumulated beneath the domed MI of a granite layer can only diffuse upward and outward along grain boundaries and micro-cracks resulting in the development of a 'halo' of alteration within the overlying country rock in which $\delta^{18}O$ values increase outwards. The Bohemia mining district, Oregon, USA (Taylor 1974), provides a good example of this phenomena where $\delta^{18}O$ contours in diorite and country rocks show a concentric pattern centred on the diorite with $\delta^{18}O$ increasing and the calculated temperature of isotopic exchange decreasing outwards (Fig. 5.20). Similar $\delta^{18}O$/temperature patterns have been found from other areas and have been attributed to interactions between meteoric groundwaters and igneous rocks (Taylor 1974).

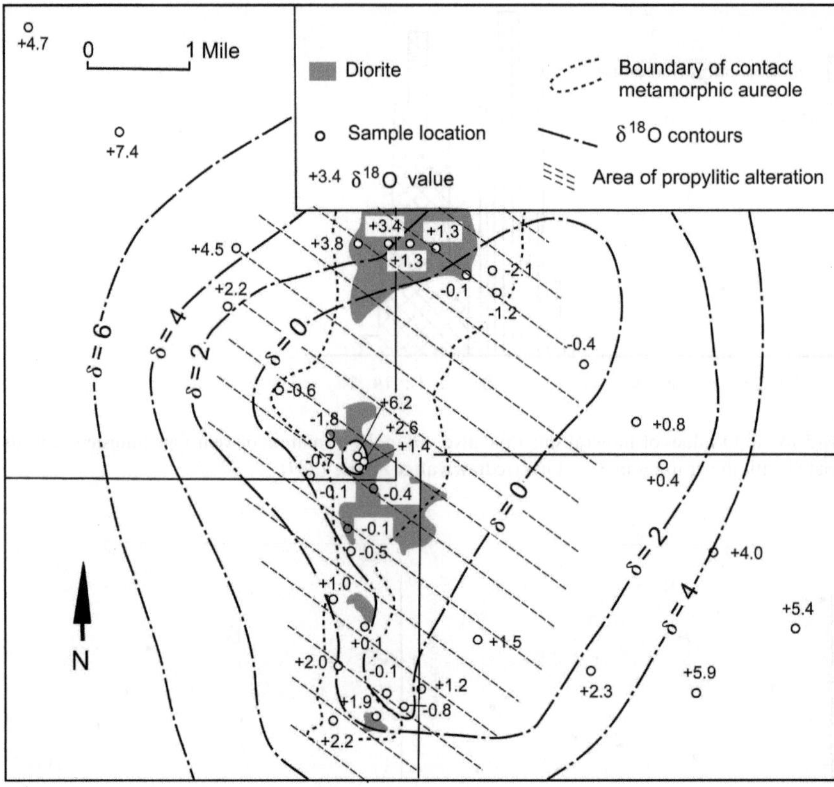

Figure 5.20. Geologic map of the Bohemia mining district, Oregon, USA, showing $\delta^{18}O$ contours in relation to outcrops of diorite and area of hydrothermal alteration (simplified from Taylor 1974).

In composite granite (CG) described above with younger granite enclosed within an older granite protolith, ^{18}O values of the young member should generally be lower than that of the old at the same erosion level. Table 5.8 gives whole rock $\delta^{18}O$ values of some Mesozoic CGs in SE China, which demonstrate that whole rock $\delta^{18}O$ values decrease with time, i.e. oldest to youngest granite in most of the examples listed.

5.5. RARE EARTH ELEMENTS

REE's whose ionic radii ranges from 0.0964 nm (Sm) to 0.1038 nm (Ce^{3+}) typically substitute for Ca (0.106 nm) in Ca-plagioclase, and accessory phases such as allanite, apatite, titanite in granitic rocks (Emmermann 1975; Liu et al 1985). Therefore, vertical variation of LREE content in a granite body is expected to follow that of calcium, i.e. increase downward. REE variation with depth through

Table 5.8. Whole rock oxygen isotopic values of some composite granites, SE China

Granite	Phase	Lithology	$\delta^{18}O$ (%)	Data source
Jiufeng	1	Biotite adamellite	11.05	GRGN (1989)
	2	Two-mica adamellite	10.18	
Lingshan	1	Hornblende biotite granite	9.7	
	2	Biotite K-feldspar granite	9.43	Chen et al. (1989)
	3	Alkali-feldspar granite	7.65	
Dadongshan	1	Coarse-grained biotite granite	10.4	
	2	Two-mica granite	10.1	Yu et al. (1987)
	3	Fine-grained granite	9.5	
Qianlishan	1	Biotite K-feldspar granite	10.8	GRGN (1989)
	2	Granite porphyry	7.8	
Shirenzhang	1	Two-mica granite	9.7	Yu et al. (1987)
	2	Muscovite adamellite	9.3	
Guidong	1	Biotite granite	8.9	Yu et al. (1987)
	2	Two-mica granite	8.2	
Xihuashan	1	Biotite adamellite	11.35	
	2	Medium-grained adamellite	10.13	Chen et al. (1989)
	3	Biotite granite	10.05	
Reshui	1	Biotite granite	10.7	Chen et al. (1989)
	2	Muscovite granite	8.7	
Dengfuxian	1	Biotite adamellite	10.9	
	2	Two-mica granite	13.95	Zhang et al. (1986)
	3	Muscovite granite	12.80	

Phase 1 = youngest granite; Phase 3 = oldest granite

the Xiaotaoyuan granite is in good agreement with this assumption (Figs.5.1 and 5.21). In the Huangsha granite, Jiangxi Province, the rocks change downward from greisenised granite, albitised granite, tourmaline two-mica granite to biotite granite (Fig. 5.22A), and the REE variation pattern with depth is the same as in the Xiaotaoyuan granite (Chen et al. 1989) (Fig. 5.22B). As described previously, the southern part of the Xingluokeng granite has been more deeply eroded than the northern part (Fig. 5.14), so that a higher LREE content and more pronounced Eu depletion of the biotite granite core with respect to the muscovite granite margin can be expected (Fig. 5.23).

According to compositional variation of Fichtelgebirge granite described above, we infer that the different granite types of this composite body formed by in situ differentiation and REE distribution patterns of the granites support this inference (Fig. 5.24).

Downward increase of LREE during crystallisation of granite should indicate that at the same erosion level, the LREE content of granite should be lower and the Eu anomaly more significant than that of adjacent country rock protolith. REE patterns of the Xihuashan granite, Jiangxi Province and its enclaves and adjacent metamorphosed country rocks (Wu et al. 1987) (Fig. 5.25) are in agreement with this, showing a weak downward trend of REE for metasediments and enclaves and a weak Eu anomaly compared with the biotite granite that exhibits a pronounced Eu anomaly and an almost flat trend.

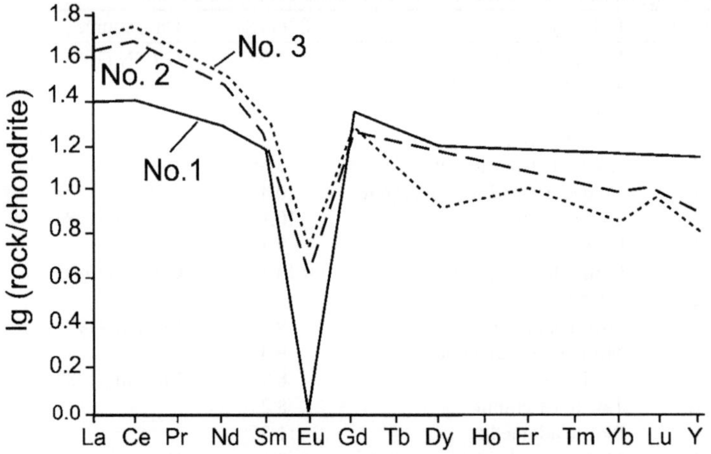

Figure 5.21. REE patterns of samples 1, 2 and 3 at different depths in borehole 169, Xiaotaoyuan granite, SE China (see Fig. 5.1).

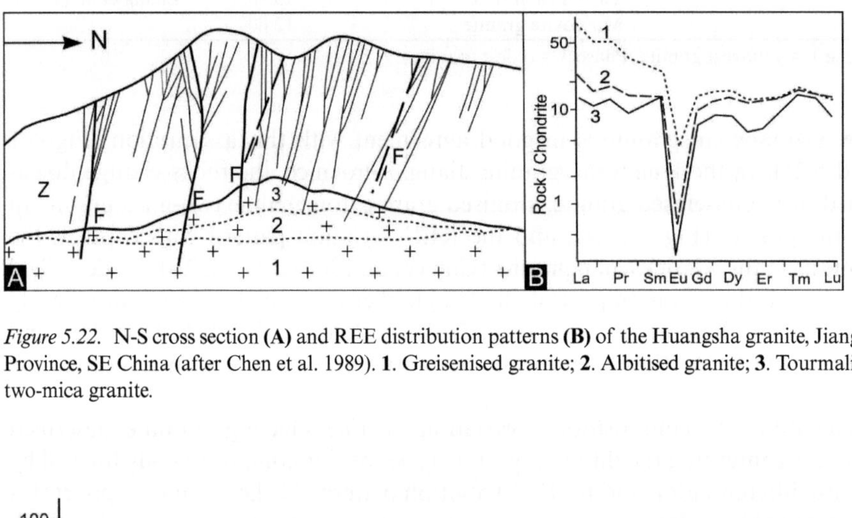

Figure 5.22. N-S cross section **(A)** and REE distribution patterns **(B)** of the Huangsha granite, Jiangxi Province, SE China (after Chen et al. 1989). **1**. Greisenised granite; **2**. Albitised granite; **3**. Tourmaline two-mica granite.

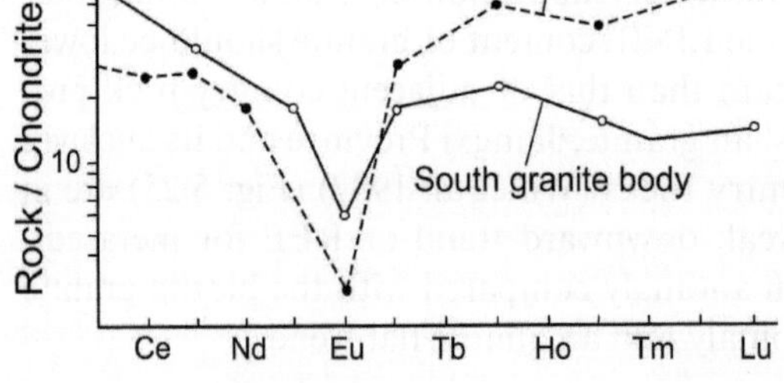

Figure 5.23. REE distribution patterns of the Xingluokeng northern muscovite granite and southern biotite granite (redrawn from Zhang and Sun 1988).

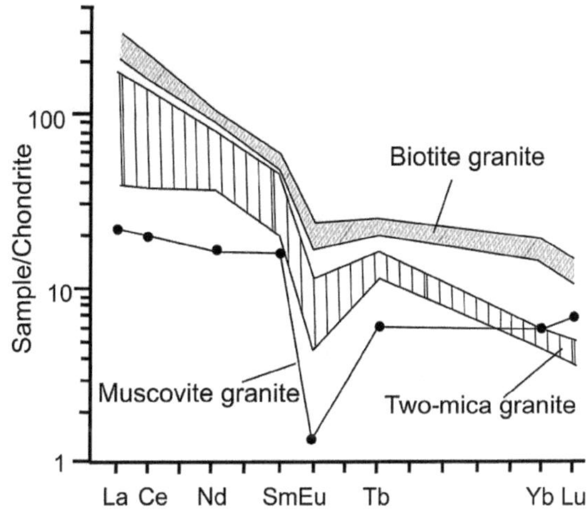

Figure 5.24. REE distribution patterns of muscovite, two-mica and biotite granite of the Fichtelgebirge granite, Central Europe (redrawn from Hecht et al. 1997).

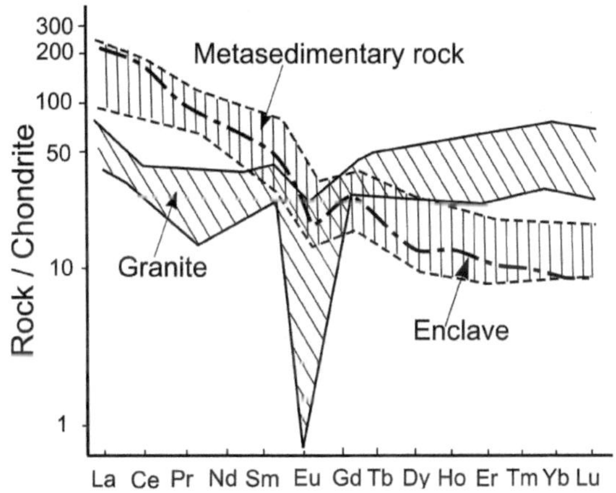

Figure 5.25. REE distribution patterns for Xihuashan granite, metasedimentary country rock and enclave (after Wu et al. 1987).

In the case of remelting of granite, remelted, younger granites should have a lower LREE/HREE ratio and a more significant Eu anomaly than the older granite. REE distribution patterns of the above described composite granites, Jiufeng, Dengfuxian, Qianlishan, Wuguishan (Fig. 5.8; Table 5.4), are illustrated in Fig. 5.26. In all four composite granite bodies, the Eu anomaly becomes more pronounced, and the REE pattern generally shifts to lower rock/chondrite normalised ratios with decreasing age. On a micro-scale it is instructive to compare by the REE patterns of partial melting of granite by trachyandesite, Sierra Nevada (Kaczor et al. 1988) (Fig. 5.27). The most silica rich melt (with 9% brown glass) has the most pronounced Eu anomaly. The more brown-coloured glass representing a greater degree of melting shows an upward shift of the REE patterns towards the granite composition but still exhibits a pronounced Eu

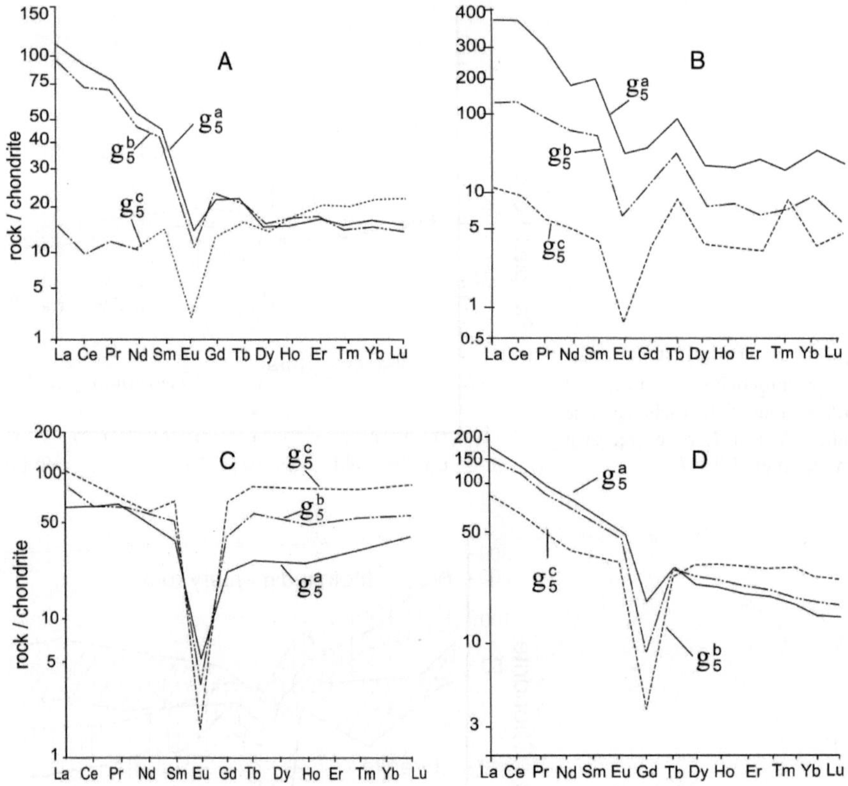

Figure 5.26. Chondrite-normalised REE distribution patterns of composite granite bodies shown in Fig. 4.33. **A.** Jiufeng (after BGMRHn 1984); **B.** Dengfuxian (after Zhang et al. 1986); **C.** Qianlishan (after Chen et al. 1989); **D.** Wuguishan (after Cai and Zheng 1997). Age notation as in Fig. 5.8.

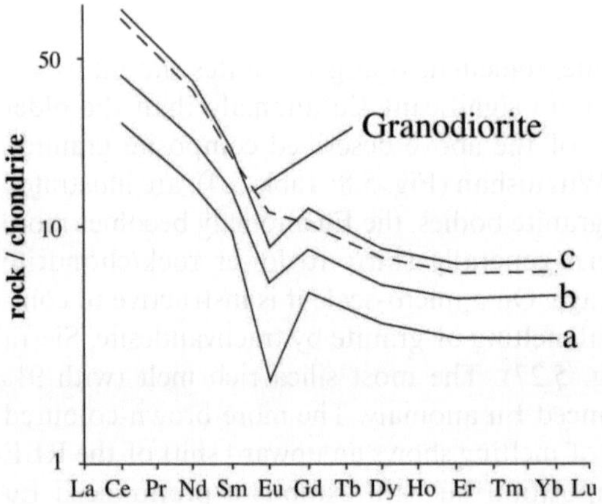

Figure 5.27. Chondrite-normalised REE distribution patterns for granodiorite, and three glass separates; **a**. 9% brown glass; **b**. 29% brown glass (25% rock melting); **c**. 100% brown glass (47% rock melting) (data from Kaczor et al. (1988), representing increasing degree of melting indicated by increasing proportion of brown verses clear glass in each melt fraction.

anomaly that is most probably related to the formation of calcic plagioclase during the melting process.

Many of the above described granites are highly evolved, i.e. they have marginal facies muscovite and two-mica granites. For those with more intermediate, i.e. granodioritic, marginal facies, such as the Criffell and Souwangfen granite bodies, a relatively higher LREE/HREE ratio with weak Eu depletion is expected, and their REE distribution curves should generally resemble that of the unmelted host country rock. Figure 5.28 shows the REE distribution patterns of four different rock types, sedimentary (A), migmatites and migmatitic granites (B), granodiorites (C), and andesitic rocks (D) in the coastal region of Fujian and Zhejiang provinces SE China. All the REE distribution curves are similar and are of the right-declination 'B' type of Xu et al. (1983). Of the four types of REE distribution curves shown in Fig. 5.28, the similarity of the first two types 'A' and 'B' is readily understandable, because migmatites have not undergone gravitational differentiation during crystallisation and so their pattern is almost identical to those of associated unmigmatised metasedimentary rocks. The fourth type

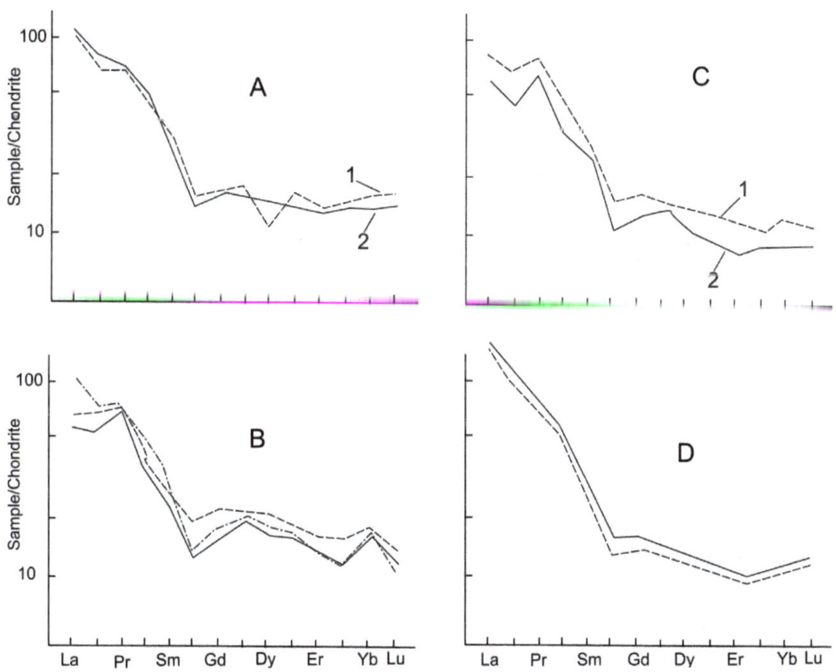

Figure 5.28. REE distribution patterns of four rock types in SE China (Chen et al. 1996). **A.** Sedimentary rocks; **1.** Metasedimentary rocks (Nanling area, SE China); **2.** Average continental crust. **B.** Migmatites from Guangzhou area; **C.** Granites; **1.** Changbu granodiorites (Guangdong Province); **2.** Adamellites (Guangning, Guangdong Province); **D.** Andesitic ignimbrite of SE Zhejiang Province.

(D) implies that andesitic–dacitic ignimbrites could have been derived from the deeper part of an extensive partial magma layer, as suggested from the typical interbedded relationship of andesite and rhyolite in the coastal area of SE China (Xue et al 1992). The third type (C) may have originated either as a separate intrusion of andesitic magma, or as an early crystallised marginal facies of a granite magma, e.g. with a composition similar to that of the Criffell and Souwangfen granites. The plots in Fig. 5.28 demonstrate that without additional knowledge of field relationships, it would difficult to distinguish different rock types solely on the basis of their REE distribution patterns. Thus, the 'genetic series (types)' of igneous rocks, especially that of granitic rocks, based on their petrological, geochemical and isotopic compositions should be called 'petrological' rather than 'genetic' (Chen et al. 1996) unless additional supporting information concerning their origin is known.

5.6. SUMMARY

The various geochemical phenomena concerning granites can be explained in terms of the model of in situ melting origin of granite, supporting the idea that, "Though there may be granites and granites, most of them are of one kind, and all of them may likely be of one connected origin" (Read 1948). Chemical composition and mineralogical variation, Rb/Sr, hydrogen and oxygen isotopic compositions, and REE distribution characteristics of granites all reflect the internal evolution of granite magma system and are not necessarily related to magma sources:

1. Petro-geochemical characteristics of granite formed through in situ melting are related not only to the composition of the protolith involved in the melting, but also to the rate at which the magma interface (MI) descends during cooling. Where FeO, MgO and CaO contents of the protolith are relatively high (e.g. mafic and volcaniclastic rocks) and the MI descends rapidly, the magma is poorly differentiated and results in early (marginal facies) formation of granodiorite–tonalite–quartz diorite, e.g. as exemplified in the domed parts of the MI. Such granites usually show a downward decrease in basicity and an increase in acidity towards adamellite and granite compositions. Where the magma of a melt layer has had sufficient time for differentiation to occur, the resultant granite will show a downward increase in basicity.
2. The in situ melting-crystallisation process of granite formation not only leads to the differentiation of rock-forming elements, but also to fractionation of Sr, H, O isotopes and REE. Mafic components, LREE, stable Sr isotopes etc. generally increase towards the lower part of a granite magma layer that should be relatively poor in silica and alkalis, HREE and $^{87}Sr/^{86}Sr$ ratio, all characteristics of intermediate dacite–andesitic/granodiorite–diorite compositions. This suggests the possibility that andesitic volcanics, particularly where they are interbedded with rhyolite, could be derived from the lower part of a crustal magma layer rather than from the mantle.

CHAPTER 6

MINERALISATION RELATED TO IN SITU GRANITE FORMATION

6.1. INTRODUCTION

Granites are typically associated with hydrothermal mineral deposits and many studies have yielded critical information about the source of metals and ligands in the ore-precipitating hydrothermal solutions that operate during formation of ore–mineral deposits. Southeast China contains perhaps the world's most extensive area of mineral resources related to granites and these have been intensively studied over the past 60 years. Large-scale, high intensity mineralisation involving Au, Ag, Cu, Pb, Zn, Pb, Sb, W, Sn, Mo, Bi, Li, Be, Ta, Nb and REE is related to the Mesozoic tectonism/granite formation (Hua 1999) (Fig. 6.1), and many of these occurrences are described below.

6.2. SOURCE OF ORE-FORMING ELEMENTS

In the granite intrusion model, ore-forming elements in granite magma are considered to have been derived from two main sources.

1. From deep source protoliths of the granite. In this case, it is expected that a large (batholithic) granite body would be associated with more mineralisation than smaller bodies because the amount of ore-forming materials ought to be proportional to the volume of magma. Coeval granites in a particular area also should be characterised by similar styles of mineralisation because of a common protolith. However, this is typically not the case; only some coeval granite exposures are associated with ore deposits, and the ore mineral assemblage varies from one granite to another. In SE China most ore deposits are almost exclusively related to smaller granite exposures (e.g. the Xingluokeng and Xihuashan granites described in Chapter 5), whereas ore mineralisation is rare or absent in larger granite bodies, e.g. the Dagongshan (Fig. 1.8A) and Fogang granites that crop out over an area of more than 1,000 km^2.

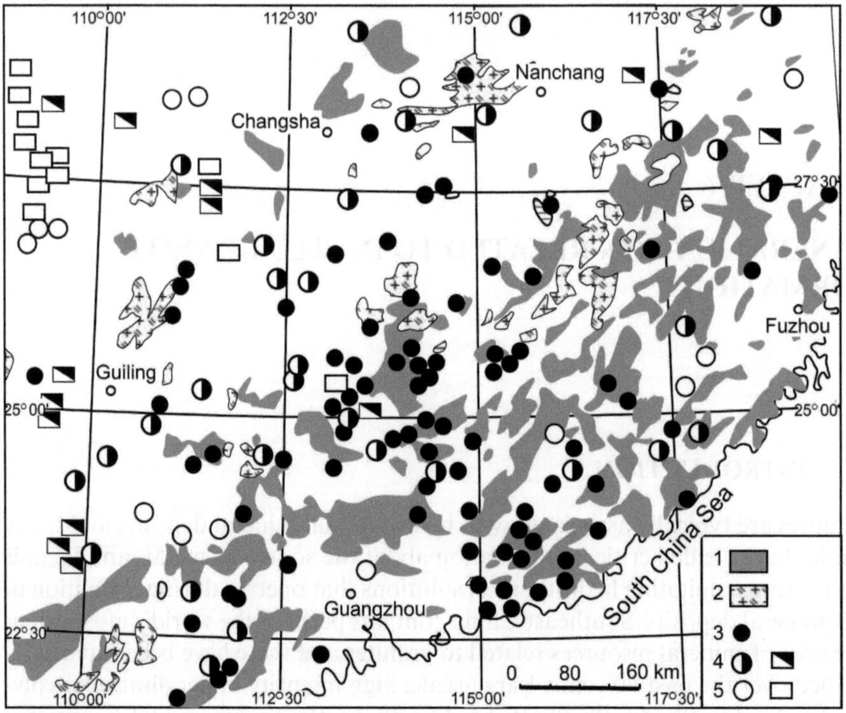

Figure 6.1. Distribution of granite and various types of hydrothermal ore deposits in SE China (modified after IGCAGS 1987). **1**. Mesozoic granite; **2**. pre-Mesozoic granite; **3–5** ore deposits; **3**. W, Mo (Bi), Fe, Sn; **4**. Cu, Pb, Zn; **5**. Au (Ag), Hg, Sb; *Open circles*: hydrothermal and contact-metasomatic type; *Open squares*: strata-bound-hydrothermal type.

2. From the country rocks of granite 'intrusions'. In this case, heating from the granite results in extensive water-country rock reaction causing a hydrothermal convection system to develop that facilitates element scavenging, concentration and deposition as ore minerals (Taylor 1978). In this case, it would be expected that batholith-size granites should be associated with abundant ore deposits because heating/cooling is of longer duration and affects a greater area than that of small granite bodies. This, however, is not the case.

The in situ melting model of granite formation implies that granites and their ore–element components are derived from the same source, i.e. the protoliths involved in melting. Figure 6.2 shows the distribution of Jurassic biotite granite and related REE-mineralisation and pre-Jurassic country rock in SE China. More than 90% of the REE mineralisation (deposits or prospects) associated with the granites are in contact with lower Paleozoic rocks that

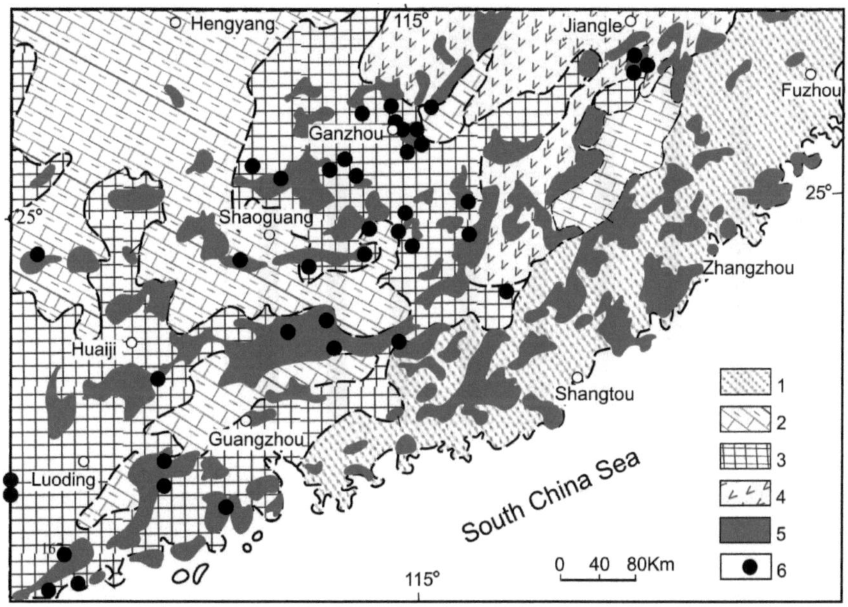

Figure 6.2. Distribution of pre-Cretaceous basement sedimentary rocks, Jurassic biotite granite and associated REE-deposits in the Nanling area, SE China. **1**. Triassic–Jurassic; **2**. Late Paleozoic; **3**. Early Paleozoic; **4**. Precambrian; **5**. Biotite granite; **6**. REE-bearing granite (deposit or prospect).

have high mean bulk REE contents (Yu et al. 1987) (Fig. 6.3). As the REE deposits all occur within the granite (Fig. 6.4) and not in the Early Paleozoic country rocks, it is concluded that they are derived internally from the granite rather than the result of hydrothermal water/rock reaction, i.e. the REE-bearing granites are derived from the melting of early Paleozoic rocks that contained REE-rich horizons.

Southeast China is the richest tungsten metallogenic region in the world and most of the tungsten is spatially related to Mesozoic granites that occur within Cambrian/Sinian (latest Proterozoic) metasedimentary rocks. According to Li (1982), there are 188 W-deposits in exocontact zones of granite exposures in Jiangxi Province, of which 138 (83%) occur in Sinian and Cambrian rocks, as are 34 (85%) of 40 large- and medium-sized W-deposits. The Sinian and Cambrian sediments have long been suspected as the 'source rocks' for the tungsten (Liu 1980). This is suggested by their high-average clastic rock abundances of ~6 ppm W (three times more than the average upper crustal W value) (Fig. 6.5). Thus, a reasonable interpretation is that these rocks were the protoliths of melting that produced the associated granites.

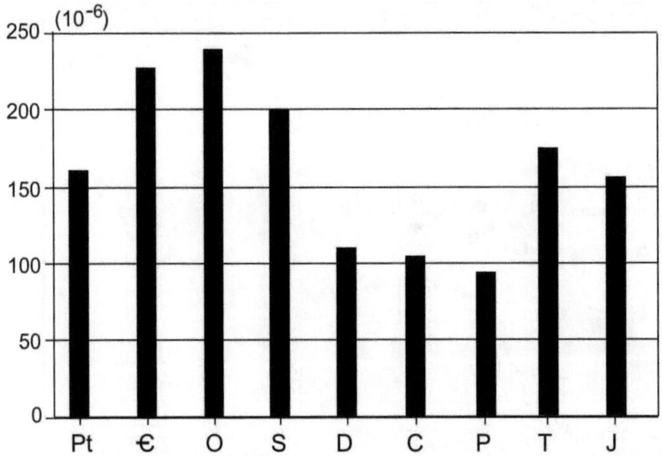

Figure 6.3. Histogram showing REE-contents of pre-Cretaceous clastic sedimentary rocks in the Nanling area, SE China (data from Yu et al. 1987).

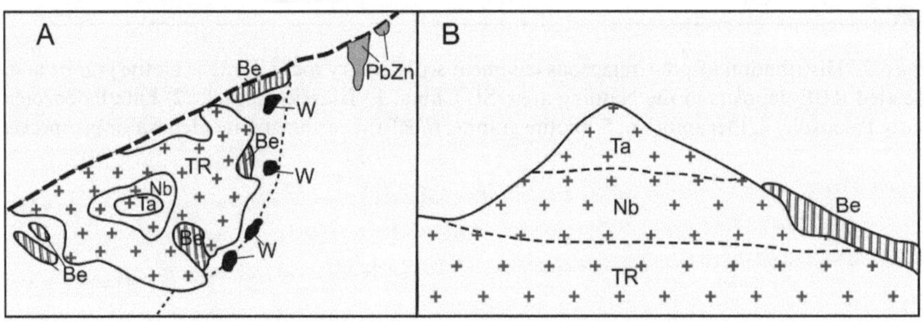

Figure 6.4. Sketch map (after Wang et al. 1988) and section (after Geological Party 935; unpublished data) of REE-bearing biotite granite (granite body No. 430), and associated Be, W, Pb–Zn deposits, SE China, showing vertical zoning of Nb and Ta enrichment within the granite. No scale is given for the original map and profile.

6.3. FORMATION AND EVOLUTION OF ORE-BEARING FLUID

The granite magma-forming process involves reactivation, reorganization and redistribution of the ore-forming elements in its source rocks. The focus of interest is the upper part of a cooling granite magma layer that is schematically illustrated in Fig. 6.6. The main control over the partitioning of ore metals (and other elements) is competition between the crystallising magmatic mineral assemblage and the magmatic volatile phase. Typically, with progressive crystallisation of silicate minerals, both anhydrous (quartz, feldspars) and less-abundant hydrous (biotite, hornblende) phases, non-compatible ore-forming metals and volatiles (e.g. H_2O, CO_2, H_2S, H_2F, HCl) become concentrated in the gas phase or in what may eventually evolve into an 'ore fluid' (Candella and Holland 1986).

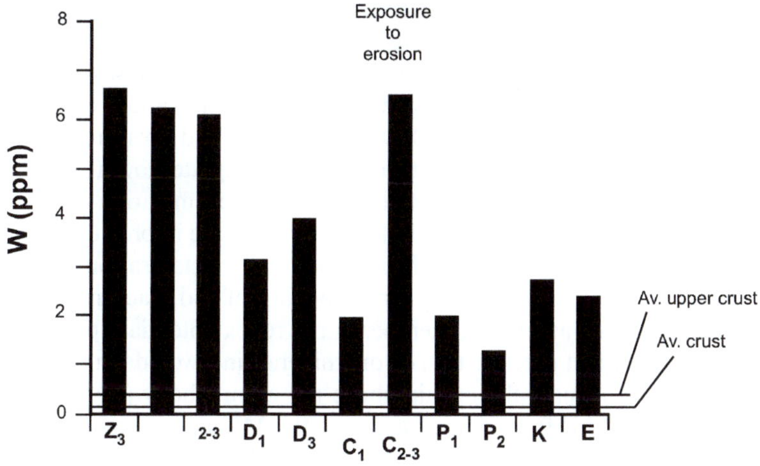

Figure 6.5. Histogram of W-contents of Proterozoic and Cambrian clastic metasediments of the Nanling area, SE China (data from Yu et al. 1987). Average upper crustal and bulk crust W-values from Rudnick and Gao (2003).

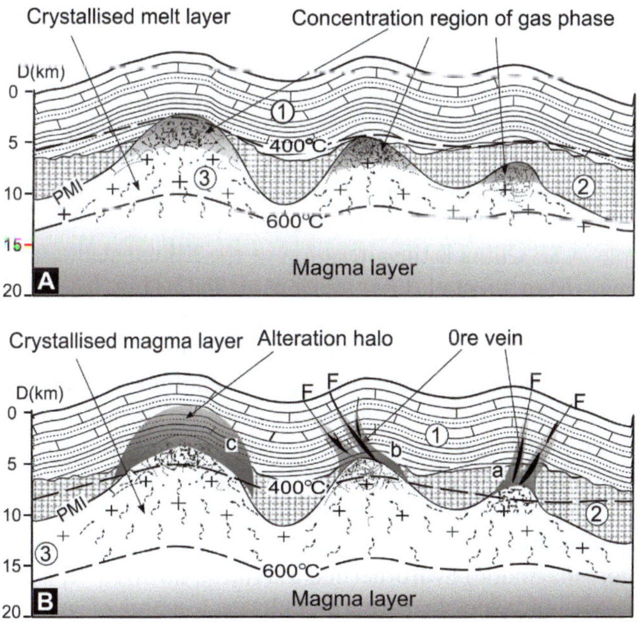

Figure 6.6. Mineralisation model of crustal melting. **A.** With decreasing temperature, the granite melt layer crystallises downwards, and exsolved ore element-bearing fluid migrates to low-P cupolas developed in the melt/magma interface (MI) thus increasing the intrinsic pressure; **B.** Fracturing of roof rocks above the MI consequent on the pressure build-up results in expulsion of ore-bearing fluid. With upward movement of the fluid, the ore-forming elements are precipitated as various ore-minerals depending on T–P and factors such as redox potential, pH, wall rock composition etc. PMI = Paleo-magma interface; 1 = sedimentary rock; 2 = basement rock 3 = crystallised upper part of magma layer (granite).

Granite magmas typically contain water concentrations of 3–6 wt % and partial pressures of magmatic gases can achieve several hundred bars (Lowenstern 1994). With an increase of vapour pressure consequent on crystallisation, the hydrostatic head of the magma may be exceeded, at which point vapour exsolves (second boiling) as an immiscible phase. Driven by thermal and pressure difference gradients, the highly fluid and buoyant vapour phase migrates towards areas of low pressure, i.e. domed parts of the MI (Fig. 6.6A). Accumulation of late-stage fluid/gas phase in these areas results in a gradual increase in pressure that, combined with cooling and contraction of roof rock and crystallisation of the magma, would likely result in fracturing, or reopening of old fractures. This results in escape of the fluid and depth-, temperature-, redox potential-, pH- and fluid composition–dependent precipitation of ore minerals in a well-documented, although often complex, zonal relationship above the granite body (Fig. 6.6B). Release of ore-bearing fluids from domed areas of the MI causes a rapid pressure decrease when a batch of hydrothermal fluid is lost. In turn, this is followed by renewed fluid/gas accumulation consequent on continuing consolidation of the granite magma, increasing pressure, renewed fracturing and ore deposition. The process can be compared with the mechanism of geyser eruption in which 'boiling' results from a reduction in pressure and it can explain the multiphase mineralisation events of most hydrothermal deposits. If fracturing of the cover rocks does not occur, upward movement of the hydrothermal fluid may take place via pores and microfractures in the roof rocks, resulting in a pervasive flow regime with the development of disseminated, as opposed to vein/fracture-type, ore deposits. Continuing permeation of fluids and vapours with decreasing temperature causes widespread alteration (e.g. sericitiation, alkali metasomatism, greisenisation, etc) of the earlier-formed minerals in the granite to form complex primary and secondary assemblages (Fig. 6.7; Table 6.1).

A well studied example in SE China is the alteration of the Xihuanshan granite, south Jiangxi Province (Mauéjol et al. 1990; Zeng et al. 2002) (Fig. 6.8). The granite consists of several grain-size facies of leucocratic porphyritic biotite granite with Qz (28–33%), Ksp (30–33%), Pl (30–50%), Fe–Bt, ± Gt (≤5%). Early formed accessory minerals include monazite, apatite, xenotime, uranothorite, uraninite, betafite and zircon. An early (late magmatic stage) hydrothermal alteration involved the breakdown of monazite to parisite ($Ce_2Ca(CO_3)_3F_2$) and new apatite with formation of abundant HREE-Y minerals (new xenotime, yttroparisite, Y-rich fluorite) and Y-rich calcite. The alteration assemblage indicates an F- and CO_2-bearing fluid that readily complexed REE's leached from the monazite. Latest (post magmatic) hydrothermal alteration involved replacement of biotite by muscovite and chlorite, garnet by chlorite, leaching of quartz, K-feldspathisation of plagioclase and albitisation of K-feldspar.

Valuable quantities of Sn, Mo, Bi, Nb, Ta, REE, Cu, Pb, Zn, Ag, Be and Li have been recovered from the Xihuashan granite. The area surrounding the granite contains hundreds of wolframite–quartz veins, that extend upwards from the Xihuashan and subsurface granite through roof-rocks of Cambrian phyllite,

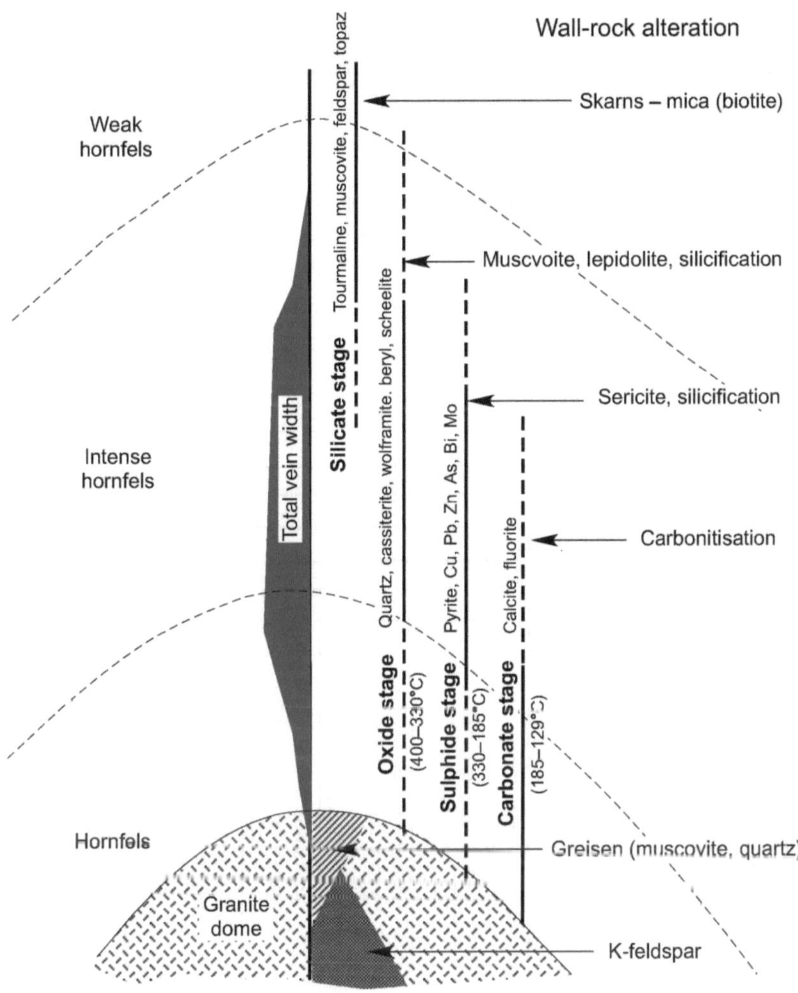

Figure 6.7. Diagram to illustrate ore mineral zoning, stages of mineralisation and associated hydrothermal alteration with a granite dome. The diagram is based on vein mineral zonation associated with tungsten mineralisation in SE China (after Yan et al. 1980). See text.

quartz sandstone and hornfels (Fig. 6.8). Around the zones of mineralisation, the country rocks are extensively altered with alkali metasomatism and greisenization affecting the granite; and tourmalinization, muscovitization, silicification, pyrophillitization affecting the metasediments. The main mineralogy and parageneses of the veins and veinlets indicate a continuous multistage process that can be summarised as follows: 1st oxide stage (wolframite, cassiterite, molybdenite, quartz, K-feldspar, beryl, fluorite, topaz); 2nd sulphide stage (chalcopyrite, galena, sphalerite, pyrite, pyrrhotite, bismuthinite, cassiterite, wolframite, quartz); 3rd carbonate stage (pyrite, scheelite, chlorite, sericite, fluorite, quartz, calcite) (Fig. 6.7).

Table 6.1. Generalised mineral paragenesis of mineral deposits of Cornwall, SW England (after Table 6.4 of Park and MacDiarmid 1975)

Gangue minerals	Zone	Ore minerals	Economically important elements	Wall-rock alteration
Feldspar, Mica, Tourmaline, Chlorite, Hematite, Fluorite, Chalcedony, Barite, Dolomite, Calcite	7	Barren, (pyrite)		Greisenised granite, Quartz-sericite hornfels, Tourmalinised granite, Tourmaline hornfels, Chloritsed granite, Chloritised country rocks, Hematised granite, Hematised country rock, Kaolinised granite, Silicified granite, Silicified country rock, Skarn
	6	Hematite, stibnite, jamesonite tetrahedrite, bournonite, pyragrite? siderite, pyrite, (marcasite)	Fe Sb	
	5b	Argentite, galena, sphalerite	Ag, Pb, Zn, U, Ni, Co, Bi	
	5a	Pitchblende, niccolite, smaltite, cobaltite, (native bishmuth, bisthmuthinite ?)		
	4	Chalcopyrite, sphalerite, wolframite, (scheelite), Arsenpyrite, pyrite	Sn, Cu, W, As	
	3	Chalcopyrite, (stannite), wolframite, (scheelite), arsenopyrite, cassiterite		
	2	Wolframite, (scheelite) arsenopyrite, (molybdenite), cassiterite		
	1	Cassiterite, specularite		
	Greisen-bordered veins	Arsenopyrite, stannite, wolframite, cassiterite, molybdenite		
	Pegmatites	Arsenopyrite, wolframite, cassiterite, molybdenite		

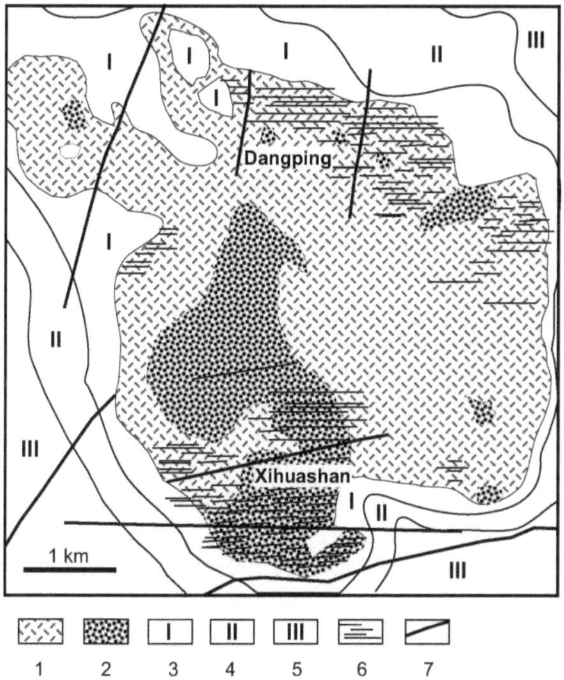

Figure 6.8. Geological sketch map of the of the Xihaushan granite, southern Jiangxi Province, SE China, showing distribution of granite grain-size facies, quartz–tungsten veins and metamorphic aureole rocks (modified from Li et al. 1993). **1**. Medium-to-medium-fine grained porphyritic biotite granite; **2**. Medium- and medium-fine-grained biotite granite; **3**. Bt–Pl hornfels; **4**. Bt–Ms hornfels **5**. Knotted slate; **6**. Quartz–tungsten veins; **7**. Fault.

A present day example that further illustrates the complex processes of mineralisation and fluid activity associated with cooling of granite is the Larderello geothermal field, Italy (see Chapter 2) (Gianelli and Ruggieri 2000). Drilling has revealed F- and B-rich muscovite-bearing granite at shallow depth with partially molten material possibly present at ~6 km depth where the geothermal gradient is ~100°C/km. The granite is considered to have resulted from melting of Hercynian gneiss and schist that also overlie the granite and have been contact metamorphosed to temperatures of 530–630°C. An early hydrothermal phase is characterised by saline fluids with magmatic signatures such as elevated F, B, Li and CO_2 associated with high-temperature (~500°C) mineral veins indicating brittle deformation (fracturing) caused by overpressuring from a large volume increase of the fluid phase during crystallisation of the granite (second boiling event). Recurrent processes of fracturing of impervious roof rocks, circulation of magmatic fluids that mixed with meteoric fluids, and closure of pore spaces with a return to impervious conditions, have been active since 4 Ma, and the process appears to be continuing today at low-to-moderate rates.

6.4. TYPES OF MINERAL DEPOSITS

6.4.1. Vein Mineralisation

During upward migration of ore-bearing hydrothermal fluid along fractures, decreasing *T* and *P* associated with variations in physico-chemical conditions of elemental solubilities, pH and Eh, the ore-forming elements in the fluid will precipitate along the fractures to form typically steeply dipping ore veins and attendant wallrock alteration extending upwards and outwards from granite (Fig. 6.6B, case a and b). Some of the numerous vein-type deposits related to the Mesozoic granites in SE China are illustrated in Figs. 6.9 and 6.10, indicating that the vein systems originate from domal structures developed at the top of an underlying extensive body of granite. This relationship is similar to many other granite-related mineral vein deposits worldwide, e.g. the well-known mining area of Cornwall and Devon, SW England, where most of the mineralisation is concentrated around and within five domes (cupolas) of an elongate Late Paleozoic granite body within a succession of folded Devonian–Carboniferous slates, shales, sandstones, limestone and volcanics. Here, a complex pattern of joints, resulting from an earlier deformation, have been reactivated to provide channels for acid porphyry dykes (called *elvans*), and Cu-, Zn-, As-, W- and Sn-bearing metalliferous veins (Fig. 6.11, Table 6.1).

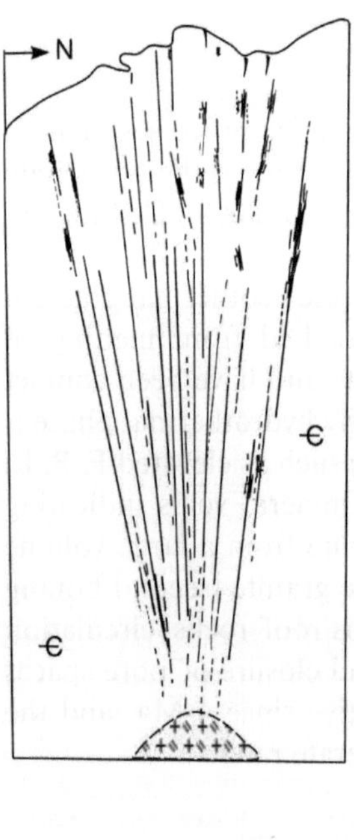

Figure 6.9. Section showing the Jubankeng tungsten deposit in Guangdong (modified after the 932 Metallurgical Geological Party, Guangdong, no scale on original map). The distribution pattern of W-veins (dark lines) suggest that W-bearing ore fluids move upwards along an inverted cone of 'explosively' formed fractures extending from the apex of a dome structure in the granite body inferred to be the result of a sudden pressure release.

6.4.2. Disseminated Mineralisation

Obviously, if the roof rocks remain unfractured, hydrothermal fluid that accumulates beneath the domed areas of the granite cannot escape via well-defined fractures and is likely to infiltrate and migrate upward and outward through pores and micro-fissures in the rock, resulting in a planar alteration of

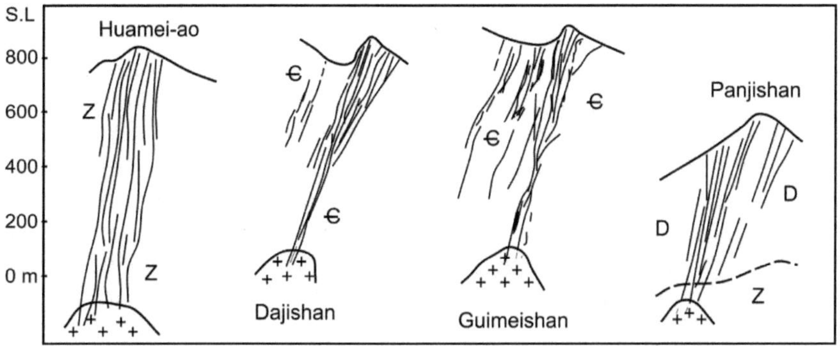

Figure 6.10. Schematic sections showing tungsten vein-type deposits associated with granite bodies in SE China (redrawn after Zhu et al. 1981, no scale on original map). For both diagrams, metasediment ages are; Z = Proterozoic; Є = Cambrian; D = Devonian.

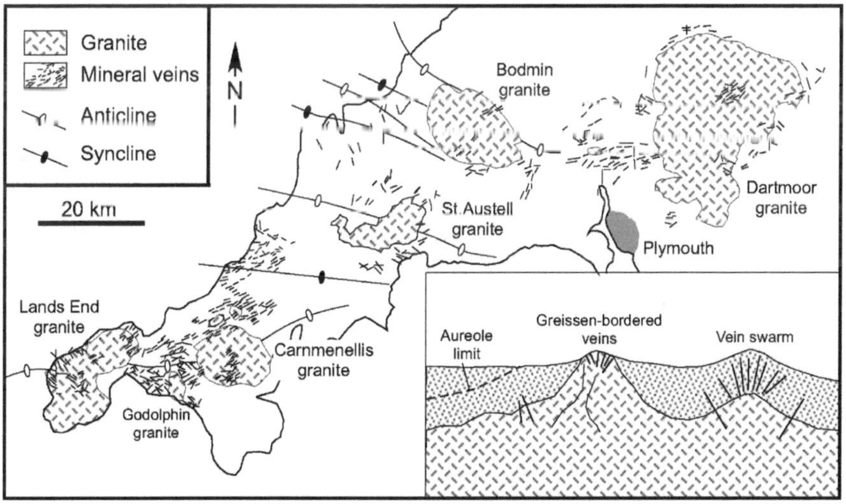

Figure 6.11. Geological map showing the distribution of mineral veins related to granites in Cornwall, SW England (after Hosking 1951; Hosking and Shrimpton 1964). An unscaled hypothetical cross section across any one of the granite exposures is also shown to indicate that the granite outcrops represent eroded dome-like structures formed in the upper surface of an extensive subsurface granite layer, possibly related to syn-crystallisation folding, and from which mineral veining is intense.

the country rock with formation of disseminated ore deposits (Fig. 6.6B: case c). The Dexing copper deposit in the northern Jiangxi Province, SE China, is an example of such a disseminated ore deposit (Fig. 6.12).

6.5. AGE RELATIONS

According to the model shown in Fig. 6.6, it can be expected that there should be an age difference between a granite body and its related hydrothermal mineral deposits, i.e. the granite should be older, because the upper part of the granite melt layer crystallizes before hydrothermal fluid escapes.

For example, the Xihuashan granite described above is dated by K–Ar at 152–131 Ma (Table 6.2), ^{39}Ar–^{40}Ar at 151–148 Ma (McKee et al. 1988) and with Rb–Sr ages at 155 ± 2 Ma (Le Bel et al. 1984). Muscovite K/Ar ages from related tungsten–quartz veins range between 141–131 Ma (Table 6.2). Rb–Sr and Sm–Nd isotopic results reported by Maruéjol et al. (1990) indicate a ~25 Ma interval between the late-magmatic (153 ± 0.2 Ma) and post-magmatic (128 ± 0.1 Ma) hydrothermal fluid alteration events described above. Another example

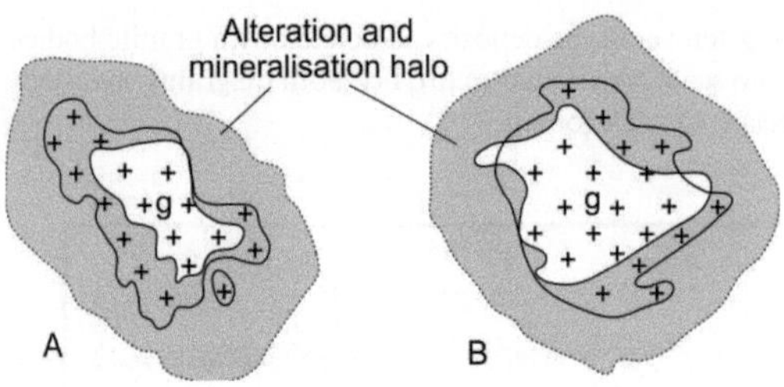

Figure 6.12. Relationship between the Dexing disseminated copper ore deposit and granite (redrawn after Zhu et al. 1983). **A**. Tongchang deposit at 5-m level; **B**. Fujiawo deposit at 140-m level; Grey shaded area = Cu ore body; **g** = granodiorite porphyry.

Table 6.2. Mineral K–Ar ages of Xihuashan granite and associated tungsten–quartz veins, SE China

Sample no.	P phase	Lithology	Mica	Granite age (Ma)	Vein age (Ma)
85	γ_5^{2a}	Coarse-grained biotite granite	Biotite	149	
86		Wolframite quartz vein	Muscovite		141
87	γ_5^{2b}	Medium-grained biotite granite	Biotite	152	
88		Wolframite quartz vein	Muscovite		132
94	γ_5^{2c}	Fine-grained biotite granite	Biotite	138	
99		Wolframite–beryl–quartz vein	Muscovite		133
92	γ_5^{2d}	Fine-grained garnet granite	Biotite	141	
93		Wolframite–molybdenite–quartz vein	Muscovite		139
101	γ_5^{2e}	Granite porphyry	Biotite	141	
102		Wolframite–quartz vein	Muscovite		131

Data from the Jiangxi Metallurgical Geological Corporation. K–Ar dating by Hubei Geological Institute.

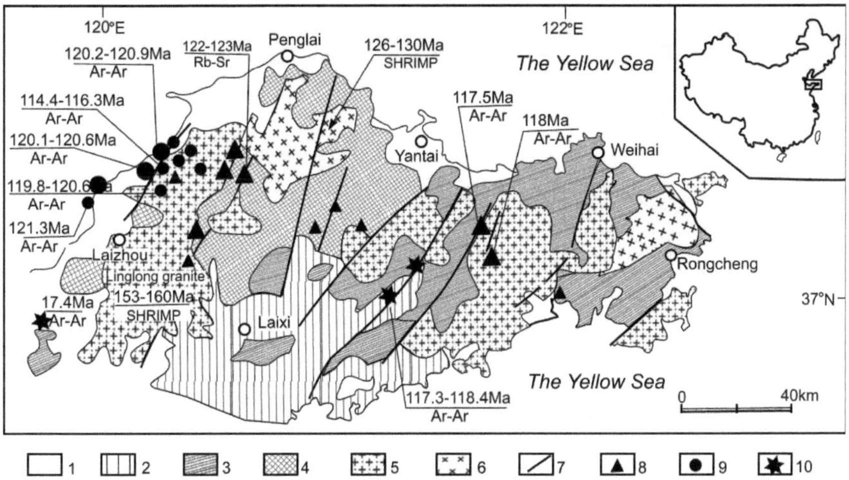

Figure 6.13. Simplified geological map of the Jiaodong Peninsula, NE China, showing age relationship between the Linglong biotite granite and related Au deposits (redrawn after Mao et al. 2005). **1**. Quaternary; **2**. Cretaceous volcanic rocks; **3**. Paleozoic metamorphic rocks; **4**. Archean metamorphic rocks; **5**. Jurassic Linglong biotite granite; **6**. Cretaceous granodiorite; **7**. Fault; **8**. Quartz vein-type Au-deposit; **9**. Fracture-alteration-type Au-deposit; **10**. Breccia-type Au-deposit.

is the Linglong biotite granite, Jiaodong Peninsula, NE China (Fig. 6.13). The granite is considered to have formed by the melting of crustal rocks (with the ISr values between 0.7086–0.7125; Qiu et al. 2002; Mao et al. 2005), and is SHRIMP dated at 153–160 Ma (Wang et al. 1998). More than ten Au deposits are known within the granite or along its periphery. Some are quartz-vein-type while others are fracture alteration-type deposits. All of the gold-bearing occurrences are considered to be genetically related to the Linglong granite (Mao et al. 2005) and yield younger ^{39}Ar–^{40}Ar and Rb–Sr ages of 121–117 Ma.

6.6. TEMPERATURE DISTRIBUTION

Corresponding to the age difference, the crystallisation temperature of granite should be higher than that of its associated mineral deposits and this is confirmed by numerous studies (among many others, Lu et al. 1974; Mao et al. 2005; Lin 2005; Zhuang et al. 2006). Figure 6.14 shows fluid inclusion temperatures of the Dajishan granite, SE China, and its related tungsten-bearing veins. From the bottom upwards, lithology and formative temperature change from biotite granite (640°C), two-mica granite (550°C) and muscovite granite (503–526°C), to pegmatite (455°C) at the top. The sequence illustrates a late-stage crystallisation sequence with Si, K and volatile enrichment within a dome structure. Temperatures of the ore-vein decrease upwards from 293 to 129°C (Lu et al. 1974).

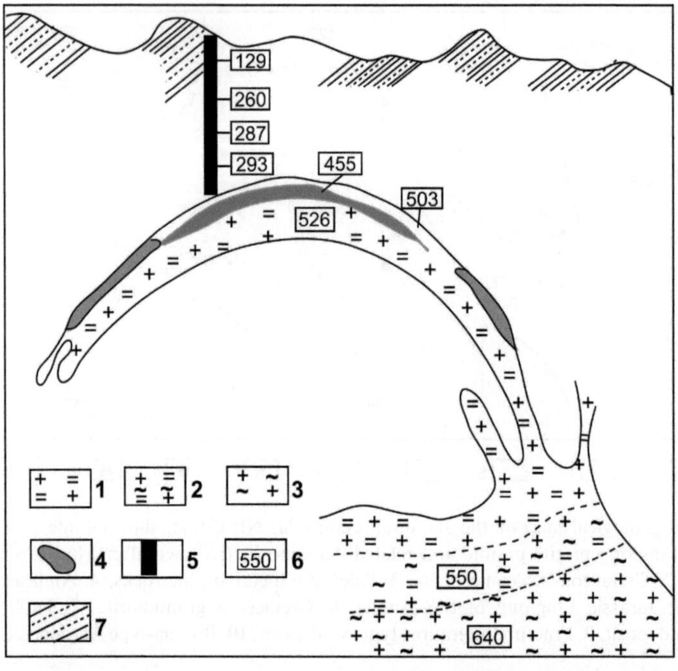

Figure 6.14. Section of the Dajishan mine, Dajishan granite, SE China, showing different lithological facies of granite and temperatures based on fluid inclusion thermometry (redrawn after Lu et al. 1974). **1**. Fine-grained muscovite granite; **2**. Medium-grained two-mica granite; **3**. Coarse-grained biotite granite; **4**. Feldspar–quartz-rich part of pegmatite; **5**. Quartz–wolframite vein; **6**. Fluid inclusion temperature; **7**. Metasandstone–slate.

If the hydrothermal fluid concentrated in domed areas of the MI is unable to escape, ore components in the fluid can only precipitate in situ or replace earlier-formed silicate minerals. This results in alteration (autometasomatism) of the upper part of the granite and produces granite-hosted ore bodies such as the example shown in Fig. 6.12.

6.7. FORMATION AND DISTRIBUTION OF HYDROTHERMAL MINERAL DEPOSITS

6.7.1. Precipitation of Ore-Forming Elements

With decreasing *T* and *P* during ascent of hydrothermal fluids, various ore-forming elements will precipitate in minerals resulting in a zoned hydrothermal mineral deposit as predicted by the zonal theory of metal arrangement in ore deposits proposed in the early twentieth century by Spurr (1907), Emmons (1924, 1926) among many others (Fig. 6.15; Table 6.1).

In SE China, W, Mo, Cu, Pb, Zn, Au, Ag and Hg minerals are commonly found in hydrothermal deposits associated with granite (Fig. 6.1). Numerous

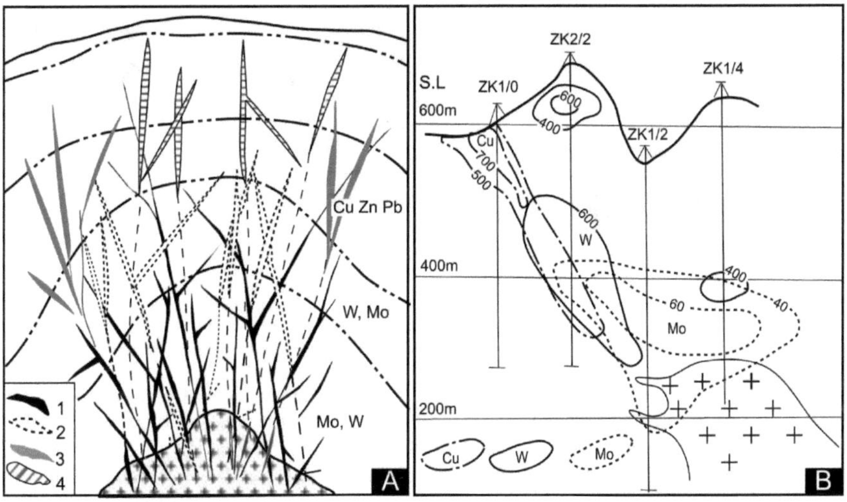

Figure 6.15. Geochemical zonation of ore-forming elements. **A**. zonation of ore-veins above the Kazakstan granite, Kazakastan (redrawn after Barnes 1975); **1**. Molybdenite vein; **2**. Wolframite vein; **3**. Sulphide vein; **4**. Carbonate vein. **B**. Contours of various ore-forming elements above the Longwangpai granite, Hunan Province, SE China (after Lin 2005).

vein-type W–Mo deposits are generally found in roof rocks close to the granite and sometimes the veining extends into the granite, indicating early, high temperature precipitation of W and Mo from the hydrothermal fluid. According to fluid inclusion thermometry, the formative temperature of W–Mo ore veins is generally between 250 and 400°C (among others, Lu et al. 1974; Mao et al. 2005; Zhuang 2006; Lin 2005). This suggests that hydrothermal mineralisation, consequent on crystallisation of granite magma, begins when the temperature of the granite contact, or just below the contact, reaches ~400°C (Fig. 6.6B).

W, Mo. Although W- and Mo-bearing phases such as wolframite and molybdenite are commonly intergrown in hydrothermal veins, maximum concentrations of Mo are usually below that of W (Fig. 6.15; Table 6.1) as documented by numerous researchers (among many others, the Nanling Tungsten Deposits Research Group 1985; Chen et al. 1996), indicating that Mo is precipitated earlier from the hydrothermal fluid at higher temperature.

Cu, Pb, Zn. In the SE China hydrothermal deposits, chalcopyrite, galena and sphalerite are commonly intergrown. As shown in Figs. 6.15 and 6.16; Table 6.1, the three elements usually occur in the order of Cu–Zn–Pb with increasing distance from the granite, and with precipitation of Cu taking place at a slightly higher level, i.e. later than W.

Au, Ag. The distribution of Au–Ag in Fig. 6.16A suggests that they precipitated later than Cu, Zn and Pb. In numerous Au–Ag deposits, the maximum concentration of Au is in the upper part of an ore body, with Ag in the lower part (Zhuang et al. 2006). An example is the Changkeng deposit, Guangdong

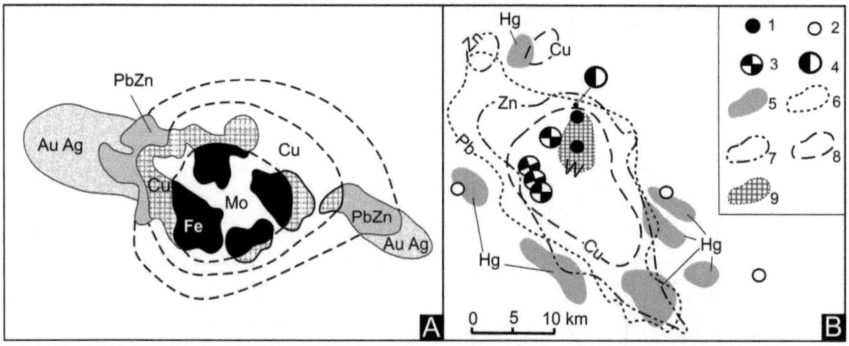

Figure 6.16. **A**. Sketch map showing metal-zoning in the Qibaoshan Mine, SE China (redrawn after Wang et al. 1988). **B**. Ore deposits and geochemical anomalies of the concealed Xianggai granite, Guangxi Province, SE China (modified after 215 Metallurgical Geological Party). **1**. W-prospect; **2**. Hg-prospect; **3** Cu–Zn–Sn deposit; **4**. Pb–Zn–Sn deposit; **5**. Areas of heavy mineral anomalies; **6–8**. Areas of geochemical anomalies; **9**. Skarn (the area corresponds to a projection of the top surface of the concealed granite).

Table 6.3. Element abundances in the Changkeng Au–Ag deposit, SE China (after Du et al. 1996)

Element	Host rock	Au	Ag	Pb	Zn	Cu
zone		ppm	ppm	wt %	wt %	wt %
Au	Silicified	11.74	10.50	0.003	0.015	0.004
	Calcsilicate	3.56	6.5	0.001	0.008	0.002
Ag	Silicified	0.32	174	0.07	0.15	0.02
	Calcsilicate	0.42	209	0.15	0.20	0.04

Province (Fig. 6.18), where Pb, Zn and Cu increase significantly downwards from the upper Au to lower Ag ore body of the deposit (Table 6.3).

Hg, Sb. In comparison with the distribution of the other elements described above, cinnabar and stibnite form at low-T in that they are typically located furthest from the granite (Fig. 6.15B). A geological map of the Shizhuyuan ore field, Hunan Province, SE China (Fig. 6.17), demonstrates a clear gradation from W–Sn–Bi–Mo deposits near the granite bodies, though Cu–Pb–Zn deposits to Hg and Sb deposits furthest from the granite.

6.7.2. Oxygen Isotope Evidence

Hydrothermal mineral deposits typically have lower $\delta^{18}O$ values than related granite because ^{16}O is preferentially partitioned into the gas phase in comparison to ^{18}O during crystallisation of granite magma. Oxygen isotopic exchange between host rock and hydrothermal fluid is controlled by temperature with ^{18}O values of the wall rock increasing away from the ore body. This is demonstrated by the Changkeng deposit where $\delta^{18}O$ increases with alteration intensity of the

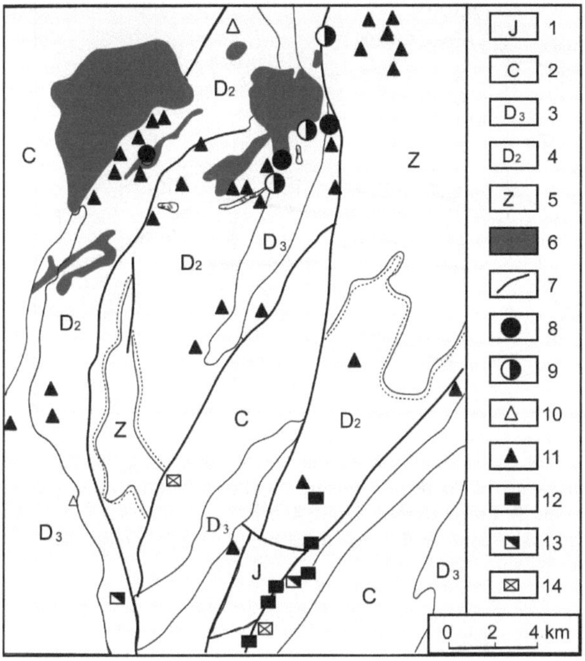

Figure 6.17. Element zonation in the Shizhuyuan ore field, Hunan Province, SE China (modified after Chen et al. 1989). **1**. Jurassic, **2**. Carboniferous; **3**. Upper Devonian; **4**. Middle Devonian; **5**. Sinian, **6**. Granite; **7**. Fault; **8–14**. Mineral deposit, **8**. W–Sn–Bi–Mo; **9**. W–Sn–Pb–Zn; **10**. Cu; **11**. Pb–Zn; **12**. Hg; **13**. Sb; **14**. Hg–Sb.

wall rock, tracking the source of the ore-bearing fluid as shown in Fig. 6.18 where geophysical exploration has revealed a subsurface granite body (Zhuang et al. 2006). Mainly, carbonate rock wall rocks of the Changkeng Au–Ag ore body have been variously silicified with the intensity of silicification decreasing upwards within and away from the ore body (Zhuang et al. 2006) reflecting decreasing temperature.

In hydrothermal processes, oxygen isotopic exchange occurs as a result of isotopic disequilibrium between hydrothermal fluid and wall rock, e.g. the fluid has a higher concentration of ^{16}O than the wall rock. ^{16}O replaces ^{18}O in silicate, oxide and oxysalt minerals in the wall rock resulting in a reduction of rock $\delta^{18}O$ thereby reducing the ^{16}O concentration (or increasing the ^{18}O content) of the fluid. From the time that hydrothermal fluid leaves its source, e.g. domed areas of a granite layer, isotopic exchange with the wall rocks occurs during upward migration of the fluid resulting in an overall increase in fluid $\delta^{18}O$. With precipitation of various ore-minerals from the fluid at different crustal levels depending on *T* etc., the source of an ore-bearing fluid cannot be reliably determined on the basis of $\delta^{18}O$ values of altered rock associated with a particular mineral assemblage.

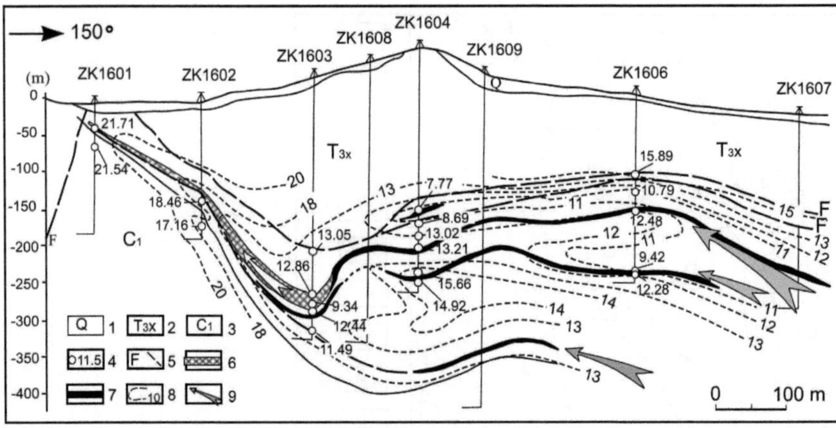

Figure 6.18. Section showing zonation of ore-forming elements and variation of $\delta^{8}O_{SMOW}$ values of the 16th prospecting section of the Changkeng Au–Ag deposit, Guangdong Province (after Zhuang et al. 2006); **1**. Quaternary; **2**. Upper Triassic; **3**. Lower Carboniferous; **4**. $\delta^{18}O_{SMOW}$ value; **5**. Fault; **6**. Au ore-body; **7**. Ag ore-body; **8**. $\delta^{18}O_{SMOW}$ value contour; **9**. Direction of ore-bearing fluid movement.

6.8. MINERALISED DEPTH HORIZONS

Since different ore-forming elements precipitate at different temperatures during the upward/outward migration of hydrothermal fluid, ore bodies enriched in a particular element in different deposits generated by the same hydrothermal event should be formed within in the same temperature interval, providing other factors such as *P*, Eh, pH, etc. have negligible effects. This implies that mineral deposits of different ore-forming elements should occur at different depths that can be related to a paleogeothermal gradient.

In the northeast part of the Xihuashan biotite granite described above, there are a series of ‘concealed granite bodies’, i.e. subsurface domal structures that are connected to a continuous layer of granite (Fig. 6.19). Within each vein system associated from each of these domes, wolframite occurs over a depth interval of ~100–200 m from above the granite contact delineating the paleotemperature range suitable for W-precipitation (Fig. 6.19).

For medium temperature ore deposits, a good example is the Hengjiang polymetallic deposit in Guangdong Province (Fig. 6.20), where polymetallic ore vein associations of Pb–Zn, Zn–Pb and Cu–Zn–Pb are restricted to well-defined horizons with increasing depth.

A longitudinal profile of the Mangchang polymetallic ore field in SW China shown in Fig. 6.21 indicates a relationship between mineralisation horizon, granite and overlying metasedimentary rocks (Fig. 6.21) as follows:

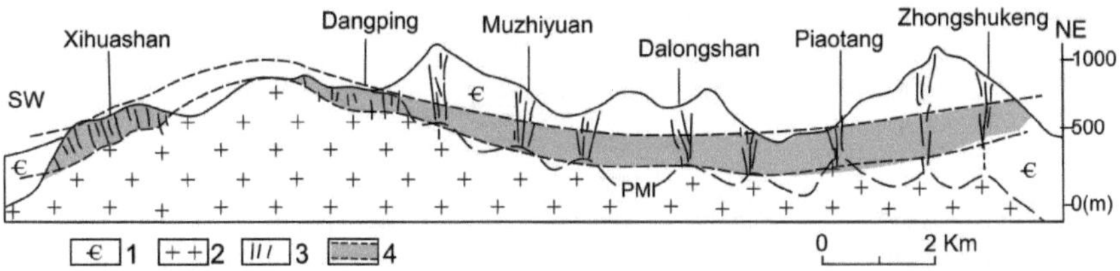

Figure 6.19. Cross section of the Xihuashan–Zongshukeng tungsten deposits, Jiangxi Province, SE China (modified after BGMRJx 1984). **1**. Cambrian metasediments; **2**. Mesozoic biotite granite; **3**. Quartz–W vein-type deposit; **4**. Horizon of W ore-deposits.

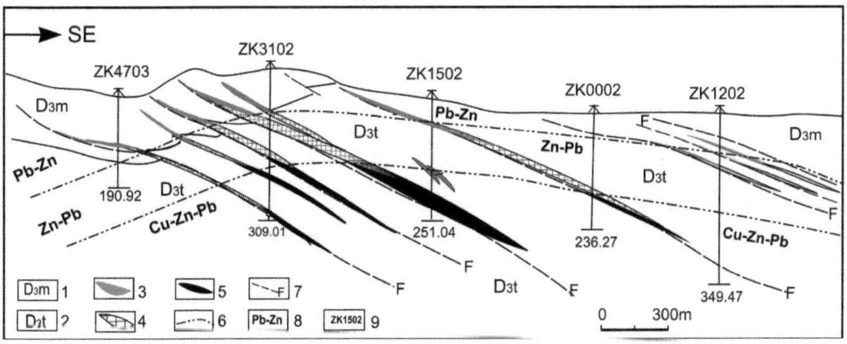

Figure 6.20. Cross section of the Hengjiang polymetallic deposit, Guangdong Province, SE China (after Zhuang 2006); **1**. Maozifeng group of upper Devonian age; **2**. Tianziling group of upper Devonian age; **3**. Pb–Zn ore body; **4**. Zn–Pb ore-body; **5**. Cu–Zn–Pb ore body; **6**. Zonation of ore-associations; **7**. Fault; **8**. Mineralised zone; **9**. Drill hole.

1. There are two Late Yanshanian granite ‘layers’ in this area, an older, upper granite dated at 91 Ma (Wang et al. 1990), and a younger, lower granite dated at 81 Ma.
2. The granite forms the core of an anticline.
3. Metamorphism decreases upwards from the upper granite contact from diopside–garnet hornfels, through biotite–plagioclase hornfels, to muscovite–chlorite hornfels.
4. Corresponding to the change in metamorphic grade, and parallel to the isograds, mineralisation sub-horizons change upwards from high-*T* (W, Mo, Nb, Ta), high-moderate-*T* (Cu, Zn, W, Sn, As), moderate-*T* (Sn, Zn, As), moderate-low-*T* (Sn, As, Zn, Pb), to low-*T* (Ag, Pb, Zn, Sb) associations over a lithological thickness of nearly 2000 m.

Regional zonation of endogenic mineralisation related to Mesozoic granites in SE China (Chen 1985) indicates increasing-*T* or ore-mineral associations towards

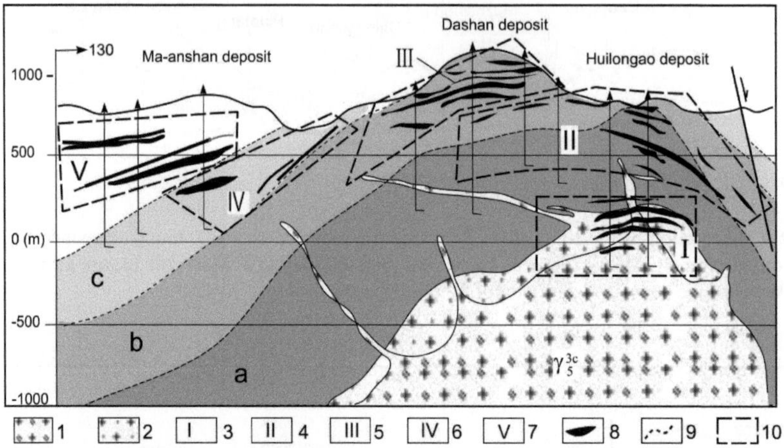

Figure 6.21. Longitudinal profile of the Mangchang ore field, Guangxi Province, SW China (modified after Wang et al. 1990). **1**. Late Yanshanian 2nd phase granite porphyry and porphyritic granite; **2**. Late Yanshanian 3rd phase granite and granite porphyry; **3**. Alteration/mineralised zone; **4**. High-*T* zone; **5**. Moderate-*T* zone; **6**. Moderate–low-*T* zone; **7**. Low-*T* zone; **8**. Ore body; **9**. Metamorphic isograd; **10**. Boundary of mineralised zone; **a**. Garnet–pyroxene hornfels; **b**. Biotite–anorthite hornfels; **c**. Muscovite–chlorite zone.

the SE (Fig. 6.22A). The section in Fig. 6.22B indicates the distribution of mineralisation horizons related to the entire inferred layer of Mesozoic granite.

6.9. MINERALISATION DURING ELEVATED CRUSTAL TEMPERATURES

If physico-chemical conditions overstep the stability of a mineral, it will either break down or react to a new stable phase and in doing so release all or some of its component elements into the environment. The process is common on the Earth's surface. For example, under supergene conditions, sulphides will break down, and sulphur and chalcophile elements such as Cu, Pb and Zn will reprecipitate in a suitable environment, while iron will stablise by combining with oxygen and hydrogen to form an iron hydroxide such as goethite. Anatexis results in mobilisation of nearly all the elements present, although prior to this, rising geotherms will have aided in the dissolution of accessory phases such as sulphides, fluorides, etc., in the rocks. For example, experimental work indicates that at 300°C, the amount of dissolved Hg in vapour can be as high as 250 mmHg, 10,000 times higher than at 50°C (Ding 1986). This indicates that with increasing temperature, Hg will be mobilised by being partitioned into a gas phase. Therefore, the formation of some low-*T*, Hg and Sb deposits could have occurred during crustal warming prior to the higher temperature hydrothermal stage of mineralisation that is associated with crystallisation of granite magma (Chen et al.

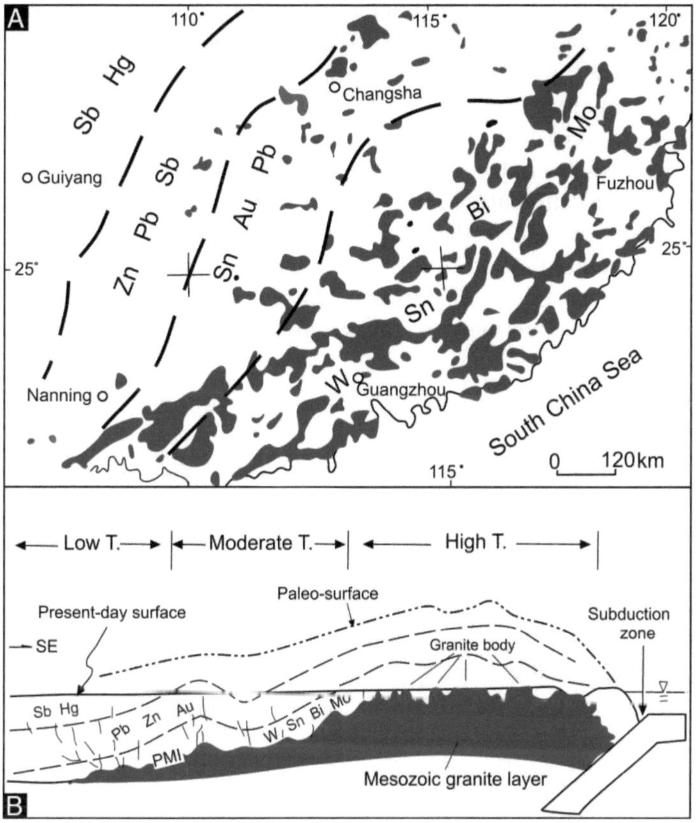

Figure 6.22. **A.** Regional mineralisation zonation in SE China and its relation to outcrops of granite (modified after Chen 1985). **B.** Schematic section showing relationship between regional mineralisation zonation and Mesozoic granite layer.

1996). These types of mineral occurrences can be referred to as hot water or thermal brine deposits. A considerable number of low temperature mineral deposits in southern China are likely to be of this type, e.g. the stratabound Pb–Zn and Hg deposits in west Hunan and Guizhou provinces (Fig. 6.1) (Wang et al. 1988; Yan et al. 1989).

Although temperature is the dominant factor controlling the formation of a particular ore-mineral assemblage, it is not the only factor responsible for mobilisation, transportation and precipitation of ore-forming elements. At least two geological parameters are also important; the composition of the rocks, and the type of conduit though which hydrothermal fluids migrate. If ore-bearing fluids encounter a 'chemically active' rock composition, such as carbonate, a significant amount of the ore-forming elements may precipitate to form an ore deposit. Figure 6.23 shows that most of the large- and medium-sized mineral deposits in

Guangdong Province, SE China, are distributed close to the interface between shallow marine carbonate and underlying clastic terrigenous strata (Lai 1981).

Fractures are important for both transportation of fluid and precipitation of ore minerals. However, in areas where fractures and joints are sparse, it is possible for the ore-bearing fluid to infiltrate and migrate upwards through pores and micro-fissures of rock in an 'evaporation' or steam mode with rising crustal temperature. When a suitable temperature interval and/or chemical environment is encountered, ore-forming elements in the 'steam' will likely precipitate (Fig. 6.24), resulting in large-scale mineralisation to form a stratiform deposit.

A map showing the distribution of Hg levels in bedrocks of Guizhou Province, SW China (Fig. 6.25) reveals high-mercury background over a vast area, and the

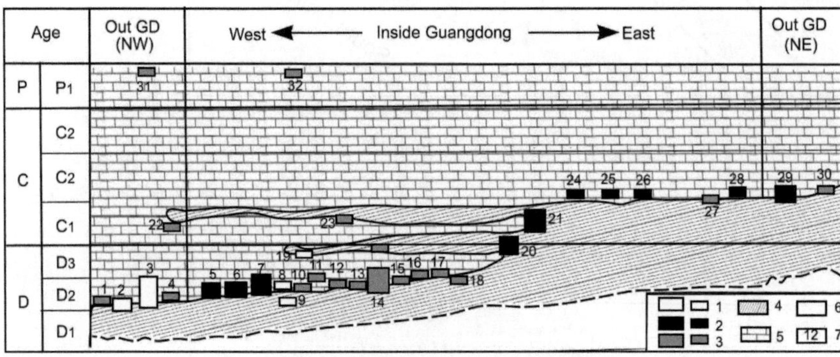

Figure 6.23. Distribution of some large- and medium-sized mineral deposits of late Paleozoic age in SE China (redrawn from Lai 1981). **1**. Fe and Fe–S deposits; **2**. W, Sn, Bi and Mo deposits; **3**. Sn, Pb, Zn, Sb and Hg deposits; **4**. Strata dominated by continental and littoral detrital rocks; **5**. Strata dominated by littoral and shallow marine carbonate rocks; **6**. Missing strata; **7**. Deposit numbers 1-32; GD = Guangdong Province.

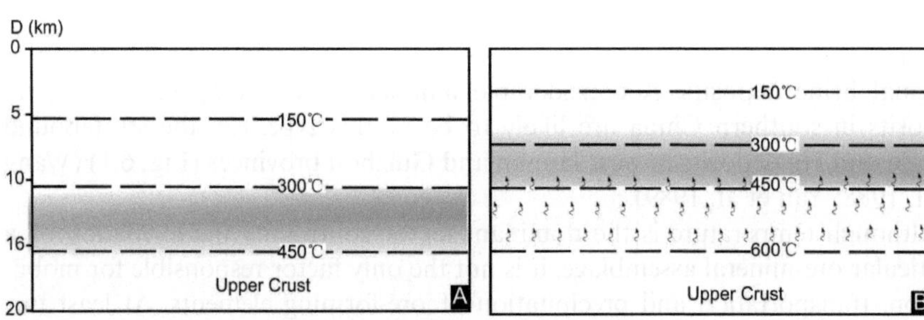

Figure 6.24. Diagram to illustrate vapour or steam migration and precipitation of ore-forming elements (in grey shaded zone) during rising crustal temperatures. **A**. Initial state assuming that 450°C is the upper temperature stability limit of a particular phase; **B**. With upward movement of the isotherms, minerals occurring below the 450°C isotherm become unstable and decompose with elements partitioning into the vapour phase that moves upwards through pores and micro-fissures in the rocks and reprecipitates them in new minerals that are stable at T <450°C.

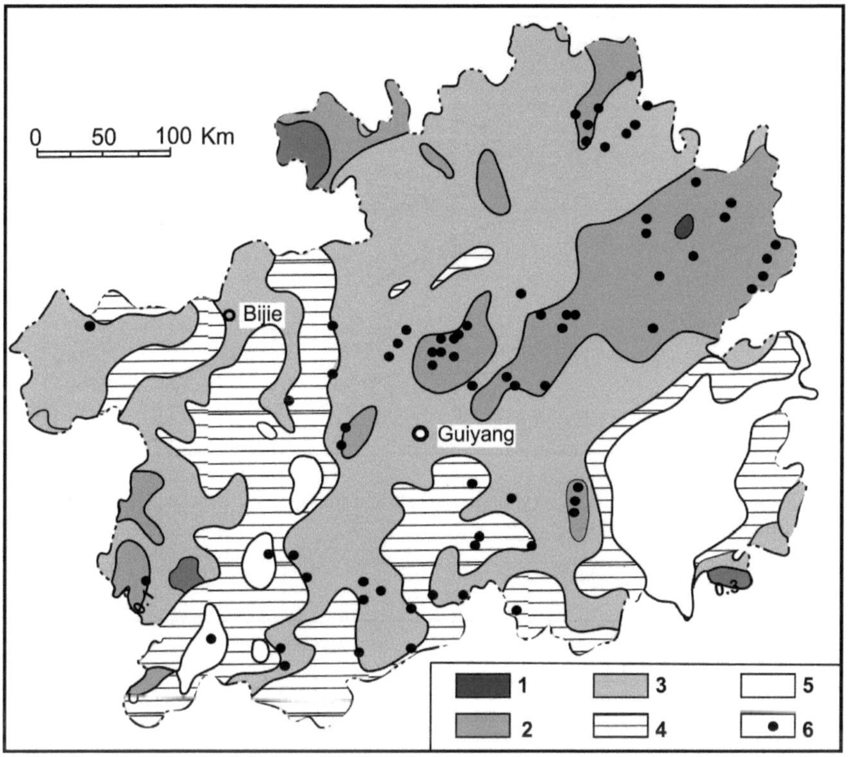

Figure 6.25. Map showing distribution of Hg (ppm) in Paleozoic bedrocks of Guizhou Province, China (modified after Xiang 1989). **1**. >0.3; **2**. >0.1; **3**. >0.03; **4**. >0.01; **5**. >0.003; **6**. Hg-deposit (prospect).

highest concentrations, as Hg-deposits (prospects), are mainly located in the central and northeastern parts of the province. According to Yan et al. (1989), the mercury deposits are not of sedimentary origin. Because Hg-mineralisation occurs over such a large area, they were probably formed by vapour phase–mode transportation/precipitation under conditions of elevated geotherms that caused remobilisation of Hg in rocks from deeper crustal levels.

The stratabound Pb–Zn deposits in western Hunan Province may be another example of 'evaporation-mode' formation. The deposits can be divided into three types using the classification scheme of Wang et al. (1988); contact type, infilling type and stratabound type (Chen et al. 1996). The contact type of Pb–Zn (polymetallic) deposits are skarns (Fig. 6.26), formed at relatively high temperatures of 200–250°C (Wen et al. 1984). Infilling-type deposits occur in veins or fracture zones with formation temperatures of 150–230°C (Wang et al. 1988). Strata-bound deposits occur within horizons of a bedded sequence, particularly in carbonates overlying clastic sediments as described above (Fig. 6.27) and have the following diagnostic features (Chen et al. 1996):

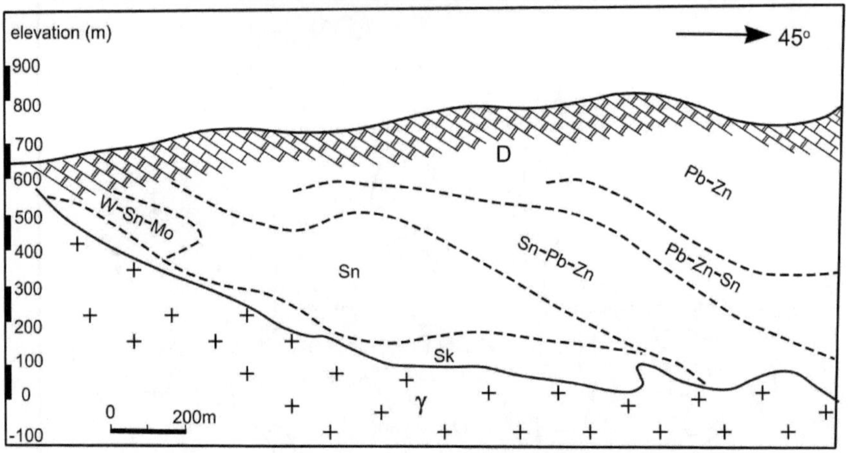

Figure 6.26. Cross section of the Xianghualing Sn–Pb–Zn polymetallic deposit, Hunan Province, SE China (after Wang et al. 1988)

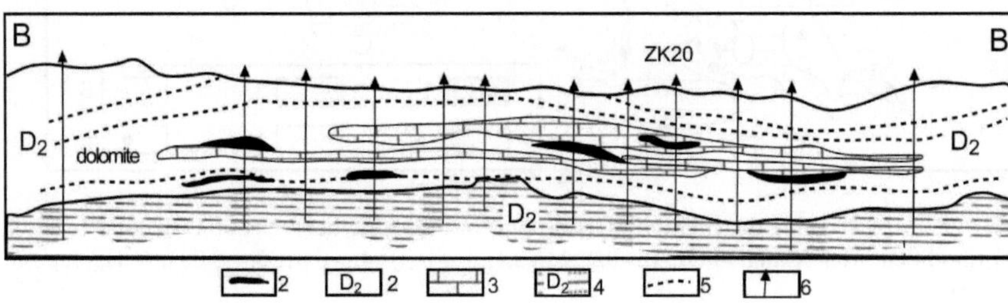

Figure 6.27. Section of the Houjianqiao Pb–Zn deposit in Hunan Province, SE China (originally unscaled; simplified after Wang et al. 1988). **1**. Ore body; **2**. Middle Devonian dolomite; **3**. Limestone lens in dolomite; **4**. Middle Devonian clastic rocks; **5**. Stratigraphic boundary; **6**. Drill hole.

1. Fluid inclusion homogenisation thermometry in sphalerite and calcite indicates formation temperatures between 120 and 165°C (Wang et al. 1988), lower than those of the contact and infilling types.
2. Associated alteration is characterised by linear bands and lenses of silicification, dolomitisation, pyritisation and discolouration (Fig. 6.28), indicating that the deposits are not the product of sedimentation.
3. The deposits occur along a definite horizon over a huge area with no spatial relationship to granite (Fig. 6.29), indicating that the ore-bearing fluid was not derived from granite magma.

In addition to the geological evidence listed above, other evidence for the origin of the strata-bound Pb–Zn deposits is from S-isotopes. Unlike oxygen, S mainly

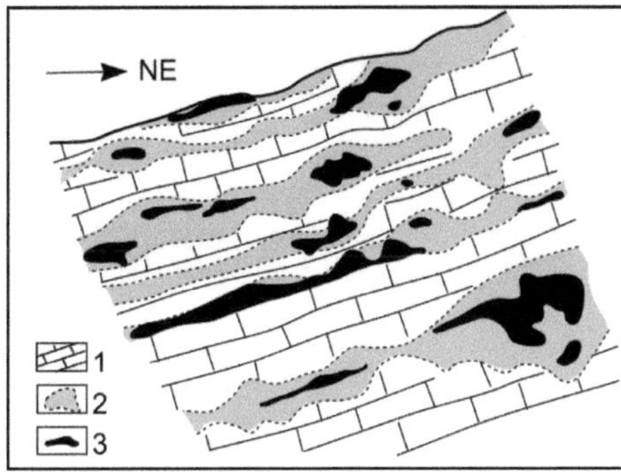

Figure 6.28. Sketch showing the occurrence of the Jiangjiaya Pb–Zn ore bodies, Hunan Province, SE China (originally unscaled; redrawn after the Hubei Regional Geological Surveying Party). **1**. Limestone; **2**. Silicified limestone; **3**. Galena-rich deposit.

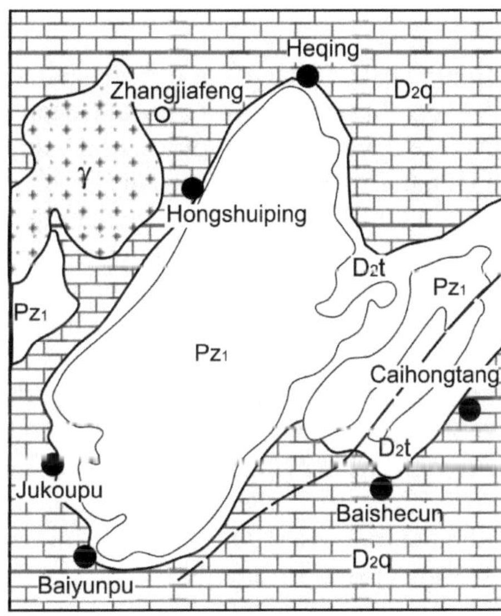

Figure 6.29. Geological sketch map showing distribution of stratabound Pb–Zn–pyrite deposits (prospects) in the Dachengshan area, central Hunan, SE China. All the deposits occur in carbonate rocks near the contact with underlying clastic sediments (redrawn after the Hunan Geological Institute). **Pz_1**. Lower Paleozoic and Precambrian (metasandstone–shale); **D_2t**. Middle Devonian sandstone–siltstone; **D_2q**. Middle Devonian carbonates; **γ**. Triassic granite; Filled circles = named Pb-Zn or Cu-pyrite deposits.

combines with metallic elements to form sulphides during hydrothermal mineralisation and does not exchange with wall rocks as they typically contain only trace amounts of S compared with S-rich hydrothermal fluids. Sulphides precipitated from ascending high-*T* hydrothermal fluids exhibit an initial preference for combining with the heavy sulphur isotope $\delta^{34}S$. A sphalerite–pyrrhotite $\delta^{34}S$ concentration contour profile of the Dachang deposit, SE China (Zhang 1989), is in good agreement with this expectation (Fig. 6.30).

Consumption of heavy sulphur with the precipitation of sulphides from the hydrothermal fluid leads to the reduction of ^{34}S concentration in the fluid,

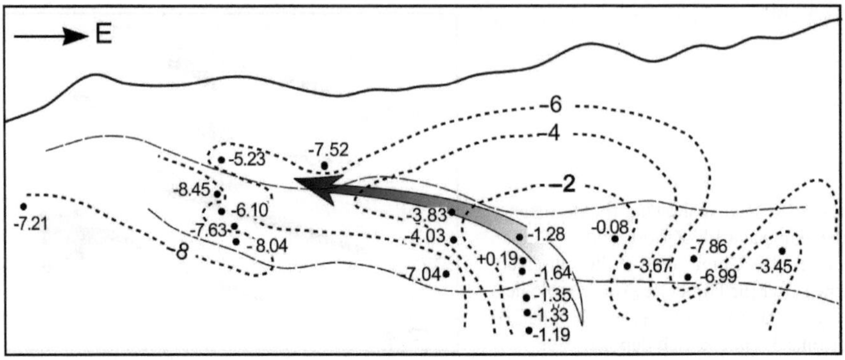

Figure 6.30. Sphalerite–pyrrhotite $\delta^{34}S$ (‰) variation in the Dachang deposit, SE China (modified after Zhang 1980).

resulting in the decrease of $\delta^{34}S$ in late-forming sulphides minerals, i.e. $\delta^{34}S$ values of hydrothermal deposits should decrease with decreasing T and increasing distance from a granite source. $\delta^{34}S$ values of the three types of Pb–Zn deposits in Hunan Province are shown in Fig. 6.31 where it can be seen that both the contact- and infilling-types have small ranges of $\delta^{34}S$, with the former having a positive mean value and the latter a negative mean value. This indicates that although they have the same granite source, the infilling-type deposits are located further from the source. Strata-bound deposits are characterised by a larger variation in $\delta^{34}S$, suggesting a different source in addition to their different mode of formation. Combining the data of the spatial distribution of Pb–Zn deposits (Fig. 6.32) with the model for regional zonation of mineralisation in SE China (Fig. 6.22), it can be concluded that most of the strata-bound Pb–Zn deposits in the western part of SE China were probably deposited from sedimentary pore water (brine) steam (evaporation mode) mobilised as a result of rising crustal isotherms.

6.10. MINERALISATION DURING GRANITE REMELTING

6.10.1. Oxidation

Removal of a considerable amount of S from the granite magma layer to its roof rocks during crystallisation indicates that the granite will be sulphide poor. Thus, if the granite undergoes remelting, the newly formed granite will be further impoverished in S. On the other hand, decreasing fS with remelting will be accompanied by an increase of the fugacities of oxygen, fluorine, chlorine, etc. This may be the reason that most of the composite granite bodies in SE China described above show increase in their oxidation ratios (Fe_2O_3/FeO) accompanied with decreasing total iron and increasing loss on ignition (LOI, representing

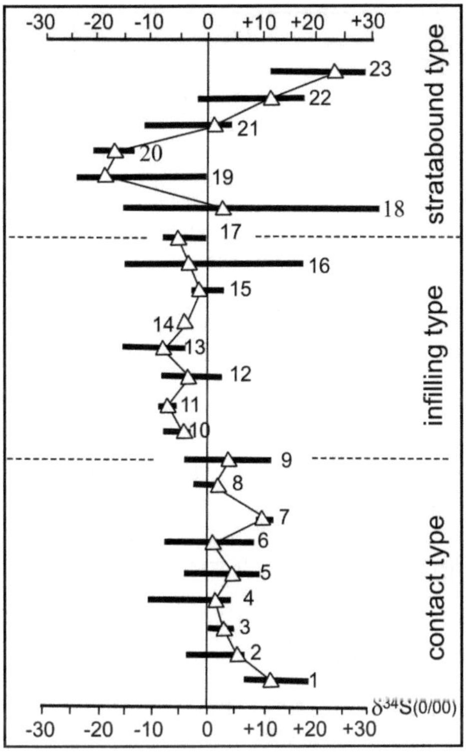

Figure 6.31. Comparison of $\delta^{34}S$ (‰) values (range and mean values) of various types of Pb–Zn deposits in Hunan Province, SE China (modified after Wang et al. 1988). **1**. Huangshaping; **2**. Baoshan; **3**. Tongshanling; **4**. Xianghualing; **5**. Dongpo; **6**. Shuikoushan, **7**. Jinshiling; **8**. Ayushan; **9**. Qibaoshan; **10**. Yaolin; **11**. Dongchong, **12**. Panjiachong; **13**. Donggangshan; **14**. Tieshikeng; **15**. Qingshuitang; **16**. Fengxijiang; **17**. Qianchangzhi; **18**. Baiyunpu; **19**. Heqing; **20**. Zhoushan; **21**. Houjiangqiao; **22**. Dongjiahe; **23**. Yutang.

total volatile content and mainly H_2O) with younging (Table 6.4; Fig. 6.33). Mean biotite granite compositions of different ages also show the same variation (Table. 6.4).

Increasing Fe_2O_3/FeO ratio and decreasing LOI implies increasing fO_2 in the remelted granite. Higher oxygen, fluorine and chlorine fugacities with remelting should facilitate complexing of lithophile ore-forming elements formerly held in rock-forming minerals such as Li, Be, Nb, Ta and REE, with O, Cl, F and CO_2 (Table 6.5) and their concentration in the upper, lower temperature part of the molten system. This provides an explanation for the occurrence of such deposits at the top of the younger granite member of many composite granite bodies (Meng 1993). Corresponding to the variation of the oxidation ratio of Mesozoic granites in Fujian Province, the concentrations of ore-forming elements in related mineral deposits evolve from sulphophile to oxygenphile as shown in Fig. 6.34.

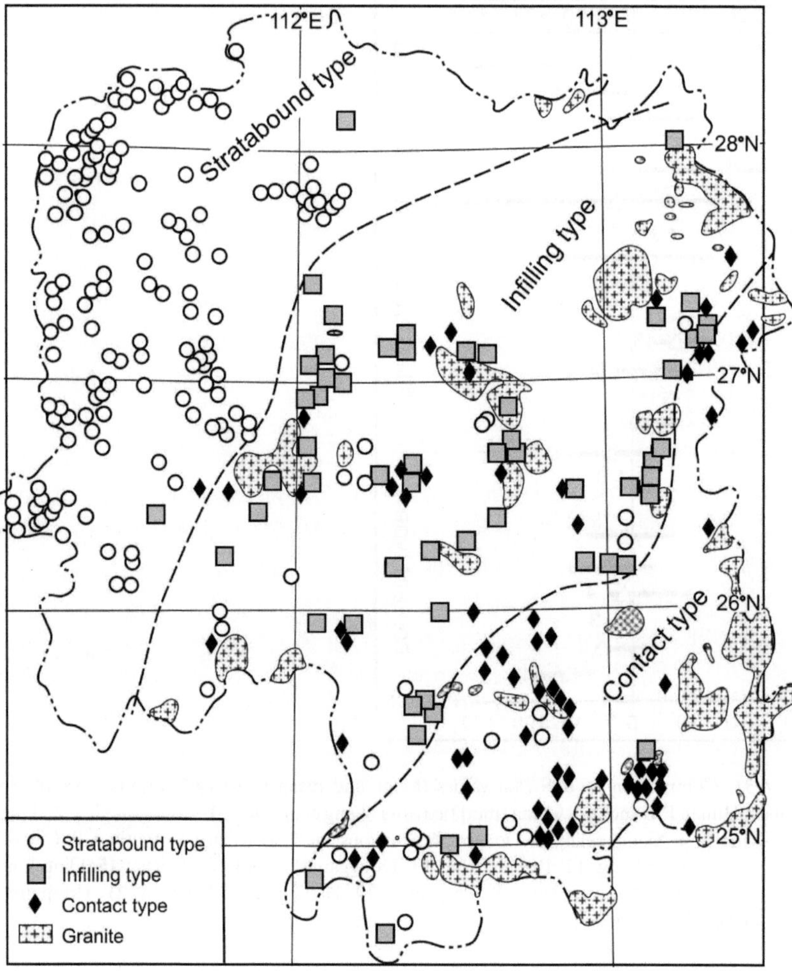

Figure 6.32. Map showing spatial distribution of the various types of Pb–Zn deposits and their relation to granite in Hunan Province, SE China (redrawn from Wang et al. 1988). Note that the number of granite exposures decreases NW, together with a SE to NW change in the type of Pb–Zn deposit from dominantly contact-type, through infilling-type to strata-bound-type. This correlates with NW deepening of the inferred PMI of the Mesozoic granite layer beneath SE China as shown in Fig. 6.22B.

6.10.2. Uranium Mineralisation

Formation of the so called 'granite type' uranium deposits in SE China may also represent an example of mineralisation resulting from the remelting of granite. This type of uranium deposit occurs within granite and typically has a greater age span than the granite host. Du (1982) divides the granite-related uranium deposits into five subtypes:

Table 6.4. Some major element and iron ratio variation in composite granites (CG), SE China

CG	Age	SiO_2	Na_2O	Fe_2O_3	FeO	Fe_2O_3/FeO	LOI
Jf	g_5^a	71.07	2.78	0.45	2.76	0.16	
	g_5^b	73.16	2.84	0.36	2.24	0.16	
	g_5^c	75.04	2.82	0.20	0.96	0.22	
Df	g_5^a	65.04	3.16	0.33	3.47	0.09	2.15
	g_5^b	73.02	3.49	0.47	1.25	0.38	1.27
	g_5^c	74.88	4.49	0.38	0.97	0.39	1.37
Ql	g_5^a	74.20	3.13	0.64	1.50	0.43	
	g_5^b	75.05	3.39	0.52	1.09	0.48	
	g_5^c	74.69	4.32	0.34	0.80	0.43	
Wg	g_5^a	71.70	3.26	0.82	2.06	0.40	0.65
	g_5^b	75.60	3.14	0.46	1.33	0.35	0.47
	g_5^c	76.08	3.57	0.5	1.07	0.48	0.61

Jf = Jiufeng; Df = Dengfuxian; Ql = Qianlishan; Wg = Wuguishan; g_5^a, g_5^b, g_5^c = oldest, younger and youngest granites, respectively.

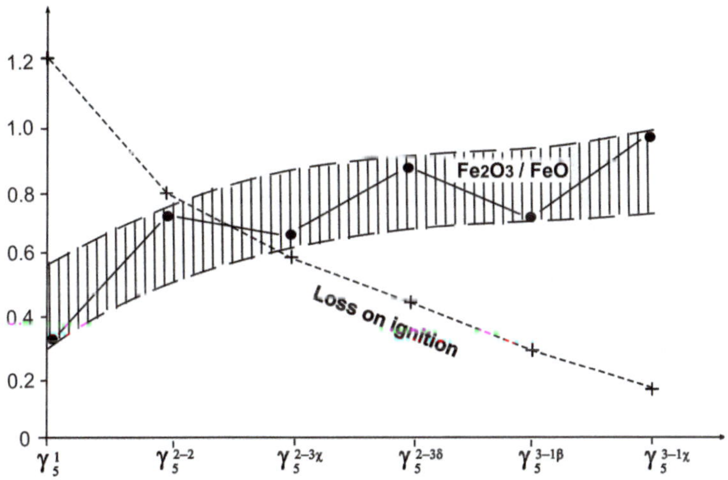

Figure 6.33. Trends of average oxidation ratio and LOI (wt %) in Mesozoic granites of different ages, Fujian Province, SE China (data source: BGMRFj 1985).

Table 6.5. Average Nb and Ta contents in mineralised and non-mineralised granites, SE China

Type	No. granite	Nb_2O_5		Ta_2O_5 (ppm)		F (ppm)	
	Body	Range	Mean	Range	Mean	Range	Mean
Mineralised granite	10	27–380	134	5–200	66	8,000–16,000	4,097
Non-mineralised granite	16	10–43	21	0–10	5	250–850	576

Data source: Institute of Geochemistry, Chinese Academy of Sciences (1970).

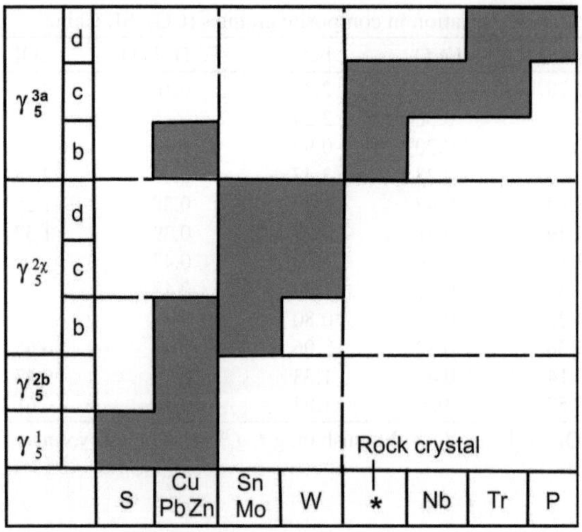

Figure 6.34. Relationship between evolution of Mesozoic granite and its ore-forming elements in Fujian Province (after Regional Geological Survey of Fujian Province, unpublished data).

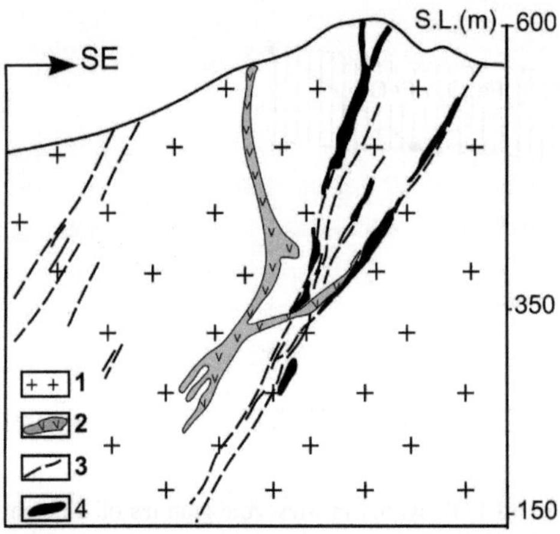

Figure 6.35. Section of the 606 U-deposit, SE China (redrawn after Du et al. 1982). **1**. Granite; **2**. Lamprophyre dike; **3**. Silicified shear zone; **4**. U-ore body.

1. Micro-quartz subtype in shear zones within the granite (Fig. 6.35) with pitchblende occurring with an aggregate of recrystallised mico-quartz associated with sporadic veinlets and disseminated grains of hematite and goethite (Fig. 6.36A).
2. Fluorite subtype: occurs as the variety of the micro-quartz subtype, and is characterised with gangue minerals dominated by fluorite (Fig. 6.36B).
3. Kaolin subtype: U-ore veins are usually 0.n–0.00n mm wide occurring mainly along micro-fissures and kaolinised zones in granite (Fig. 6.36C).

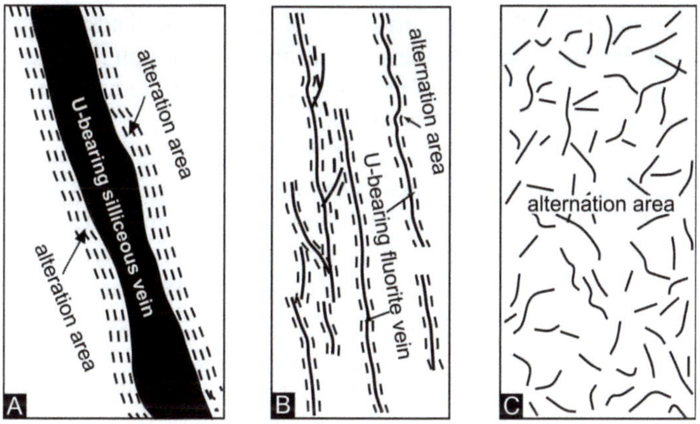

Figure 6.36. Examples of subtypes of U-ore veins described in text (after Jin et al. 1991). **A**. micro-quartz subtype; **B**. Fluorite-subtype; **C**. Kaolin-subtype.

4. Alkali metasomatic subtype: alkali metasomatism (mainly Na and sometimes K metasomatism) along shear zones is associated with U mineralisation and concentration of Y, Ce, Nb, Ta, Zr or Hf.
5. Carbonate subtype: pitchblende occurs in veins of calcite and dolomite in granite, skarn or marble.

A feature shared by the above subtypes is the almost complete absence of sulphides and an abundance of hematite in the vein wall rocks (Wang 1985).

According to the above data, diagnostic features of the SE China U-deposits can be summarised as

- Granite-type uranium deposits were formed by the injection of ore-bearing fluid along fractures.
- Compared with other hydrothermal deposits, U-deposits are impoverished in S and enriched in O.
- The U-ore fluid is rich in Si, Na and K together with F and CO_2.

It is well known that uranium has two valency states in nature, i.e. U^{4+} and U^{6+}. Under reduction conditions, uranium largely occurs as U^{4+} that can replace Zr^{4+}, Hf^{4+} and REE^{3+} in zircon, titanite, apatite, xenotime, allanite, monazite, thorite, pyrochlore and other accessory minerals during crystallisation of granite magma (Table 6.6). This means that although uranium distribution is similar to REE, Zr, Hf, Ta, Nb, etc., which become concentrated in the domed areas of the granite

Table 6.6. Uranium contents (ppm) in various minerals in granite (after Liu et al. 1984)

Biotite, hornblende, magnetite, garnet	3–10
Apatite, titanate	n* 10
Allanite, pyrochlore, monazite, xenotime, zircon	n*100~ n*10,000

MI during crystallisation, when the concentrations of the above elements are high, U is low (Liu et al. 1984), or possibly inert. For example, although the background concentration of uranium in the Xihuashan granite can be as high as 20 ppm, no U-deposits have been found (Wang 1982).

Under oxidising conditions, U^{4+} changes into U^{6+} which typically occurs in uranyl ($(UO_2)^{2+}$), and unable to enter accessory minerals because of its large ionic radius. It is therefore likely that $(UO_2)^{2+}$ will be partitioned into late-stage hydrothermal fluids. Fission track studies have revealed that in altered rock-forming and accessory minerals such as quartz, feldspar, biotite, zircon and allanite, fission tracks are sparse within the grains but dense between grains, which has been explained as a 'U-purification' process of the minerals during their alteration (Du et al. 1982; Zhang and Zhang 1992). However, another possibility cannot be excluded, i.e. that the high concentration of fission tracks in alteration surfaces on hydrothermal minerals may contain higher concentrations of U derived from the hydrothermal fluid.

Although increasing of O, F, Cl and CO_2 fugacities results in higher oxidation potential of remelted granite, partitioning of U into the fluid phase requires a higher ambient redox potential than that required for the solution of REE etc., in the domal structures of the granite MI. This implies that several remelting events appear to be necessary for uranium to become concentrated in the hydrothermal fluid. Hydrothermal uranium deposits in SE China formed between 87 and 47 Ma (Li 1982), considerably later than the formation of W, Mo, Nb and Ta deposits between 180 and 120 Ma (Chen et al. 1989), and usually with a greater age span than their respective host granite (Du et al. 1982).

Fluid inclusion thermometry indicates that the formative temperature of most uranium deposits in SE China was ~250–150°C (Zhang and Zhang 1992), and this defines the temperature–depth interval of a U-mineralisation horizon (Du et al. 1982; Chen et al. 1996). Because the remelted granite is located below older granite layer(s), the U-mineralisation horizon occurs within older granite as shown in Fig. 6.37.

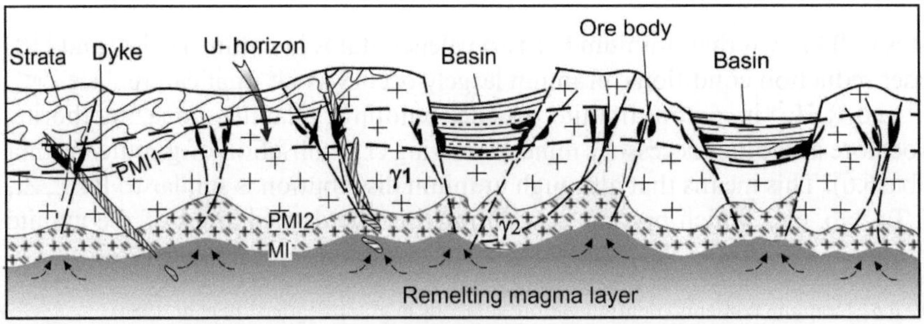

Figure 6.37. Schematic diagram showing relationship between repeated granite melting and distribution of a U-mineralisation horizon (between the parallel dashed lines) in the older granite layer (redrawn after Chen et al. 1996). γ1 = older granite; PMI1, PMI2 = older and younger paleo-magma interfaces; MI = youngest magma interface.

Table 6.7. Coefficient of oxygen[a] and U-content of pitchblende from hydrothermal uranium deposits in China (after Cai 1982)

Sample no.	UO_2(%)	UO_3(%)	Coef. of oxygen content	Occurrence
1	44.49	28.07	2.46	Granite
2	36.21	47.43	2.54	Granite
3	34.61	49.23	2.59	Granite
4	50.41	31.98	2.39	Exocontact zone
5	39.07	25.02	2.39	Exocontact zone
6	47.80	32.94	2.41	Granite
7	22.16	51.29	2.70	Rhyolite
8	35.92	46.37	2.56	Granite
9	41.36	37.05	2.47	Granite
10	39.35	38.94	2.50	Granite
11	38.25	39.74	2.51	Granite
12	49.40	29.00	2.37	Granite
13	45.40	28.30	2.38	Granite
14	43.70	29.60	2.40	Granite
15	36.20	20.20	2.36	Granite
16	33.00	17.80	2.35	Granite
17	51.30	26.90	2.37	Granite
18	50.10	24320	2.32	Granite

[a]Coef. of oxygen = x in UO_x, and $2<x<3$ (after Du et al. 1982).

To emphasize the importance of T in the precipitation of uranium does not mean to minimise the influence of chemical factors. As stated above, in hydrothermal fluids U is most likely in the form U^{6+}. However, both U^{6+} and U^{4+} occur in pitchblende (Table 6.7) in approximately equal amounts (Cai 1982). If it was only falling temperature that caused precipitation of U, then there should be less U^{4+} and it is considered that the U^{4+} in pitchblende is likely to have been derived by an oxidation–reduction reaction between hydrothermal fluid and wall rock. Evidence for this is that hematitisation (also known as red alteration) is commonly developed in the wall rock of U-bearing veins. This can be explained by U^{6+} capturing electrons from Fe^{2+} and being reducing to U^{4+}, with oxidation of Fe^{2+} to Fe^{3+} during precipitation of pitchblende (Du et al. 1982; Jin and Huang 1991).

When a U-bearing vein intersects an intermediate or mafic dyke, there is a significant increase in the ore-grade of the vein (Figs. 6.35 and 6.38). As U-contents of such dykes typically range between 1 and 2 ppm (Du et al. 1982), much lower than that of granite, the elevated U of the vein could not have been derived by leaching from the dyke. The most likely explanation is that electrons from the dissolution of iron-bearing minerals in the dyke cause the reduction of U^{6+} to U^{4+} in the infiltrating hydrothermal causing rapid precipitation of pitchblende. At the same time, oxidation in the dyke results in reddening of the wall rock (Fig. 6.38).

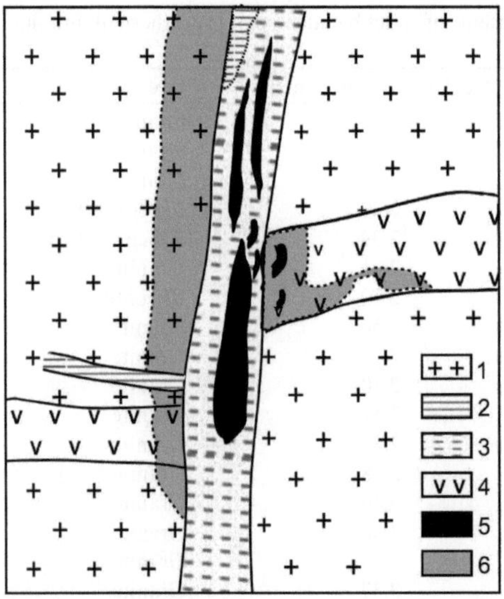

Figure 6.38. Relationship between U-bearing vein and a mafic dyke (redrawn from Jin and Huang 1991; unscaled). **1**. Kaolinised biotite granite; **2**. White micro-quartz; **3**. red micro-quartz (with hematite); **4**. Mafic dyke; **5**. Massive pitchblende; **6**. Disseminated pitchblende and rock alteration.

6.11. PATTERNS OF ELEMENT REDISTRIBUTION AND ELEMENT FIELDS

In situ melting involves different protoliths that contain different element concentrations. During crystallisation of a granite magma layer, however, the various elements are reorganized at different temperatures and redistributed in new minerals at different depths within the crust. Most of lithophile elements are fixed on solidification when the temperature of the granite solidus is crossed and those elements are distributed in the rock beneath the position of the MI (Fig. 6.39). The sulphophile elements and those not substituted into silicates are expelled from the system with exsolved H_2O during the latest stage of crystallisation. During upward movement of hydrothermal fluid, sulphophile elements precipitate as sulphide assemblages with decreasing temperature in the roof areas of a granite magma layer (Fig. 6.39). Following precipitation of sulphophile elements, the hydrothermal fluid will be dominated by H, C, N, O, B and halogens, and is likely to mix with meteoric water. During the hydrothermal process, the main function of these elements is to act as complexing agents of the ore-forming elements, and they will ultimately be expelled into the hydrosphere (Fig. 6.39). For example, brine from the KTB deep drill hole in southern Germany has a high $^{36}Cl/Cl$ ratio believed to be largely derived from the subsurface continuation of the Falkenberg granite that crops out about 2 km east of the KTB hole (Manuela et al. 1998). Inert gases do not take part in the hydrothermal process described and they are directly or indirectly liberated into the atmosphere after being exsolved from the granite magma (Fig. 6.39).

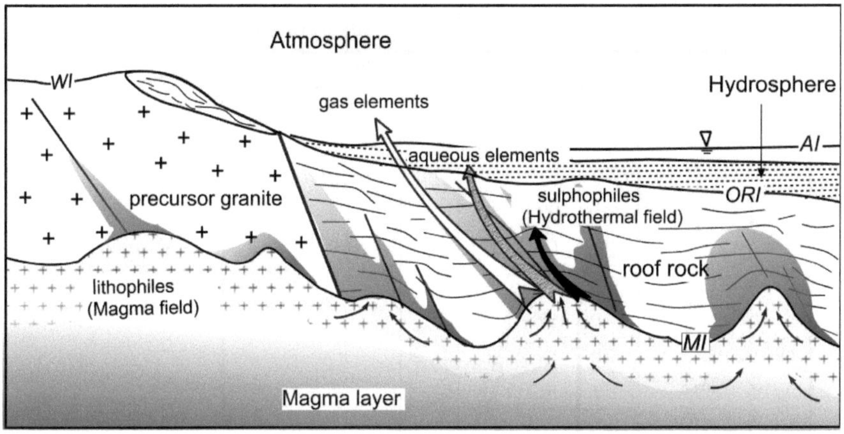

Figure 6.39. Diagram to illustrate migration and distribution of elements during crustal melting–crystallisation. Lithophile elements are concentrated below the MI; sulphopile elements are precipitated in rocks overlying the MI; aqueous elements migrate to the hydrosphere and gases are ultimately expelled into the atmosphere. Black, stippled and white arrows indicate migration paths of sulphophile, aqueous and gas elements, respectively. MI = magma interface; WI = weathering interface between rock and atmosphere that develops in areas of erosion; AI = accumulation interface (upper limit of areas of deposition); ORI = oxidation–reduction interface (interface between sediments and unweathered basement rock).

The behaviour of the various elements mentioned above is evidently related to their texture, a relationship explored by Goldschmidt, Zavaritsky, among many others, Rosler and Lange (1972) and most recently Railsback (2003). As textural variations of the elements are explained from the periodic table, the problem is to decipher the role of the elements in geological processes. The in situ melting model of granite genesis provides a framework for better understanding element behaviour during the crustal melting–crystallisation process (Fig. 6.39), and on this basis a concept referred to as *Geochemical Fields of Elements* has been proposed by Chen et al. (1996) and Chen (1998a), which divides the periodic table into four geochemical fields, e.g. magmatic, hydrothermal, aqueous and gas. The magmatic field includes the lithophile, siderophile, rare and radioactive element groups of the Zavaritsky classification. Siderophile elements in the magmatic field used to be considered to be mainly concentrated in the Earth's interior, yet in the in situ melting–crystallisation process, they enter silicates or form accessory phases such as Fe–Ti oxides. The gas field equates with Zavaritsky's H and inert gas groups and the aqueous field relates the magmatic–pneumatolytic and hydrothermal fields to the others. From Fig. 6.40, the geochemical fields mirror the layered structure of the crust and show the distribution of the various elements during crustal melting–crystallisation (Fig. 6.39).

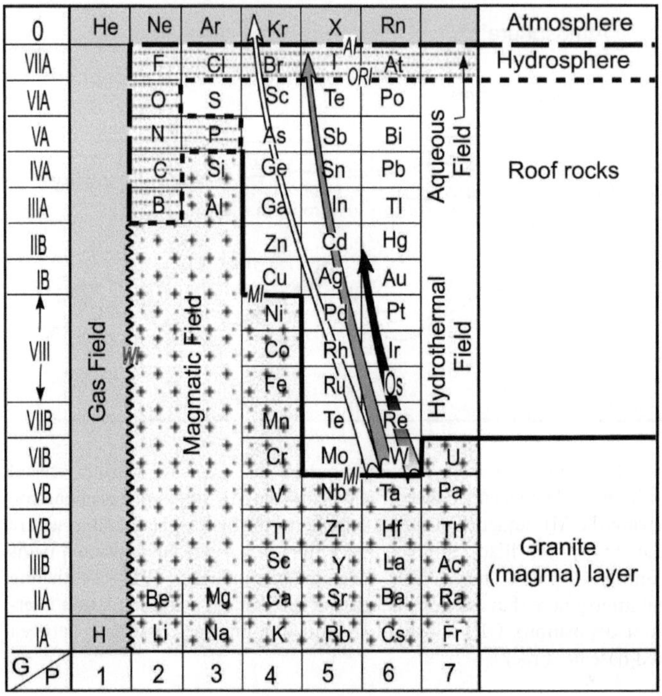

Figure 6.40. The geochemical fields of elements. Different fields of the upper crust are characterised by different elements and represent; the granite layer, roof rocks, sedimentation sphere (part of the hydrosphere), and the atmosphere. The arrows indicate directions of element movement and destinations in the three upper crust fields during melting–crystallisation to from the granite magma layer (compare with Fig. 6.39). Dark grey, stippled and white arrows = siderophile, aqueous and gas elements, respectively.

The dividing line between the magmatic and hydrothermal fields marks the position of the MI, i.e. typically between Group VB (Nb, Ta) and Group VIB (Mo, W). On both sides of the dividing line, the aqueous ore-forming elements are arranged in the same way as mineral zonation from the MI downwards, i.e. in the order Ta, Nb and REE (Fig. 6.4). The Zr and Hf concentration zones are generally associated with Ta, with Hf above and Zr below (Liu et al. 1975; Li et al. 1975). Above the MI, upward/outward zonation of the ore-forming elements is Mo, W, Cu, Zn, Ag, Au, Hg as detailed above (Figs. 6.15–6.21).

The above mineralisation zonation is represented in the geochemical fields of elements as a zigzag path (Fig. 6.41). In other words, the positions of the above-mentioned elements in the geochemical fields essentially reflect their positions in nature.

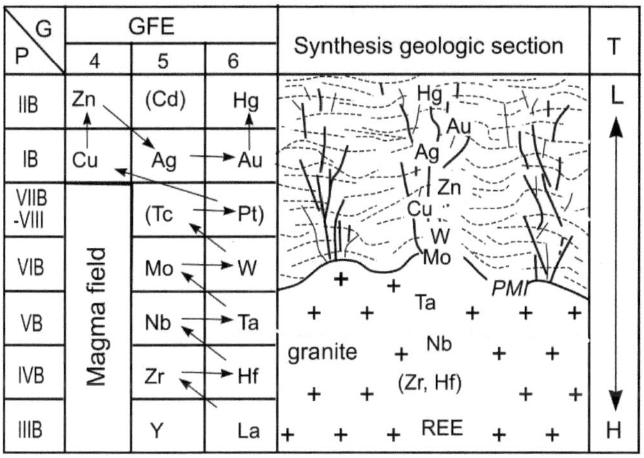

Figure 6.41. Comparison between mineralisation zonation above and below the MI of a granite layer and positions of aqueous ore-forming elements in the geochemical field of the elements (after Chen 1999). Bracketed elements are dispersed elements. See text.

6.12. SUMMARY

Granite genesis concerns not only the granite itself, but also the numerous related ore deposits. In the in situ melting framework of granite genesis, rock-, ore- and fluid-forming elements are all derived from the melted crust. The melting–crystallisation process results in reorganization and redistribution of elements with lithophile elements redistributed to form a new rock (granite) below the magma interface (MI), and those unable to be accommodated in silicate minerals are expelled from the melt system together with mainly H_2O and other volatiles exsolved on crystallisation. During the upward movement of hydrothermal fluid, most of the ore-forming elements are again fixed in the lithosphere by precipitation of appropriate minerals from the fluid under various physicochemical conditions resulting in mineralisation zoning at different crustal levels. Aqueous elements responsible for the transportation of ore-forming elements are ultimately released into the hydrosphere and gas elements to the atmosphere. The main conclusions are as follows:

1. The ore-forming elements tend to concentrate in the gas phase and accumulate within domed areas of the MI to be expelled during the late stage of magma crystallisation, probably when temperatures within the domed areas decrease to ~400°C.
2. During upward movement of the hydrothermal fluid, various ore-forming elements are precipitated within different temperature intervals to form mineral assemblage zones in the order, Mo, W, Cu, Zn, Pb, Ag, Au and Hg (Sb)

with decreasing temperature, resulting in maximum element/ore mineral concentrations at different distances (map and section) from the domed parts of granite bodies. Maximum element concentration may not necessarily result in the formation of an ore body as this will depend on the concentrations of ore-forming elements in the metasedimentary protoliths involved in melting to form the granite.

3. Enhancement of crustal temperatures before the rock melting begins can result in the mobilisation of some ore-forming elements. The ore-bearing fluid migrates upwards as a result of the *T–P* gradient and ore-forming elements precipitate as appropriate minerals under suitable conditions. In many instances, mineral deposits formed in this way occur within well-defined horizons. The 'strata-bound' appearance of deposits should thus represent the 'optimal' physico-chemical regime for the precipitation of ore-forming elements from ore-bearing fluids mobilised in rocks.
4. With the upward migration of hydrothermal fluid, the concentration of ^{18}O in the fluid should increase as ^{34}S decreases, resulting in ore-deposits formed at different levels, corresponding to different temperature intervals, having different oxygen and sulphur isotopic compositions. Thus, it is questionable whether the source of an ore-bearing fluid can be determined from the isotopic signature of the deposit compared with those where fluid sources are dominantly meteoric water or the mantle.
5. The physico-chemical state of the melt determines the accumulation and dispersion of the ore-forming elements. Repeated melting–crystallisation of a granite crust results in a decrease of fS and increase in the fugacities of O, F, Cl and CO_2 in the remelted granite magma, favouring concentration of lithophile elements in residual and gas–liquid phases.
6. Most so-called granite type U-deposits are of hydrothermal origin. If fO_2 of the magma is relatively low, U tends to partition into both accessory and rock-forming minerals during crystallisation. Decreasing fS and increase in O, F, Cl and CO_2 fugacities of the magma, results in oxidation of U^{4+} to U^{6+} that becomes concentrated in the hydrothermal fluid. Thus, 'granite type' U-deposits can be significantly younger than their host granite.

CHAPTER 7

HEAT SOURCE FOR CRUSTAL GRANITE MAGMA LAYERS: TECTONIC MODELS

7.1. INTRODUCTION

In Chapter 2, a number of ways of generating sufficient heat to cause crustal melting and the formation of partial melt layers were described, i.e. thickening of radiogenic heat-producing rocks, lithospheric extension, injection of magma into the continental crust, shear heating and isothermal decompression. Yet, as outlined in Chapter 3, partial melting of rocks cannot produce granite magma that forms batholiths if there is no convection within the region of partial melting. For this to occur, constant energy input into the system is essential for the formation and evolution of a crustal granite magma layer. Such a requirement requires the large dynamic system of plate convergence with which granite belts of 'hot orogens' formed in different geological times are mostly related to coeval plate convergences, e.g. the circum-Pacific and Tethys Meso Cenozoic granite belts, the Paleozoic granite belts of the North American Appalachians, Europe, the north and south Tianshan Mountains, NW China. Surface heat flow measurements indicate that crustal temperatures of continental margins are strongly perturbed by plate convergence. In this Chapter, two examples, i.e. SE China and Tibet–Himalayas from the western continental margin of the Pacific and the Tibet–Tethys, respectively, representing different convergent types, i.e. oceanic plate subduction and continent collision, are used to model the relationship between plate convergence and formation/evolution of crustal convection layers of granite magma.

7.2. CRUSTAL TEMPERATURE DISTURBANCE RELATED TO PLATE CONVERGENCE

Analysis of subduction zones (Toksöz et al. 1971; Molnar and England 1990; Davies and Stevenson 1992; Peacock et al. 1994) have shown that their thermal structure is controlled by rate of convergence, geometry and thermal structure of the subducting slab, rate of shear heating and the geometry and rate of flow in the mantle wedge above the subduction zone. The typical thermal structure of a subduction zone setting given by Oxburgh and Turcotte (1970), Minear and Toksöz (1970), Turcotte and Schubert (1982) and subsequently confirmed by many other studies (Fig. 7.1), suggests a domed area of isotherms in the continental margin where the temperature is much higher than that of the subducting oceanic plate at same depth level. An interpretation of this phenomenon is given in Fig. 7.1. Heat from movement on the subduction slip zone is conducted to both the continental and the oceanic crustal rocks. Assuming at time t_1, the continental and the oceanic crusts have the same temperature at a particular depth (represented by the 600°C isotherm), the isotherm beneath the subduction interface will be deflected downwards with oceanic crust moving from t_1 to t_2 because of subduction, and the same isotherm in the continental crust will move upward where the energy input/output >1 (Fig. 7.1). It can be expected that with a faster the rate of subduction, the stronger the deformation of isotherms on the both sides of the subduction plane.

An example of the temperature disturbance of a continental margin is the northern Cascadia subduction zone, North America, where surface heat flow inboard of the volcanic arc averages ~80 mW/m^2. This value is similar to background heat flow along the arc away from the volcanic centres and is consistent with petrological and geophysical constraints that indicate temperatures in excess

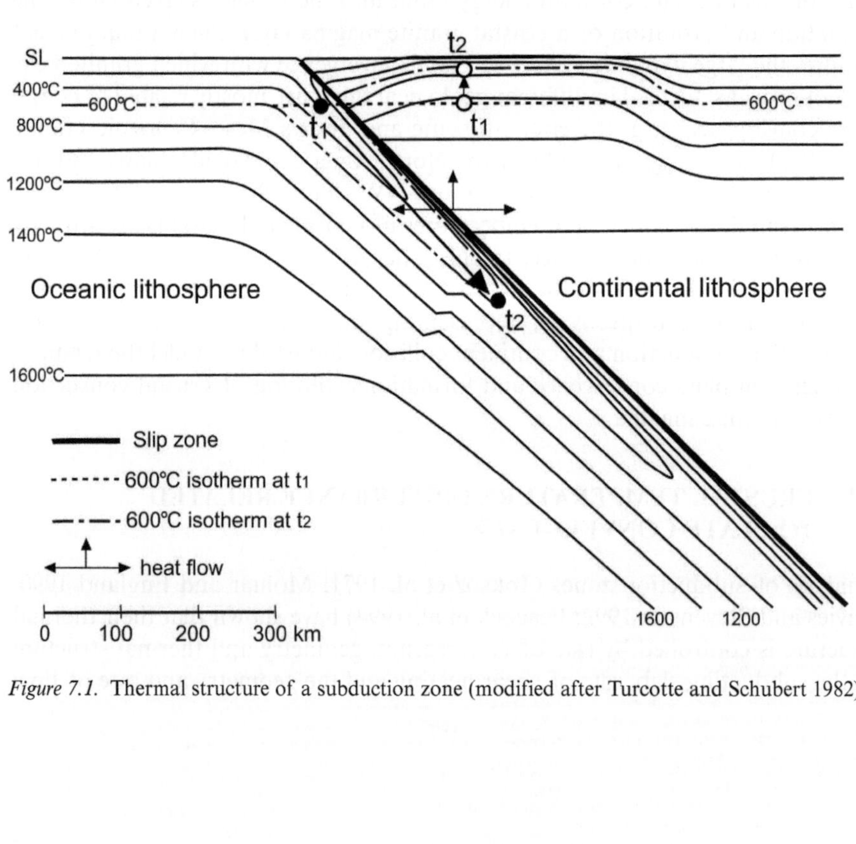

Figure 7.1. Thermal structure of a subduction zone (modified after Turcotte and Schubert 1982).

of 1200°C at 40–70 km depth below the arc (Elkins-Tanton et al. 2001). In the Cascadia example, the 'hot' area extends for some 800 km from the subduction zone to the 'cool' North American craton. Similarly, high heat flow of 80–100 mW/m² is observed for over 400 m across the backarc of the Central Andes subduction zone (Springer and Forster 1998) (Fig. 7.2) and values are typically >60 mW/m² across backarc regions of western Pacific subduction zones (Watanabe et al. 1977). In northern Cascadia, calculated geotherms give temperatures of 800–900°C for the crust-Moho boundary at 35 km (Hyndman and Lewis 1999). Below the Central Andes the crust is 60–70 km thick and the crust/Moho boundary has an estimated temperature of 1200–1300°C (Springer and Förster 1998) (Fig. 7.2).

Figure 7.3 shows three different geotherms, cool, warm and hot, related to surface heat flow values of 42, 75 and 90 mW/m², respectively, as prescribed to the landward boundary of the Cascadia backarc (Currie et al. 2004). With a crustal thickness of 35 km, only the warm and hot geotherms will induce the beginning of crustal melting at 20 km (hot geotherm) and 25 km (warm geotherm) with crust/Moho boundary temperatures of ~1050 and 810°C, respectively.

The model and situations described above are related to present-day surface heat flow. However, crustal temperatures vary with time depending on the energy input/output of the system. A surface heat flow map of China (Fig. 7.4) clearly indicates that the values of surface heat flow are closely related to the age of convergent orogens: maximum heat flow (>80 mW/m²) occurs in the Himalayas formed in the Cenozoic; lesser heat flow values of mainly 70–80 mW/m² with isolated highs of >80 mW/m² are located within the Mesozoic orogenic belt of eastern China; and the two E–W-trending Late Paleozoic orogenic belts in northern (Tianshan–Yinshang Belt) and central China (Qinling Belt) have relatively low

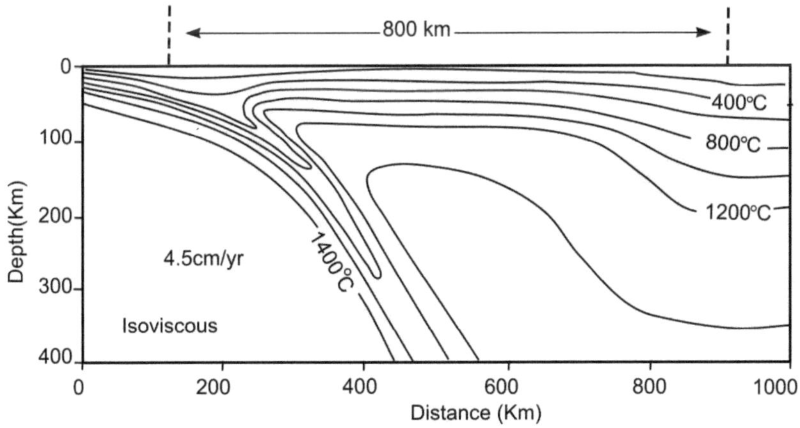

Figure 7.2. Pattern of geotherms related to subduction beneath the central Andes, South America (modified after Springer 1999).

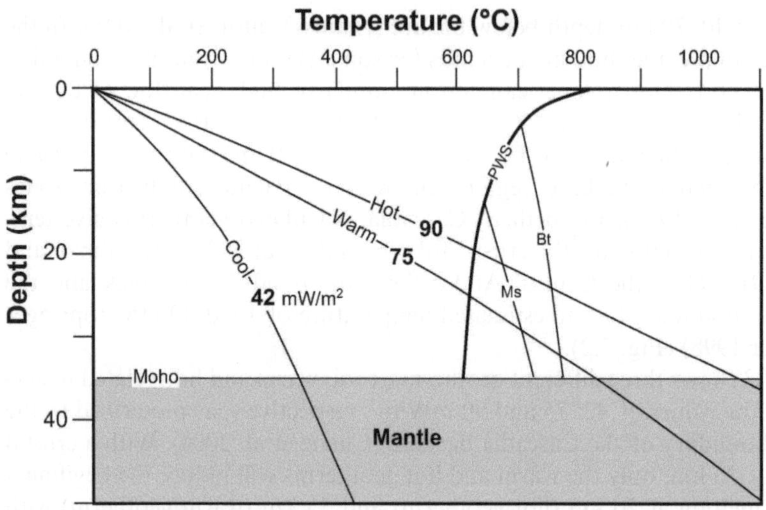

Figure 7.3. Plot of hot, warm and cool geotherms derived from heat flow data related to North Cascadia, SW Canada (after Hyndman and Lewis 1999; Lewis et al. 2003). PWS = Pelite wet solidus; Ms and Bt = muscovite and biotite dehydration melting curves, respectively.

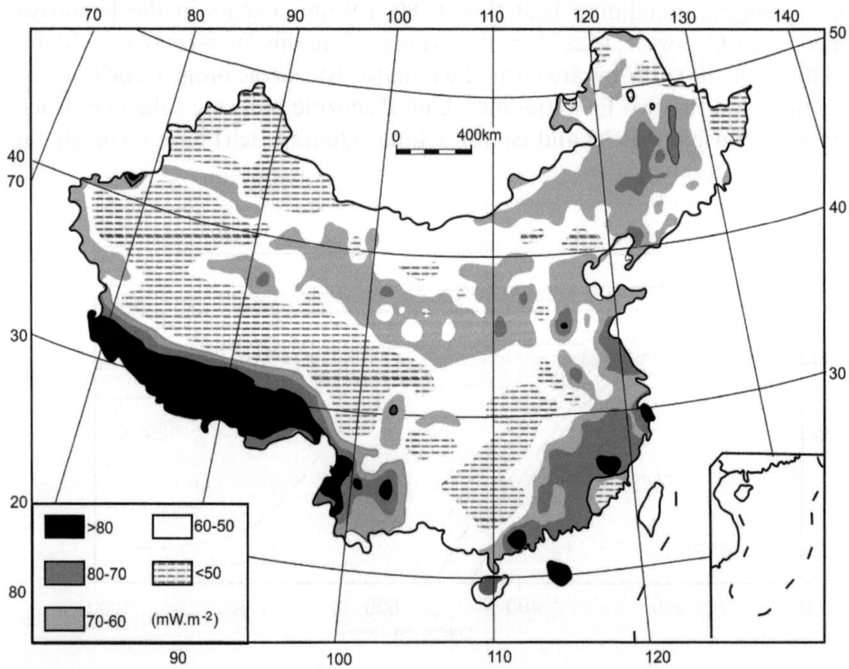

Figure 7.4. Map of surface heat flow in China (redrawn from Wang 2001).

values of 60–70 mW/m^2. Yet, even in the Cenozoic orogen of the Himalayas, the present-day crustal temperature must be much lower than it was 10 Ma ago before crystallisation of the Himalaya granite belts at 10–20 Ma (see below) because there would have been numerous magma reservoirs located at relatively shallow depth at that time.

7.3. SUBDUCTION AND GRANITE FORMATION: WESTERN PACIFIC CONTINENTAL MARGIN

7.3.1. Introduction

Southeast China, including the provinces of Guangdong, Jiangxi, Fujian, Zhejiang, Hunan and Guangxi (Appendix 1), is located in the southwest part of the Mesozoic tectonomagmatic belt of the western Pacific continental margin, and the Mesozoic granite and rhyolite that crops out over an area of some 240,000 km^2 have been studied by Chinese geologists for more than 50 years. Compositions of the granites are plotted in terms of normative ratios involving Qz An Ab Or in Fig. 7.5 indicating that the great majority are granite (biotite granite) and adamellite with extensions into the granodiorite and alkali-feldspar granite fields. Comparison with plots of experimental granite melt compositions

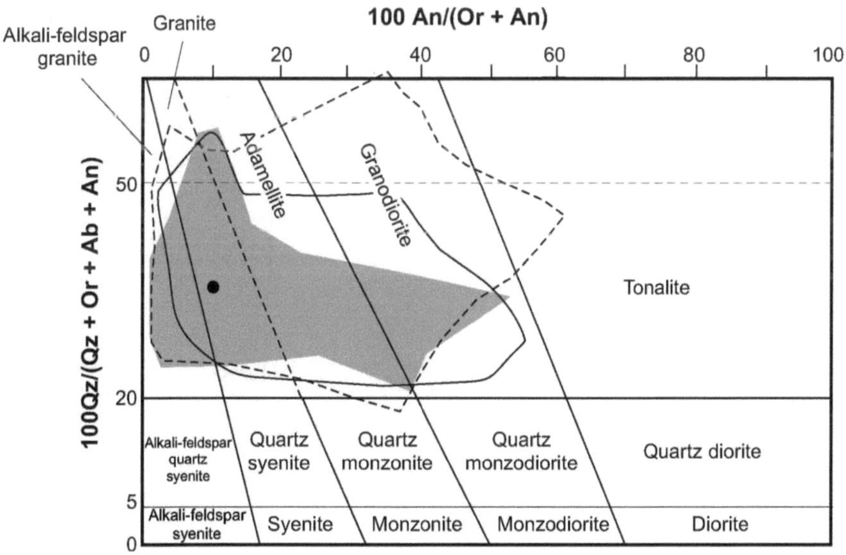

Figure 7.5. Field of Mesozoic (Yanshanian) granite compositions in SE China (grey shaded area) in terms of normative ratios. Filled circle = mean composition. Areas delineated by dashed and solid lines = experimental melt compositions of mica-bearing metasedimentary and granodiorite–tonalite compositions, respectively.

with various protolith compositions (Figs. A2a–d; B2a; C2a–d; and D2a in Appendix 2), indicates that the SE China granites are derived almost exclusively from the melting of mica-bearing quartzofeldspathic metasediments and granitic rocks rather than mafic protoliths.

7.3.2. Tectonic Framework of SE China and Granite Formation

In the Late Paleozoic, SE China was characterised by platform sedimentation (Fig. 7.6) that underwent weak deformation, metamorphism and igneous activity. The Indosinian deformation event occurred at the end of the middle Triassic time (Huang et al. 1985; Yang et al. 1989; Chen 2001) was followed by Yanshanian (Ys) deformation that has been subdivided into five tectonic episodes occurred at the end of early (Ys_1), middle (Ys_2) and latest Jurassic (Ys_3), and of early (Ys_4) and latest Cretaceous (Ys_5), respectively (Fig. 7.7A). All the Mesozoic deformation episodes are recorded by stratigraphic unconformities (Fig. 7.7B), which strongly changed the Paleozoic framework of SE China (BGMRJx 1984, 1988; BGMRGx 1985; BGMRFj 1985; BGMRGd 1986; Chen 1987; Xu 1987; BGMRHn 1988; BGMRZj 1989).

Mesozoic deformation resulted in the development of the dominant NNE–SSW trending structural grain of SE China (referred to as the Cathaysian Fold belt) that parallels the continental margin and the trend of the coeval granite belts. With the continent-ward decrease of granite exposure, the main structural elements change from major faults in the coastal area to folds inland

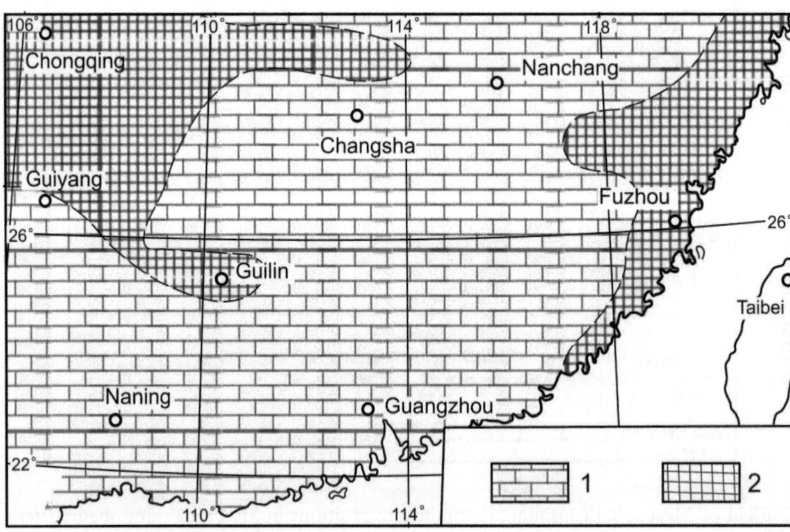

Figure 7.6. Lithofacies paleogeography of SE China in the late Carboniferous (after Feng et. al.1999). **1**. carbonate platform sedimentation; **2**. denudation domain.

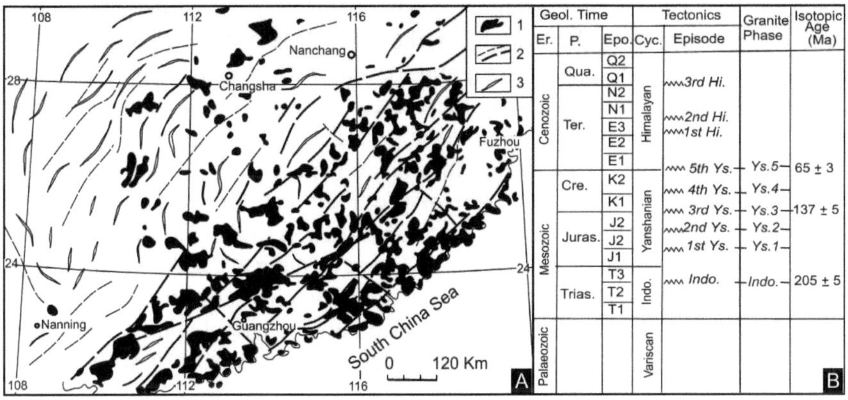

Figure 7.7. **A**. Map showing spatial relation between Mesozoic granites and main tectonic structures in SE China. The structural elements strike parallel to the continental margin and granite outcrops decrease NW. Structures vary from regional faults developed in the coastal area to folds developed inland; **1**. = Mesozoic granite; **2**. = Fault; **3**. = Fold. **B**. Temporal relationship between Mesozoic deformation episodes and granite activity in SE China.

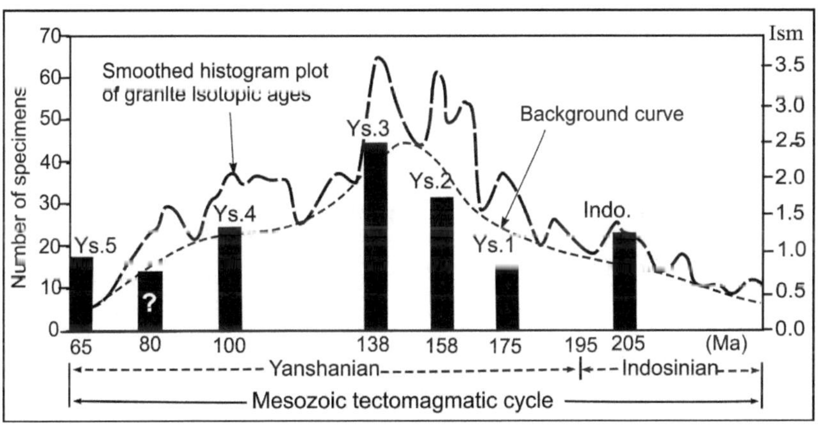

Figure 7.8. Smoothed histogram plot of 1,552 isotopic ages of SE China granites (heavy dashed line) compared with ages of Mesozoic deformation episodes (black columns). The height of the columns is determined by their respective *Ism* (intensity of structural movement) value determined as the area of unconformable region of a deformation phase/area of unconformable region produced during the 6th Yanshanian deformation phase as shown in Fig. 7.10. The thin dashed line is inferred to represent the Mesozoic tectono-magmatic event of SE China. Question mark = uncertain *Ism*. One unit = 3 Ma (after Chen 1987).

(Fig. 7.7A). Detailed geological mapping and radiometric dating has established that Mesozoic granite formation consisted of six discrete phases that are correlated with the six tectonic episodes mentioned above (Fig. 7.7A). The relationship is demonstrated in Fig. 7.8 as a smoothed histogram plot of 1,552 K–Ar granite whole rock and mineral ages (excluding volcanic rocks) (Chen 1987). The background curve connecting the troughs of the smoothed histogram is considered to reflect one Mesozoic tectono-magmatic event with maxima approximating the timing and relative intensity of the corresponding tectonic episode (Chen et al. 2003). The relationship between each period of granite formation and tectonism implies that they were related in time and space.

Unlike faults and folds that can be reformed, reinforced or destroyed by subsequent deformation, stratigraphic unconformities such as those recorded in Fig. 7.8 preserve the best evidence of tectonic disturbance (Huang et al. 1985; Chen 1987) (Fig. 7.9). The significance of a regional stratigraphic unconformity is not limited to recording a particular tectonic event. Presuming that lateral contractual deformation is exerted on the crust, deformation intensity should decrease with increasing distance from the compressive boundary as shown in Fig. 7.10A. Following erosion of the deformed rocks and subsequent deposition on the eroded surface,

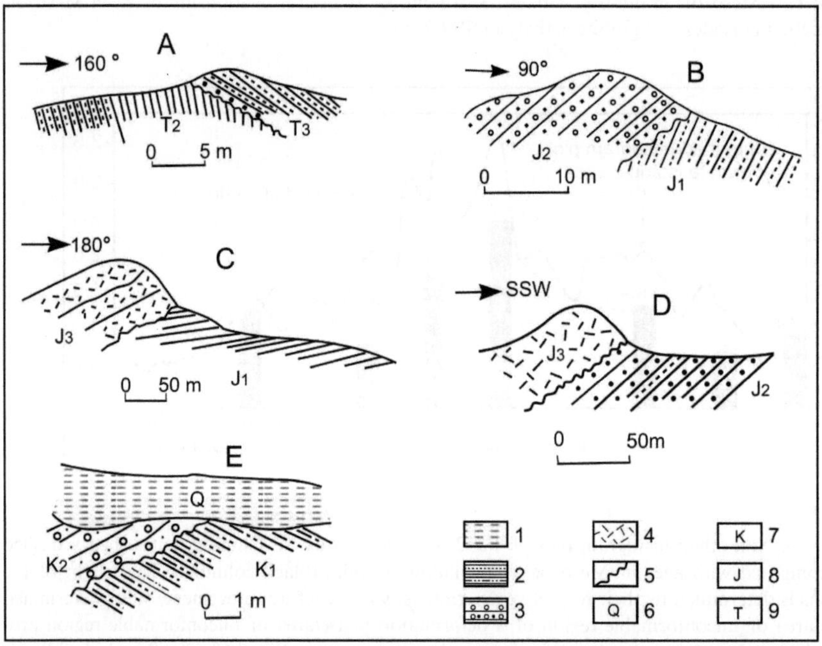

Figure 7.9. Sections illustrating unconformities that indicate episodes of Yanshanian deformation in SE China (data sources: Jiangxi, Zhejiang, Guangdong and Fujian Provincial Geological Surveys). **1**. Recent sediments; **2**. Shale/sandstone; **3**. Conglomerate; **4**. Volcanic sediments; **5**. Unconformity; **6**. Quaternary; **7**. Cretaceous; **8**. Jurassic; **9**.Triassic. **A**. Unconformity between the upper and middle Triassic in Gao'an County Jiangxi Province. **B**. Unconformity between the middle and lower Jurassic in Jishui County, Jiangxi Province. **C**. Unconformity between the upper and lower Jurassic in Boluo County, Guangdong Province. **D**. Unconformity between the middle and lower Jurassic in Zhuji County, Zhejiang Province. **E**. Unconformity between the upper and lower Cretaceous in Cong-an County, Fujian Province. See inset map (Appendix 1) for localities.

the degree of angularity of the resulting unconformable relationship is expected to increase towards the compressive boundary (Fig. 7.10B).

According to stratigraphic investigations of Provincial Geological Bureaus, the SE China area is divided into 172 stratigraphic regions. On the basis of 172 section localities, the intensity of the tectonic episodes delineated in Fig. 7.11

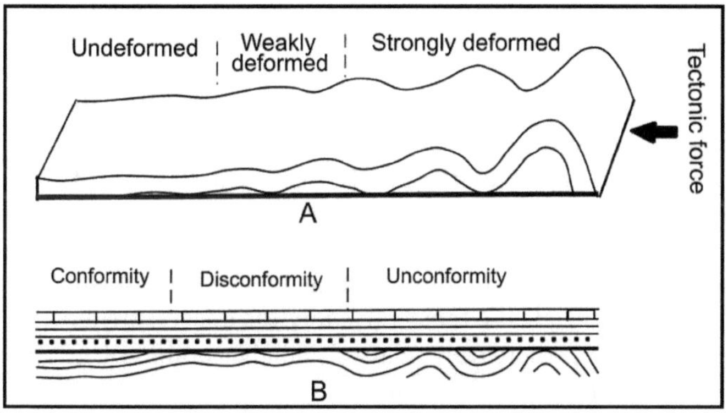

Figure 7.10. Diagram showing the relationship between direction of compression, intensity of crustal deformation and change in the degree of angular unconformity.

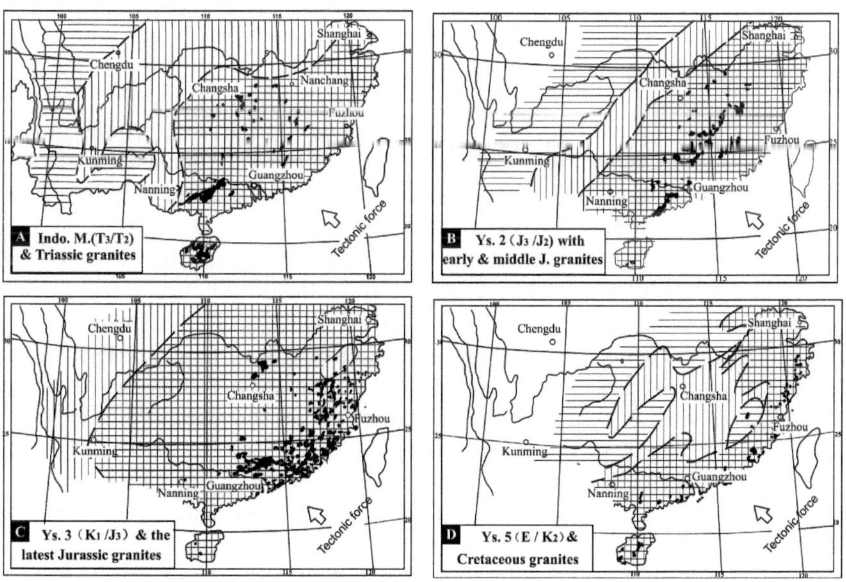

Figure 7.11. Areas affected by Mesozoic deformation episodes and distribution of coeval granites in SE China; Cross hatched = area of unconformable contacts; vertical lined = area of disconformable contacts; horizontal lined = area of conformable contacts. Black = granite. Arrow indicates vector of NW Mesozoic subduction of the paleo-Pacific (Kula) plate.

vary from markedly unconformable in the coastal area through moderately unconformable to conformable towards the continental interior (Chen 1987; Chen et al. 2003). The regional extent of the mapped unconformable–conformable relationships inwards from the SE China coast can be broadly related to a SE-directed increase of tectonic intensity and strongly supports the idea that deformation episodes were related to Mesozoic subduction of the paleo-Pacific (Kula) Plate beneath the eastern continental margin of the Eurasian Plate (Hilde 1977; Dickinson 1978; Guo et al. 1980; Li 1992; Northrup et al. 1995; Li and Zhou 1999).

Of the six Mesozoic tectonic episodes, the 3rd episode of the Yanshanian event (Ys.3) in the latest Jurassic represents the area of maximum angular unconformity (Fig. 7.11C). The age maxima of Mesozoic granite in SE China coincides with the timing of this deformation episode (Fig. 7.8). Following the 3rd Yanshanian episode, there was a decrease in the intensity of tectonic activity and the areal extent of granite formation, so that by the 5th Yanshanian deformation (Ys.5) at the end of latest Cretaceous, only the coastal strip of SE China was affected (Fig. 7.11D), where only relatively smaller bodies of granite are exposed. There is no known Cenozoic granite.

Granite generated during a particular deformation event is invariably limited to the area of stratal unconformity that relates to the deformation episode, and the overall trend of coeval granite outcrops is parallel to the continental margin and to the areas of temporal tectonic disturbance. In each case, the area of granite outcrop decreases towards the continental interior (Fig. 7.11).

7.3.3. Tectonic Model

The correlation between Mesozoic tectonic episodes and granite production in SE China, and their inferred relationship to the direction of paleo-Pacific (Kula) plate subduction as discussed above, constrains the source of heat required for large-scale crustal melting in the eastern Asian continental margin in the Mesozoic in relation to perturbation of the isotherm pattern illustrated in Fig. 7.1. This provides a model to explain the relationship between tectonism, granite formation and plate subduction based on the data from SE China.

The mid–upper crust of the late Paleozoic platform of SE China is designated with thickness of 25 km and an original temperature of 800°C at its base, the same as that shown in Fig. 3.5. Subduction of the paleo-Pacific plate, which may have begun in the latest Paleozoic in response to breakup of the Pangaea supercontinent-enhanced temperatures of the continental margin resulting in partial melting of rocks at the base of the crustal section being considered (Fig. 7.12A). With increasing temperature caused by accelerating subduction (likely accompanied by shallowing of the subduction angle) (Chen et al. 2003), a convecting granite magma layer began to develop in the lower part of the partial melting region as discussed in Chapter 3 (Fig. 3.12). Heat and isotherms were advected upwards as the convection layer thickened (Fig. 3.13 and 7.12B).

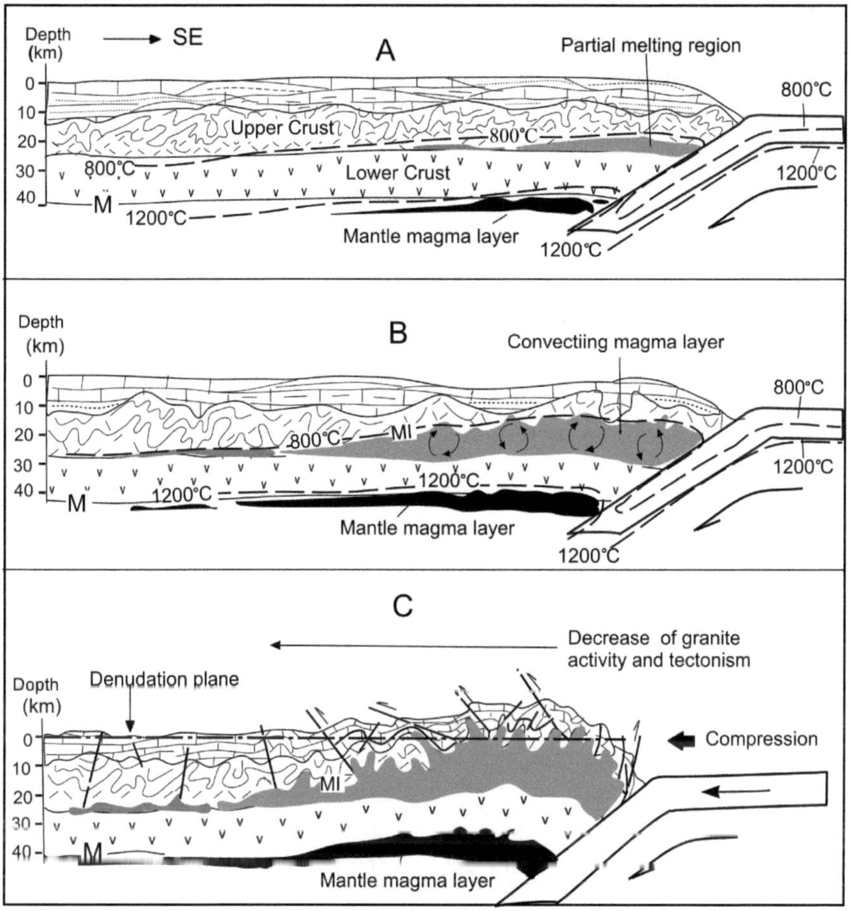

Figure 7.12. Schematic diagrams to illustrate the relationship between paleo-Pacific plate subduction, the intracrustal melting and magma layers and deformation of continental crust. **A**. 'Fertile' rocks at the base of the upper crust begin to melt when crustal temperatures rise to values above their solidi. Underplating of basalt occurs at the same time at the top of the mantle due to melting caused by convective thinning above the subduction zone and rise of asthenosphere. This was the source for Cenozoic basalt distributed along the coastal area of SE China and inferred to have been derived from a depth of 40–80 km (Zou 1994). Melting to produce granitic magma does not occur in the lower crust because of its refractory granulitic composition. **B**. The region of lower upper crustal melting enlarges due to convection in a thickening growing magma layer (MI moves upward) advecting heat and isotherms upward concomitant with rising crustal temperatures. **C**. Deformation of the crust (tectonism) occurs as a result of a weakened crustal rheology due to presence of the granite magma layer (waterbed effect) resulting in uplift and deformation together with simultaneous uplift/deformation of the magma interface (MI) and Moho

Rheological weakening of the crust with upward thickening of the convecting magma layer dependent on constant energy (heat) input due to continuous subduction reduced crustal resistance to compressive forces exerted by the subducting plate resulting in a relatively rapid episode of deformation (orogeny) (Fig. 7.12C).

Overall continent-ward inclination of the geotherms during this process is evidenced by decreasing outcrops of Mesozoic granite and diminishing tectonic intensity (Fig. 7.7B) as well as mineralisation temperature (Fig. 6.22).

7.3.4. Multiple Melting (Remelting) and Granite Belts

As described in Chapters 3 and 4, fluctuation of the MI results in remelting of the first formed granite layer, to produce a new generation(s) of downward-younging granite layers that are commonly exposed at any one locality as composite granite bodies, and with different-aged granite layers exposed as regional granite belts (Chen et al. 2003). Figure 7.13 shows the distribution of Mesozoic granitoids in Zhejiang Province of SE China; a western belt of essentially Late Jurassic age (K–Ar ages between 153 and 138 Ma), a central belt of between 100 and <138 Ma, and an eastern belt where the granitoids are generally less

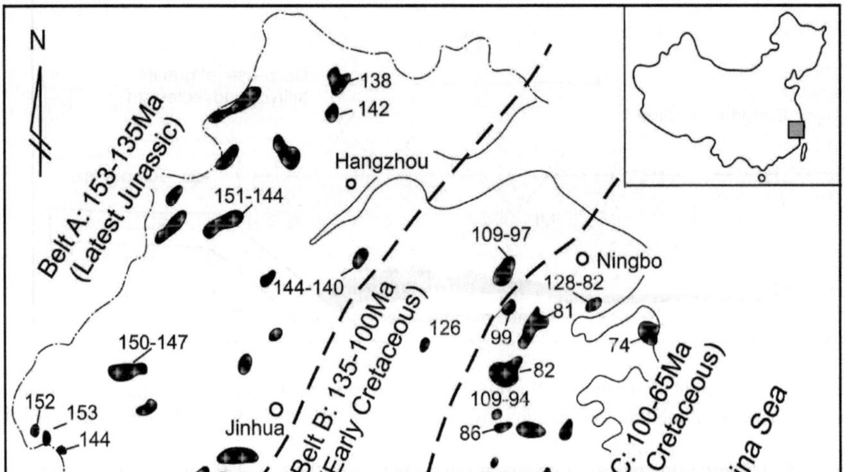

Figure 7.13. Map showing different-aged granite belts in Zhejiang Province, SE China.

than 100 Ma. Over the whole area of SE China, the same SE-younging tendency of granite is apparent (Fig. 7.14A). This pattern is also characteristic of the eastern Pacific continental magmatic arcs such as the Coast Plutonic Complex (Brown and McClelland 2000) and Sierra Nevada batholith (Bateman 1992), where discrete granite belts have migrated normal to the trend of the arc with time possibly in response to changes in the dip of the subduction slab or slab roll-back.

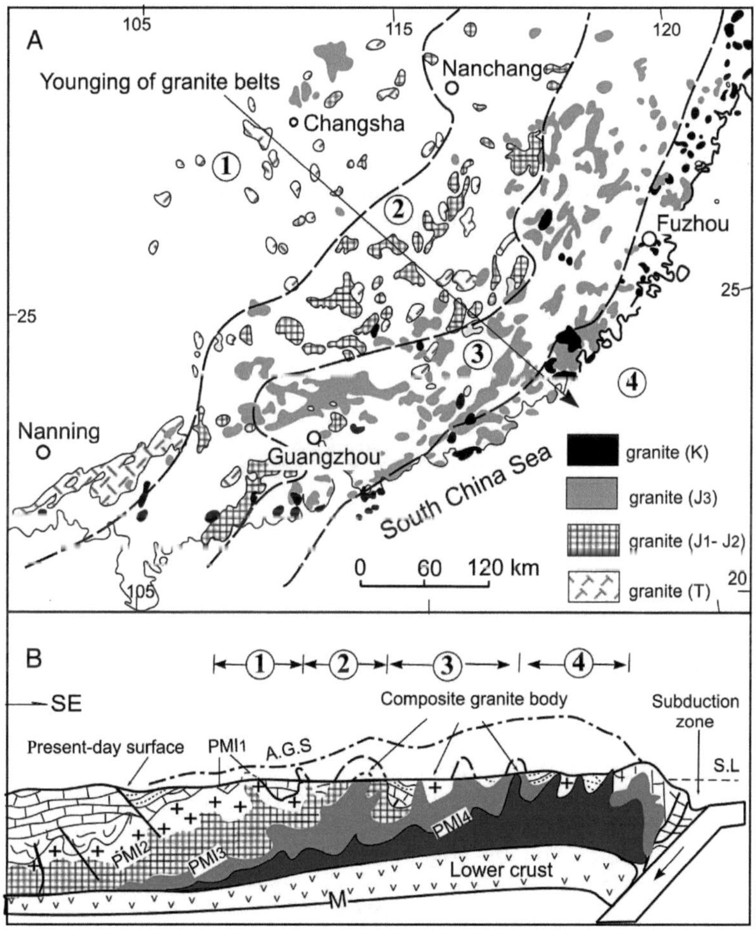

Figure 7.14. **A**. Map showing Mesozoic granite belts of SE China and their oceanward younging. Dashed lines delineate boundaries of the granite belts; numbered 1 to 4. **B**. Schematic cross-section showing relationship between the 'migration' of granite belts and their relationship to positions of granite layers. PMI_1–PMI_4= different aged paleo-magma interfaces. Circled numbers correspond to those in A.

Remelting of granite results in renewed differentiation of the magma system, as described in Chapter 5 with individual composite granite bodies and regional granite belts showing trends of increasing acidity and decreasing basicity with younging. An interpretation of the relationship between the SE-younging tendency of granite belts in SE China and the downward-younging granite layers resulting in a series of overlapping temporal- and spatially separated thermal events in the Mesozoic is shown in Fig. 7.14B. This is supported by a geological interpretation of the SE–NW Taiwan–Sichuan geophysical traverse given in Fig. 7.15 indicating that from Xiamen to Shaoyang, there are five different aged Mesozoic granite layers, all with their inferred PMI's dipping northwestwards (Chen et al. 2003).

7.3.5. Summary

The Mesozoic tectono-magmatism of SE China reflects the response of the lithosphere of the western Pacific continental margin to subduction of the paleo-Pacific (Kula) plate. Varying rates of subduction, accompanied with changes in

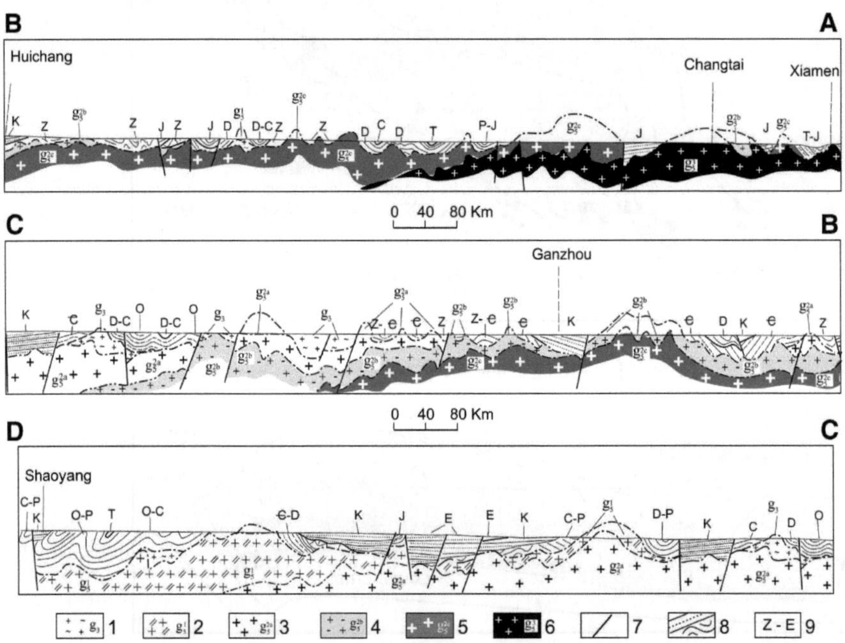

Figure 7.15. Geological re-explanation of the GET geophysical traverse from Xiamen (Fujiang Province) to Shaoyan (Hunan Province), SE China. **1**. Pre-Mesozoic granite; **2**. Triassic granite; **3**. Early Jurassic granite; **4**. Middle Jurassic granite; **5**. Latest Jurassic granite; **6**. Cretaceous granite; **7**. Fault; **8**. Unmelted country rock; **9**. Stratal age notation (E = Eogene; K = Cretaceous; J = Jurassic; T = Triassic; P = Permian; C = Carboniferous; D = Devonian; O = Ordovician; € = Cambrian).

the subduction angle, increases and decreases the crustal temperature, the former resulting in extensive partial melting of the rocks in the lower part of the 'fertile' upper crust that ultimately results in formation of a convecting crustal granite magma layer. Constant energy (heat) input as a result of long-term subduction enables the magma layer to thicken upward (rising MI) causing rheological instability of the upper crust resulting in compressive deformation of unmelted rocks and granite layer. This produces structural and granite belts parallel to the continental margin with decreasing intensity and granite outcrop (decreasing isotherms) towards the continental interior.

7.4. CONTINENTAL COLLISION AND GRANITE FORMATION: TETHYS BELT

7.4.1. Tectonic Framework and Granite Distribution of Tibet Plateau

The Tibet plateau is located in the eastern part of the Tethys belt (Sengor 1984) and the following analysis is based on a systematic geological and geophysical investigation of the Tibet plateau–Himalaya region carried out in 1980s by Academy of China and the Chinese Academy of Geological Sciences. As shown in Fig. 7.16, two E–W trending suture zones that mark the sites of subducted oceanic crust, i.e. the Bangong–Amdo (BA) zone in the north and the Yarlung–Zangbo (Tsangpo) (YZ) zone in the south, divide the Tibetan region into three terranes named from

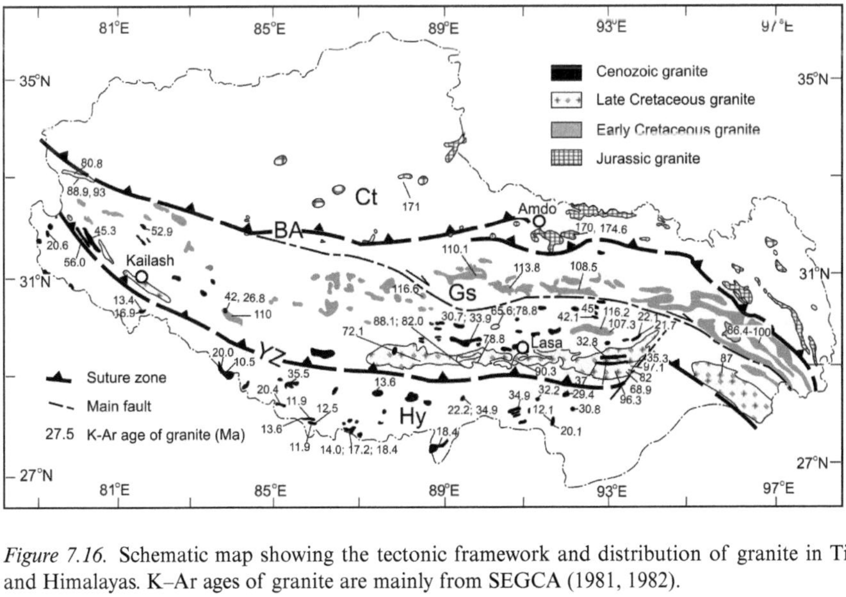

Figure 7.16. Schematic map showing the tectonic framework and distribution of granite in Tibet and Himalayas. K–Ar ages of granite are mainly from SEGCA (1981, 1982).

north to south as the Chang Tang (Ct), Gangdise (Gs) and Himalaya (Hy). The Meso–Cenozoic granites in the Tibet–Himalaya region are predominantly formed during four periods, i.e. Early Jurassic with the radiometric ages of 170–175 Ma, Early Cretaceous (100–116 Ma), Late Cretaceous (65–88 Ma) and Cenozoic (10–50 Ma) (Fig. 7.16), which are distributed in different areas, constituting different aged granite belts younging southwards (Fig. 7.16).

Exposures of the Early Jurassic granites are limited in the north Ct terrane of the BA suture zone, and are mainly of granodiorite, biotite granite and quartz monzonite (SEGCA 1981, 1982). The Cretaceous granites are predominantly distributed in the Gs terrane between the two suture zones (YZ and BA) and those formed in the Early Cretaceous are mainly exposed in the north part of the terrane with decreasing occurrences from north to south and from east to west (Fig. 7.16).

The Gangdise batholith, one of the largest granite batholiths in the world, crops out over 2500 km from east to west, is located in the southernmost part of the Gs terrane near the YZ suture zone and consists of diorite, granodiorite, quartz diorite and quartz monzonite. The main body of the batholith formed in the Late Cretaceous between 65 and 95 Ma, and numerous associated small leucogranite occurrences are dated at between 20 and 50 Ma.

Though all the granites located south of the YZ suture zone are formed in the Cenozoic, they can be divided into two sub-belts: the Lhagri-Kongri in the north and the Himalaya in the south. Granites in the north sub-belt are dated at 30–40 Ma and those in the Himalaya between 10 and 20 Ma. Granites in both belts are mostly leucogranites (two-mica granite and muscovite granite) (SEGCA 1981; Liu 1984).

7.4.2. Tectonic Phases in Relation to Subduction and Collision

It has been documented that the three terranes (Ct, Gs, Hy) divided by the two suture zones (YZ, BA) have different histories of tectono-magmatism, especially in the Meso–Cenozoic (Chang et al 1982; GQXP 1983; Sengor 1984). The classification of strata in Fig. 7.17 is mainly according to CGQXP (1983) and Wang (1983, 1984), showing that there are six tectonic phases (compressive deformation) recorded by unconformities of the Meso–Cenozoic strata in this area (designated as Y1, Y2, Y3, Y4, H1 and H2) and that the different tectonic phases occur in different terranes (Fig. 7.17). Deformation of Y1 represented by the unconformity between the Zamunaqu and Tanggula groups appears only in the Ct terrane , and does not affect the Gs terrane, indicating that it originated from subduction of oceanic crust towards the Ct terrane along the BA zone. Deformation of Y2 represented by the unconformity between the Tanggula and Gangba groups in the Ct terrane and between the Guoqu and Xiaqiong groups in the Gs terrane indicates collision of the two terranes resulting in deformation on both sides of the BA zone (Fig. 7.17).

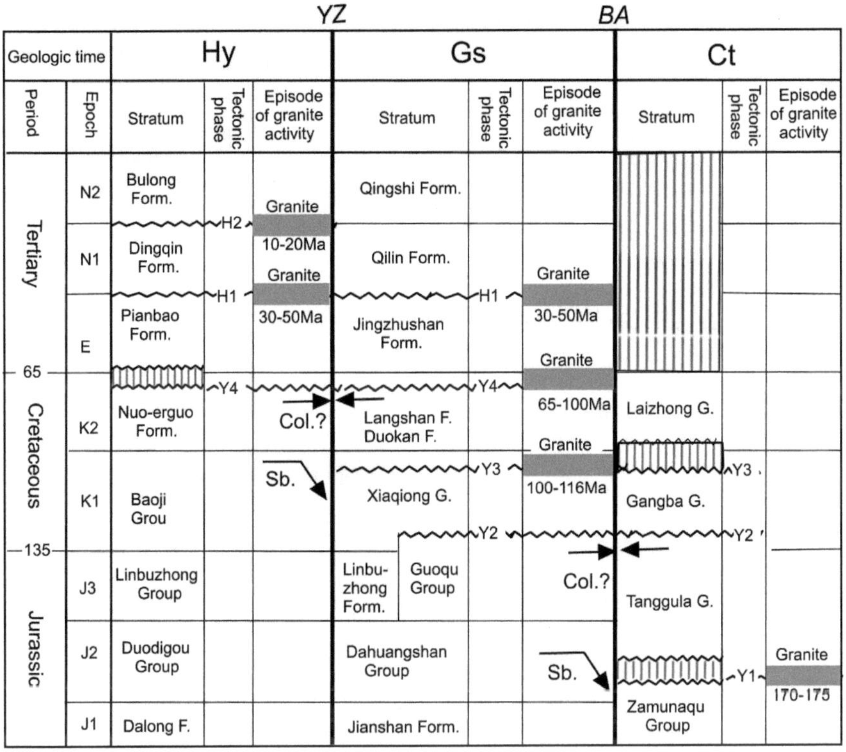

Figure 7.17. Diagram showing the stratigraphic sequence, tectonic phases and episodes of granite formation in different terranes of Tibet-Himalaya. Arrows indicate plate subduction or collision. See text.

The lack of evidence for significant compressive deformation in the Gs terrane in Jurassic time (Fig. 7.17) suggests that subduction of the oceanic crust toward the Gs terrane along the YZ zone began after latest Jurassic. Collision with the Eurasian Plate ended movement of the Gs terrane and formed the BA suture zone. This resulted in breakoff of the oceanic lithosphere behind the Gs terrane, resulting in the formation of the present-day MBT due to the continuing northward movement of the Indian Plate. This is consistent with the fact that the Y3 deformation phrase at the end of Early Cretaceous occurred in the Gs terrane and did not affect the Hy terrane (Fig. 7.17).

At the end of latest Cretaceous time, the Hy terrane may have arrived at its present-day position colliding with the Gs terrane to form the Yz suture zone. Evidence for this is provided from crustal deformation indicated by the Y4 unconformity (Fig. 7.17) that occurs on both sides of the YZ zone, and post–latest Cretaceous marine regression that occurred in this area (SEGCA 1981).

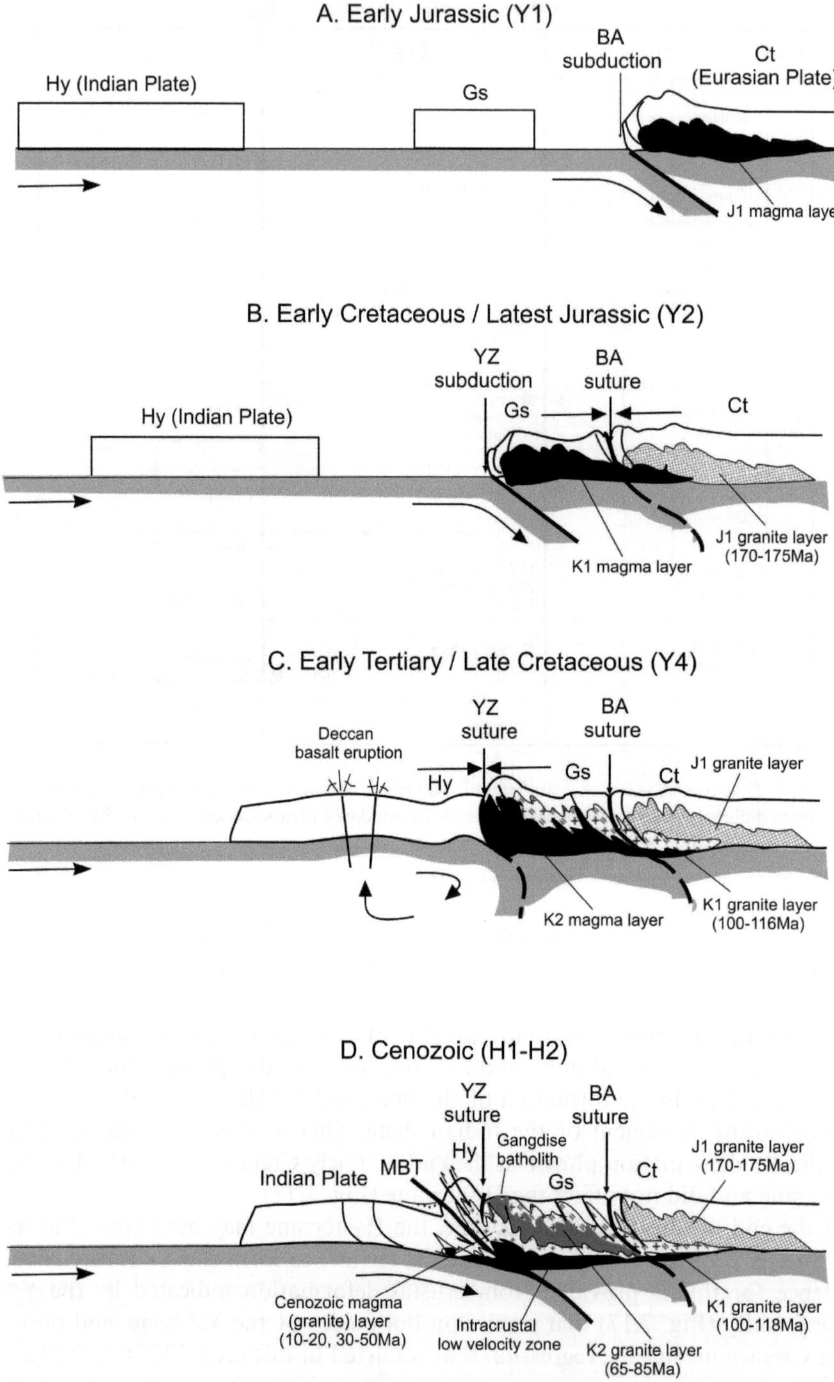

Figure 7.18. Schematic diagrams to illustrate the relationship between the formation of Meso–Cenozoic granite and plate convergence in Tibet-Himalaya. See text.

The H1 and H2 tectonic phases in the Cenozoic are considered to be related to compression of the Indian continent.

7.4.3. Magma Layers and Plate Convergence

According to the tectonic framework, distribution and ages of granites, and characteristics of the tectonic phases described above, a model for Meso–Cenozoic crustal deformation and granite formation of Tibet–Tethys and their relationship to plate subduction and continental collision are shown in Fig. 7.18.

Early Mesozoic subduction along the BA zone resulted in the formation of a granite magma layer within the upper crust of the continental margin of the Eurasian Plate, i.e. the Ct terrane, and ultimately during Late–Early Jurassic compressive crustal deformation between 170 and 175 Ma (Fig. 7.18A). This situation is similar to SE China. For this reason, the distribution of Early Jurassic granite in the Ct terrane occurs north of the BA zone.

Collision of the Gs terrane with the Eurasian Plate (Ct terrane) at the end of Jurassic or beginning of the Cretaceous marked the transformation of the BA zone from subduction to suture. This terminated movement of the Gs terrane and caused breakoff of the oceanic lithosphere behind the terrane (Fig. 7.18B). Subduction was subsequently initiated along the YZ zone towards the Gs terrane resulting in the formation of a granite magma layer within the Gs with deformation at the end of Early Cretaceous (Fig. 7.17), forming the 100–118 Ma granite belt (Fig. 7.16).

Collision of the relatively rigid Indian continent with hot, rheologically weak Gs terrane rocks at the end of latest Cretaceous caused strong deformation in the leading edge of the Gs terrane (Fig. 7.18C) as indicated by the Gangdise batholith (Fig. 7.16). Contact of the hot Gs terrane resulting in heating the frontal part of the Indian Plate, in addition to the shear heating generated by subduction of the Indian Plate along the BMT (Fig. 7.18D), may be the reason for formation of the Cenozoic granite magma layer (with crystallisation ages between 10 and 50 Ma) within both the Hy and Gs terranes (Fig. 7.15). This could also explain the great crustal thickness beneath Tibet and why uplift to form the Himalayas occurred in the Late Cenozoic rather than Early Cenozoic (Fig. 7.18D).

7.5. CONCLUDING STATEMENT

Like SE China, the Tibet–Tethys belt provides an example of the relationship between plate convergence (subduction and collision) and formation of a crustal granite magma layer, further suggesting that constant energy input is essential for the formation of granite magma layers. Plate convergence is probably responsible for the origin of most granites.

CHAPTER 8

GEOLOGICAL EFFECTS OF CRYSTALLISATION OF A CRUSTAL GRANITE MAGMA LAYER: SE CHINA

Increasing crustal temperature results in formation of a granite magma layer and compressive deformation of the crust. Decreasing crustal temperature leads to crystallisation of the magma layer and generates various geological effects such as fault-bounded basins and volcanic activity that are introduced in this chapter with reference to Mesozoic–Cenozoic evolution of SE China.

8.1. FAULT-BLOCK BASINS

8.1.1. Characteristics and Distribution of Mesozoic Basins

Mesozoic–Early Tertiary basins in SE China can be classified in to three types according to their deposits, i.e. coal-bearing sequences, intermediate-acid volcanics or volcanogenic sediments, and red beds. Coal-bearing sediments are mostly distributed in the central part, volcanic filled basins in the coastal region, and red-bed basins in the central–western area of SE China (Fig. 8.1). Different types of basins are formed at different times. Coal-bearing sedimentary basins are Late Triassic to middle Jurassic in age; red-bed basins without coal formed during the Late Cretaceous and Early Tertiary, and those formed in the Late Jurassic–Early Cretaceous are filled by volcanics and volcaniclastic sediments in the coastal region and by red-beds inland (Fig. 8.1).

Stratigraphic sections from localities in different tectonic belts from NW to SE across Fujian Province, SE China, illustrate the variation of basin deposits with time (Fig. 8.2) .

1. In coal-bearing basins, coal-bearing strata are dominantly of Late Triassic–Early Jurassic age. From about the middle Jurassic, volcanic sediments dominate the sequence.
2. Large-scale eruption of intermediate-acid volcanic rocks took place in Late Jurassic–Early Cretaceous times, and their thickness decreases from East to West.

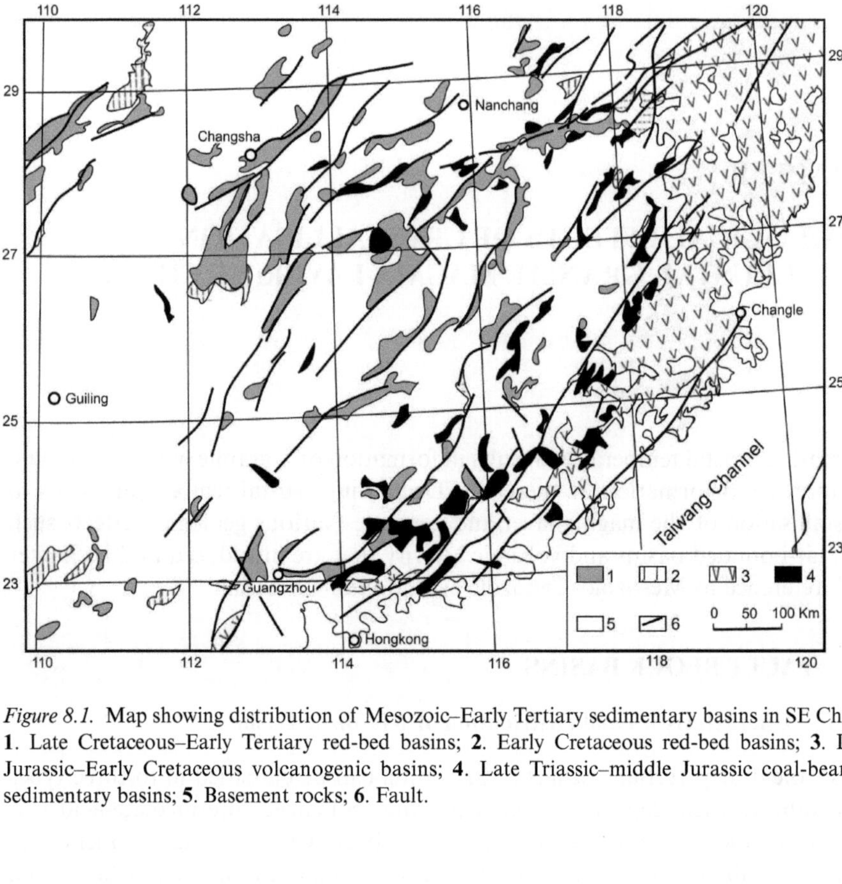

Figure 8.1. Map showing distribution of Mesozoic–Early Tertiary sedimentary basins in SE China. **1**. Late Cretaceous–Early Tertiary red-bed basins; **2**. Early Cretaceous red-bed basins; **3**. Late Jurassic–Early Cretaceous volcanogenic basins; **4**. Late Triassic–middle Jurassic coal-bearing sedimentary basins; **5**. Basement rocks; **6**. Fault.

3. In the Early Cretaceous, basins in western Fujian Province are filled by continental red beds; those in the central and eastern parts of the province contain volcanics.

8.1.2. Basin Formation

Regional paleogeography and structure of the sedimentary basins indicate that most of the Early Mesozoic coal-bearing basins are synclinal depressions (BGMRGd 1986 BGMRJx 1984), e.g. the Late Triassic–Early Jurassic basin in the Pingxiang area, Jiangxi Province (Fig. 8.3), and that the Early Jurassic basins were evidently produced by large-scale crustal flexures rather than fault-block movement (Fig. 8.4) .

Increasing crustal temperatures during the Early Mesozoic in SE China led to the development of an extensive mid-upper crustal magma layer as described in Chapter 7. Rheological instability caused by this large area of magma layer accentuated far-field deformation related to NW-directed subduction of the

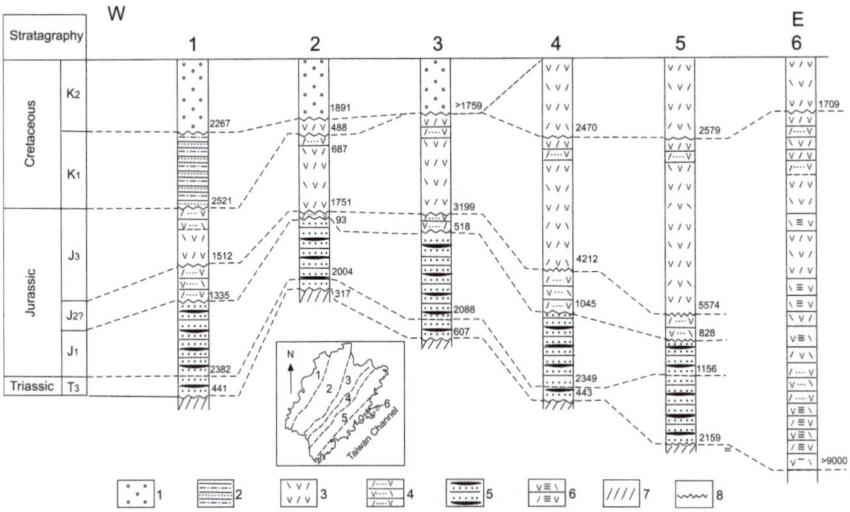

Figure 8.2. Stratigraphic sections (1–6) from various localities in Fujian Province, SE China (data from BGMRFj 1985). **1**. Coarse sandstone and conglomerate (red beds); **2**. Fine sandstone and mudstone (red beds); **3**. Intermediate-acid volcanics; **4**. Volcanigenic sediments; **5**. Coal-bearing sediments; **6**. Volcanogenic flysch; **7**. Basement rocks; **8**. Unconformable contact. Numbers on the map of Fujian Province indicate locations of the stratigraphic columns.

Kula Plate, is evidenced by intense folding and thrusting of pre-Cretaceous rocks (Figs. 8.5 and 8.6) , including those of the coal-bearing basin sediments, e.g. as shown from the Tan Hushan coal mine, Fujian Province (Fig. 8.5).

In contrast to fold/thrust tectonics of the coal-bearing basins, the Cretaceous–Early Tertiary red-bed basins are predominantly controlled by normal faulting, and are referred to as 'fault-block basins' (FBB) (Guan 1988). Most of the Mesozoic–Cenozoic red-bed basins in SE China are associated with regional NE-trending faults (Fig. 8.1) considered to have been initially formed as thrust faults in the Early Mesozoic (Li 1973). From the end of Late Jurassic time, thrust faulting changed to normal faulting resulting in the formation of numerous FBB as indicated from sections of some typical FBB in SE China (Figs. 8.7 and 8.8) . Numerous models have been advanced to explain the formation of these FBB (Zhang 1986). Most assume that the basins developed on a rigid undermass that raises the 'room problem' of how block subsidence by normal faulting can be accommodated.

The formation of FBB began from the end of Early Cretaceous time (Fig. 8.9), suggesting they may be related to cooling of the crustal magma layer beneath SE China, and lowering of the crustal geothermal gradient. This may have been in response to a slowing in the rate of Kula Plate subduction from ~120–140 mm/year in the Late Cretaceous to ~30–40 mm/year in Eocene

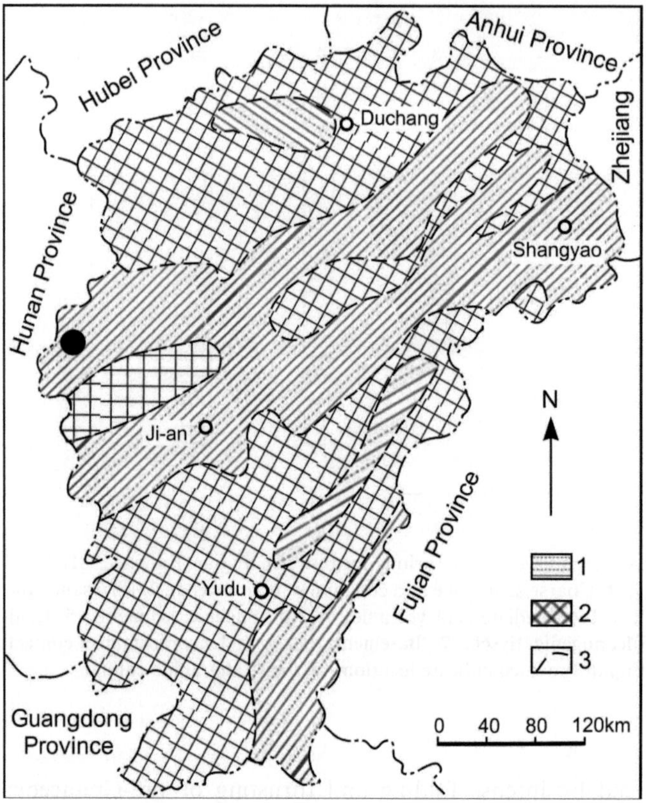

Figure 8.3. Early Jurassic paleogeography map of Jiangxi Province, SE China (simplified after BGMRJx 1984), indicating that the sedimentary basins are mainly synclinal. **1**. Sedimentary basin; **2**. Region of uplift; **3**. Tectonic unit boundary. Section localities depicted in Fig. 8.4 are indicated by filled circles.

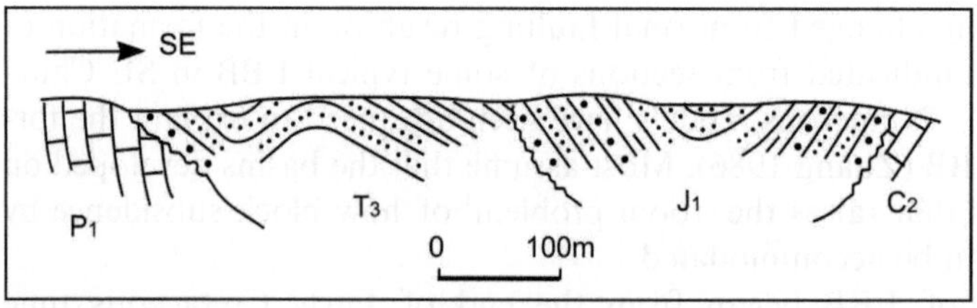

Figure 8.4. Cross section of Late Triassic–Early Jurassic strata in the Pingxiang area, Jiangxi Province, SE China (for location see Fig. 8.3), showing the synclinal structure (modified after BGMRJx 1984). **J1**, lower Jurassic; **T3,** upper Triassic; **P1**, lower Permian; **C2,** middle Carboniferous.

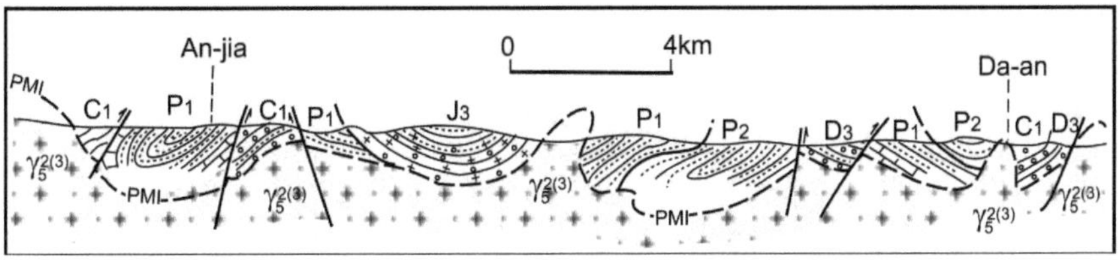

Figure 8.5. Cross section from An-jia to Da-an in NW Fujian Province, SE China (modified after BGMRFj, 1985), suggesting that the rocks below the granite interface have been melted in the Late Jurassic to form the third Yanshanian granite layer. 'Syn-magma' deformation of roof rocks took place above the magma layer. **J3**, upper Jurassic; **P1, P2,** lower and upper Permian, respectively; **C1,** lower Carboniferous; **D3,** upper Devonian; $Y_5^{2(3)}$, third Yanshanian (Late Jurassic) granite; PMI, position of paleo-magma interface at end of Late Jurassic.

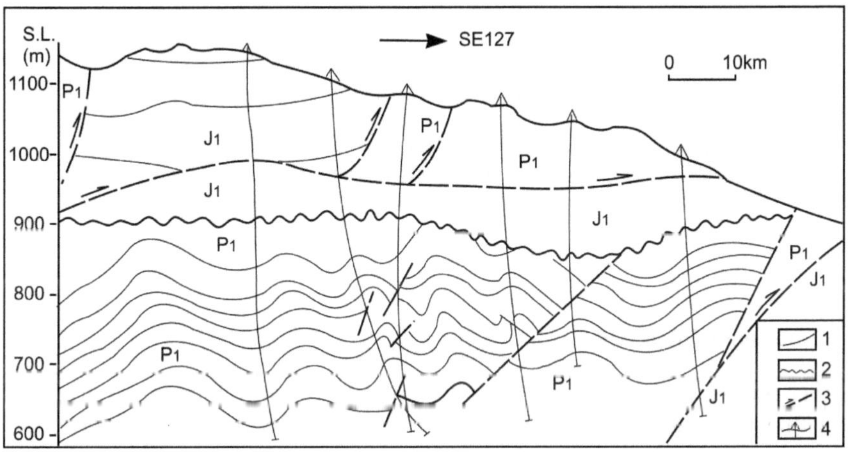

Figure 8.6. Prospecting section of the Tan Hushan coal mine, Yongcun County, Fujian Province, SE China, showing folding and thrusting of the Early Permian (P1) and Early Jurassic (J1) strata (modified after BGMRFj 1985). **1**. Stratigraphic boundary; **2**. Unconformity; **3**. Fault; **4**. Drill hole.

time (Northrup et al. 1995). Contraction of roof rocks overlying the downward crystallising granite layer facilitated differential movement on boundary faults to create FBB (Fig. 8.10). Magma beneath sinking ('heavy') fault-blocks (with a larger density or located in 'gravity high') flows to relatively low-pressure areas and rising ('light') blocks provide room for the downward-moving fault-blocks and create surface 'highs' fault-block mountain (FBM) with erosion supplying detritus into the FBB. The onset of fault-block tectonics is one of gravitational adjustment following the cessation of compression in many orogens (Mattauer 1980).

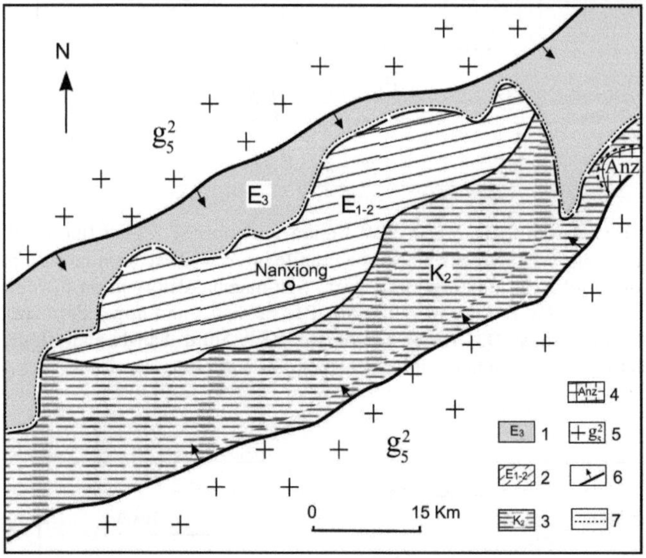

Figure 8.7. Geological map of the Nanxiong basin, northern Guangdong Province, SE China, bordered by normal faults and filled by Late Cretaceous–Early Tertiary red beds (simplified after BGMRGd 1986). **1**, **2**. Early Tertiary; **3**. Late Cretaceous; **4**. Pre-Paleozoic; **5**. Jurassic granite; **6**. Normal fault; **7**. Unconformity.

8.1.3. Origin of Red Beds

Another problem related to the formation of FBB in SE China is the origin of the distinct red colouration of the basin strata. This is commonly considered to be related to paleoclimate, i.e. arid conditions during deposition, or to erosion and redeposition of red lateritic soils produced under tropical conditions (e.g. Wu and Yang 1978; Turner 1980).

A vast amount of poorly-sorted conglomerate/sandstone (Fig. 8.11A) exhibiting rapid facies variation (Fig. 8.11B) characterise the Mesozoic–Cenozoic red-bed basins in SE China (Fig. 8.11B). These data indicate strong physical weathering in the source regions during the accumulation of the red beds and conflicts with the idea that they originated from redeposition of red lateritic soils. The idea that red beds were formed in arid environment appears to be in consistent with the following observations:

1. Sporamorphes in the red beds contain ephedripites that reflect an arid climate.
2. Red beds contain gypsum and halite deposits.
3. Glyptoliths or eolian deposits occur in the red beds.

However, if reddening of the sediments is due to weathering in an arid climate during their accumulation, it is difficult to explain the following:

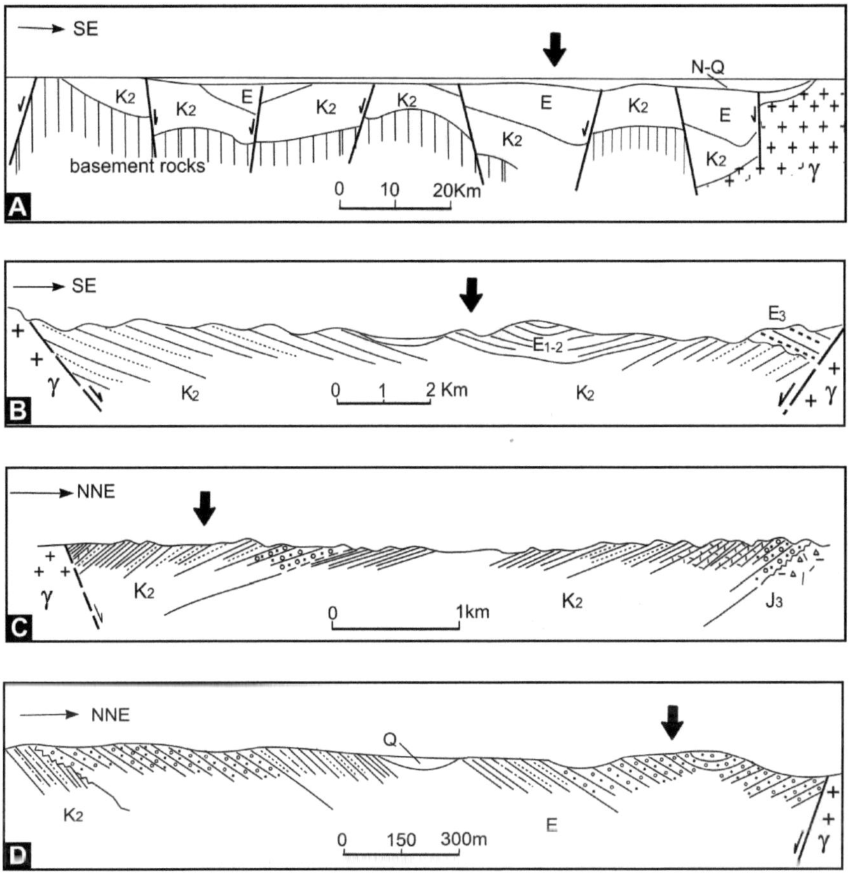

Figure 8.8. Cross sections of Cretaceous–Early Tertiary red-bed basins in SE China. **A**. Dongting basin, Hunan Province; **B**. Nanxiong basin, Guangdong Province; **C**. Shanghang basin, Fujian Province; **D**. Liu basin, Jiangxi Province.

1. Analysis of Sporamorphes of from numerous red-bed basins indicates that the Early Cretaceous in SE China was characterized by gymnosperms and ferns that imply a humid tropical–subtropical climate (Wu and Yang 1978). The lower Cretaceous rocks of the basins, however, are typical red bed (Fig. 8.11B).
2. Dinosaur eggs have been found in many red-bed basins in SE China. If climate was arid or even producing desert conditions, how could an abundance of such large-sized animals survive during the Mesozoic?
3. Although eolian (desert) deposits can be found in some basins, present-day desert sediments are not red. The most abundant sediments in the red-bed basins are freshwater lacustrine, alluvial and deltaic deposits. Present-day freshwater sediments are typically dark grey, but in the Mesozoic basins, these sediments are also red.

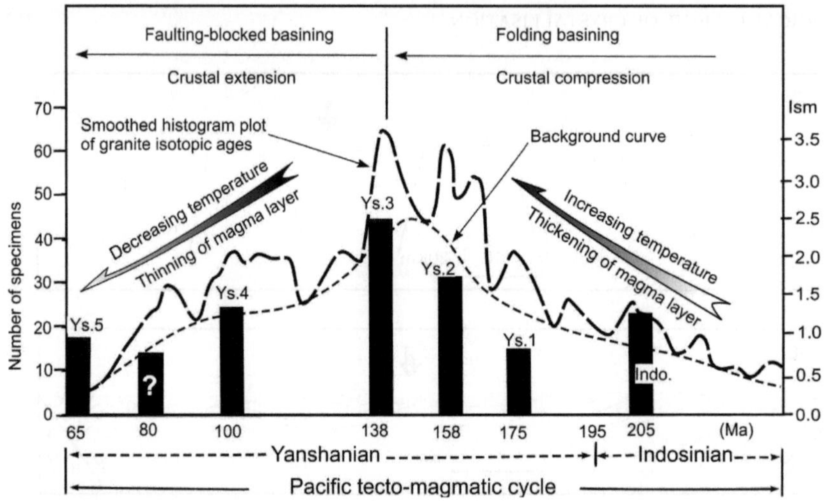

Figure 8.9. Diagram to illustrate the relationship between fold/fault basins and crustal compression /extension, respectively, in SE China.

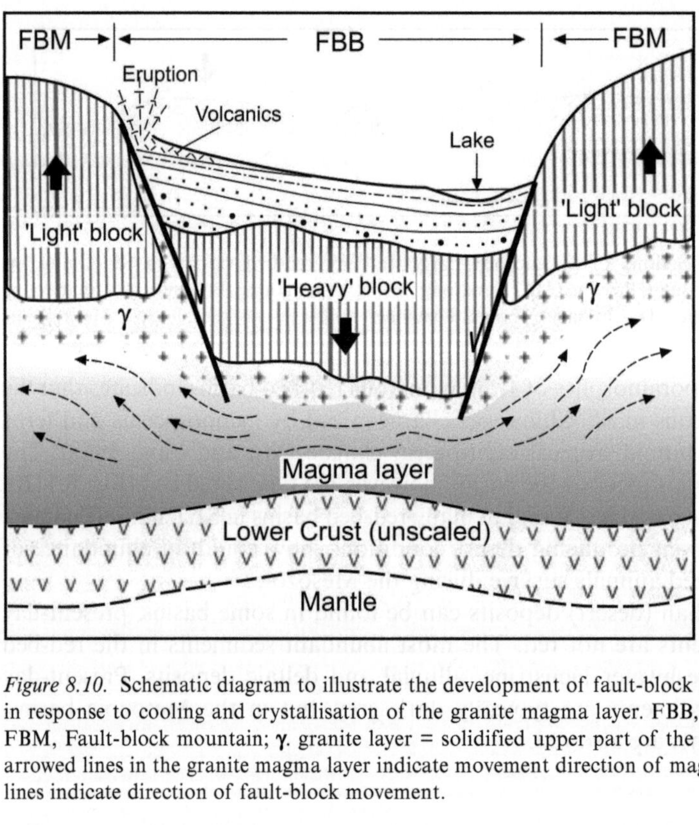

Figure 8.10. Schematic diagram to illustrate the development of fault-block basins in SE China in response to cooling and crystallisation of the granite magma layer. FBB, Fault-block basin; FBM, Fault-block mountain; **γ**. granite layer = solidified upper part of the magma layer. Thin arrowed lines in the granite magma layer indicate movement direction of magma; thick arrowed lines indicate direction of fault-block movement.

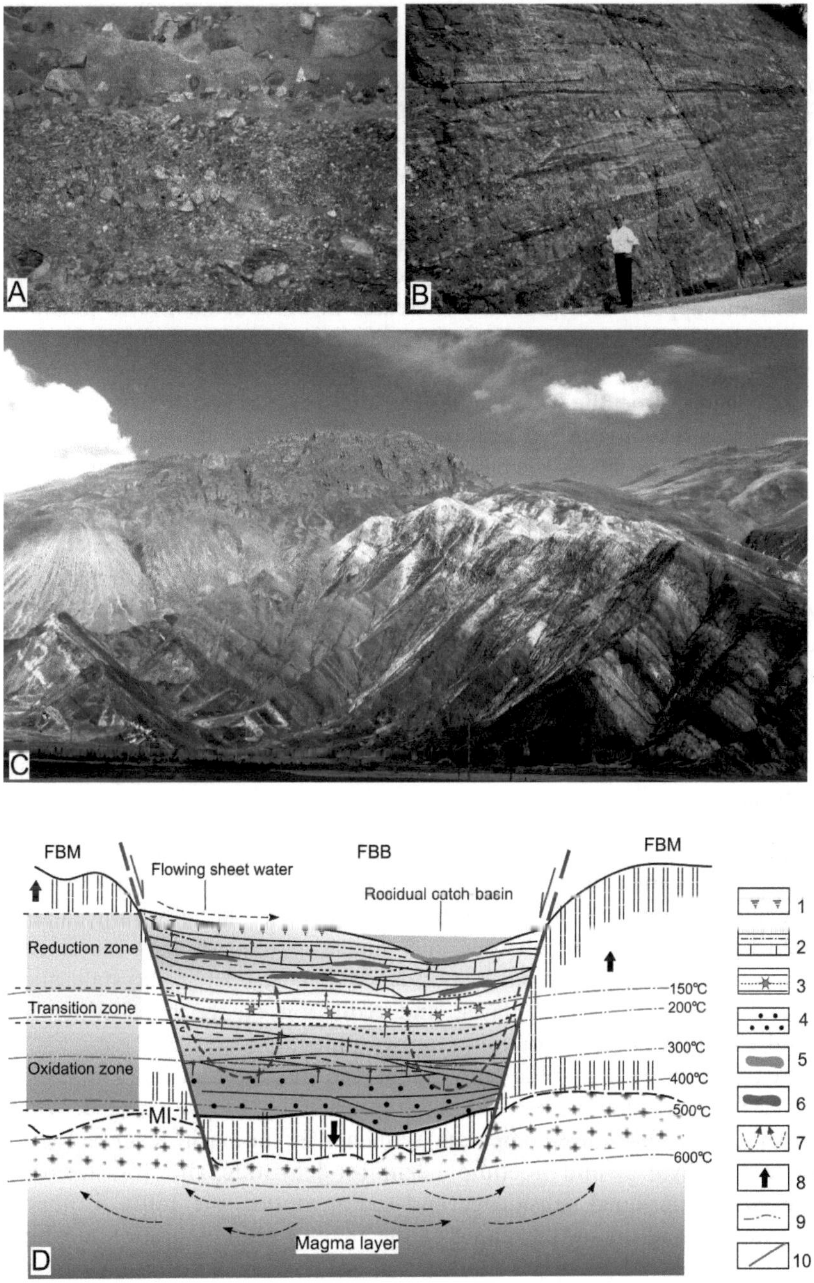

Figure 8.11. **A**. Poorly-sorted and abraded Lower Tertiary conglomerates in the Danxia red-bed basin, SE China; **B**. Lower Cretaceous alternating lenticular conglomerate/sandstone in the Pingyuan red-bed basin, SE China. **C**. Red bed sequence in northern Tibet showing a vertical colour change from red in the lower part, white in the middle part and gray to gray–black in the upper part. **D**. Model to explain the formation of Mesozoic–Cenozoic red-bed sequences in fault-block basins of SE China. **1**. Saline; **2**. Mudstone, siltstone and limestone; **3**. Cu-mineralisation; **4**. Conglomerate; **5**. Halite; **6**. Oil or gas deposit; **7**. Direction of fluid movement during diagenesis; **8**. Direction of fault-block movement; **9**. Inferred isotherms with temperatures indicated; **10**. Fault. For other symbols see Fig. 8.10.

Another explanation for the red colouration is diagenesis. The red pigment of continental red beds principally consists of fine authigenic hematite (van Houten 1973; Walker 1976; Turner 1980; Weibel and Grobety 1999), whereas ferryhydrite and goethite are common in younger sediments (van Houten 1973). Studies of the Skagerrak Formation, Denmark by Weibel and Grobety (1999) provide an instructive example for the transformation of goethite to hematite with increasing depth (temperature). At depths down to 550 m, goethite is the only Fe-oxide. At greater depth hematite appears and gradually increases in amount relative to goethite. Hematite crystal in the deeper levels of the Skagerrak Formation hematite has the same crystal morphology and size as goethite at shallow levels due to pseudomorphic replacement of goethite by hematite. The study also indicates that goethite is the precursor of hematite in red bed formation. The range of temperature over which goethite transforms to hematite has been experimentally determined at between ~150 and 400°C (Dekkers 1990; DeBoer and Dekkers 2001).

The above observations and experimental results suggest:

1. Formation of red pigment of rocks requires a diagenetic temperature not lower than 150°C.
2. Red beds formed in a tectono-sedimentary cycle are variegated with gule or *Amaranthus* in the lower and middle parts of the sequences, and are gray–black in the upper part (Fig. 8.11C). This indicates that diagenetic temperatures of the upper gray–black part of the red beds was <150°C and suggests an approximate position for the 150°C paleo-isotherm at the red–gray–black transition.

Considering the above data, the origin of red beds in the Mesozoic–Cenozoic FBB in SE China can be modeled as shown in Fig. 8.11D. The lower part of red-bed sequences is commonly dominated by conglomerate, gradually fining changing upward into sandstones, siltstone and mudstone (occasionally limestone), indicating a decreasing rate of basin subsidence that we relate to crystallisation of the underlying granite melt layer, and reduction of the relief difference between the basin and the surrounding topographic highs.

In the early stage of fault-block movement, the sinking fault-block impinges on the melt layer or solidified but still hot granite, so that they are heated. When the temperature within the sedimentary sequence in the basin reaches 150°C, authigenic ferryhydrite and goethite begin to dehydrate and transform to hematite. With continued subsidence and sedimentation the 150°C isotherm moves upwards resulting in thickening of the zone of red colouration in which hematite is formed. Oxidation conditions coupled with dehydration, oxygen consumption with the formation of hematite and the release H_2O precludes formation or stable persistence of sulphides that are not found in red beds. Downward infiltration of cool meteoric water helps to establish a fluid cycling system of fluid within the basin sediments (Fig. 8.11D). Elevated temperature and an abundance of free oxygen brought in by groundwater accelerate oxidation of any sulphides present and remove resultant sulphate.

Scavenging of alkalis such as K^+, Na^+ and Ca^{2+} by the upward moving fluid results in a reduction in the redox potential of the fluid resulting in reduction of sulphate and precipitation of sulphides, e.g. CuS, over a certain depth/temperature/redox potential interval in the sequence. In the red-bed basins, Cu-mineralisation is usually associated with a horizon of yellow–green coloured sandstone (Li and Pu 2000) located between the lower red sequence (mainly conglomerates-sandstones) and the upper gray–black sequence (mainly siltstones and mudstones) (Fig. 8.11C) that marks the oxidation–reduction transition.

Continued upward movement of fluid carries Na and K to the uppermost level of the basin sequence to form saline deposits at or near the surface with evaporation of the fluid. Resolution of the saline deposits by meteoric water or by springs increases salinity of shallow lakes within the basin (Fig. 8.11D) to a level at which halite is precipitated during periods of evaporation. In the red-bed basins, halite deposits occur in the gray–black sediments at the top of the sequence.

The absence of fossils in the red-bed sequences suggests that organic material in the sediments has been completely decomposed and removed in solution because of higher than normal temperature during diagenesis. Where temperatures were <150°C in the upper part of the sequence and within the stability temperature range of kerogen, organic material may have been deposited (Fig. 8.11D) to eventually form the oil and gas deposits within the gray–black sediments of the red-bed basins.

8.1.4. Summary

Decreasing crustal temperatures from the end of the Late Jurassic in SE China resulted in the formation of numerous FBB. Subsiding fault-blocks penetrated the upper part of the underlying granite melt layer or hot, newly crystallised granite enhancing the temperature of basin sediments leading to the formation of red beds in the lower part, overlain by sediments hosting Cu-mineralisation, with eventual formation of petroleum and gas concentrations at higher levels.

8.2. VOLCANISM

Crustal relaxation with decreasing temperature following compression from the Late Jurassic not only created FBB, but provided channels for large-scale eruptions of acid-intermediate magma. In SE China the middle Yanshanian (Late Jurassic and Early Cretaceous) volcanic activity (Fig. 8.2) led to the eruption of rhyolite, dacite, trachyite, trachy–andesite and trachy–basalt and formation of the coastal volcanic belt (Fig. 8.1) (Weng et al. 1987, Yin et al. 1991). A second, Late Yanshanian (Late Cretaceous) volcanic episode resulted in eruption of rhyolite, dacite, andesite and basalt (Nanling Institute of Geology and Mineral Resources 1987; BGMRZj 1989; Lapierre et al. 1997). Mesozoic volcanicity, however, is not limited in these two periods, as most Mesozoic sedimentary sequences usually contain volcanic horizons. The spatial distribution of the Mesozoic volcanics indicates that acid (-intermediate) volcanics occur in the coastal region and basalt inland (Fig. 8.12) .

The acid (-intermediate) volcanics are considered to have been derived from the mid-upper crustal granite magma layer and basalt from mantle melting (e.g. Xu et al. 1999; Zhou and Li 2000) (Fig. 8.13) . Numerous well-defined faults occur in the coastal area of SE China (Fig. 8.1) and are presumed to extend down to the top of the granite melt layer and provided avenues for eruption of dominant rhyolite and subordinate dacite, acid-andesite magma (Fig. 8.13).

Comparison of Figs. 8.14–8.17 indicates an oceanward migration of volcanic belts since the Late Jurassic, with the eventual cessation of acid (-intermediate) eruptions and a dominance of Cenozoic basalt volcanism in the coastal area.

With downward inclination of isotherms and resultant thinning and termination of the granite magma layer towards the continental interior, faults west of the crustal magma layer could be expected to transect the entire crust to tap underplated basalt magma at the crust–mantle boundary giving rise to the basalt dykes and volcanics inland of the acid-intermediate volcanic rocks (Fig. 8.13).

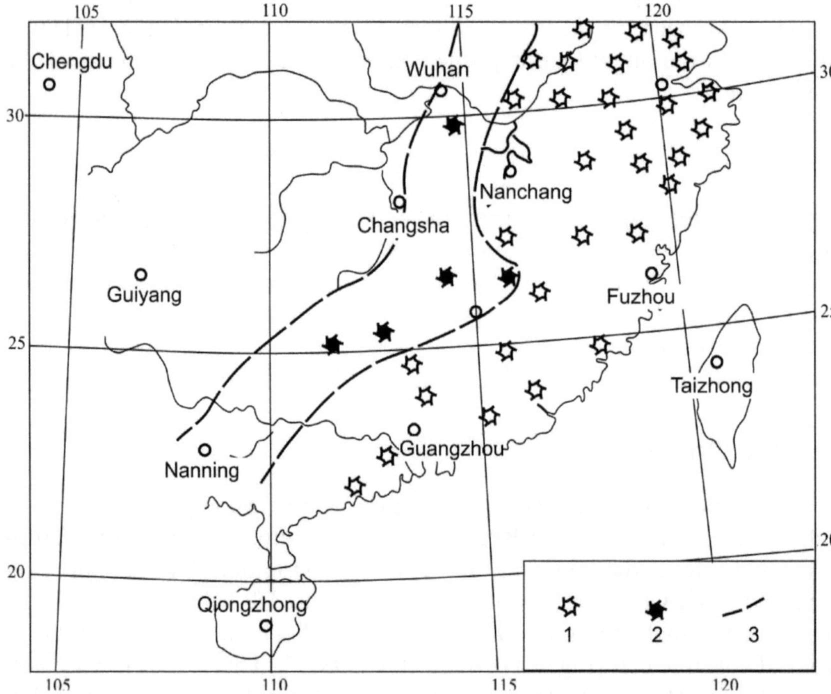

Figure 8.12. Map showing distribution of latest Jurassic volcanics/volcanic-bearing sediments, with acid (-intermediate) volcanics in the coastal region and basalt inland. **1**. Rhyolite (andesite); **2**. Basalt; **3**. Boundary of volcanic compositions.

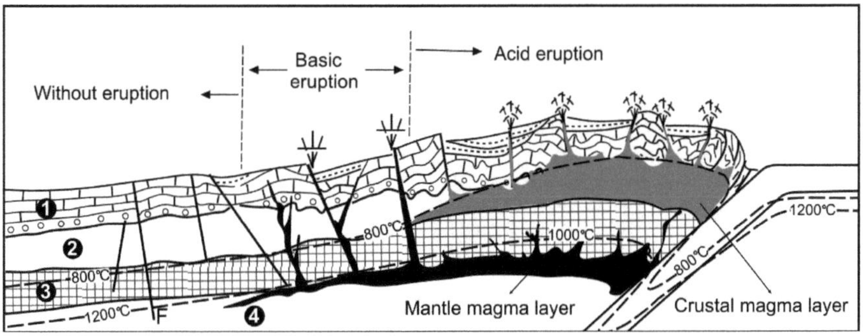

Figure 8.13. Cartoon to illustrate the essential eruption of bimodal (acid-intermediate and basic) volcanic rocks in SE China, derived from a mid-upper crustal granite magma layer and the mantle, respectively. **1**. Upper crust; **2**. Middle crust; **3**. Lower crust; **4**. Mantle.

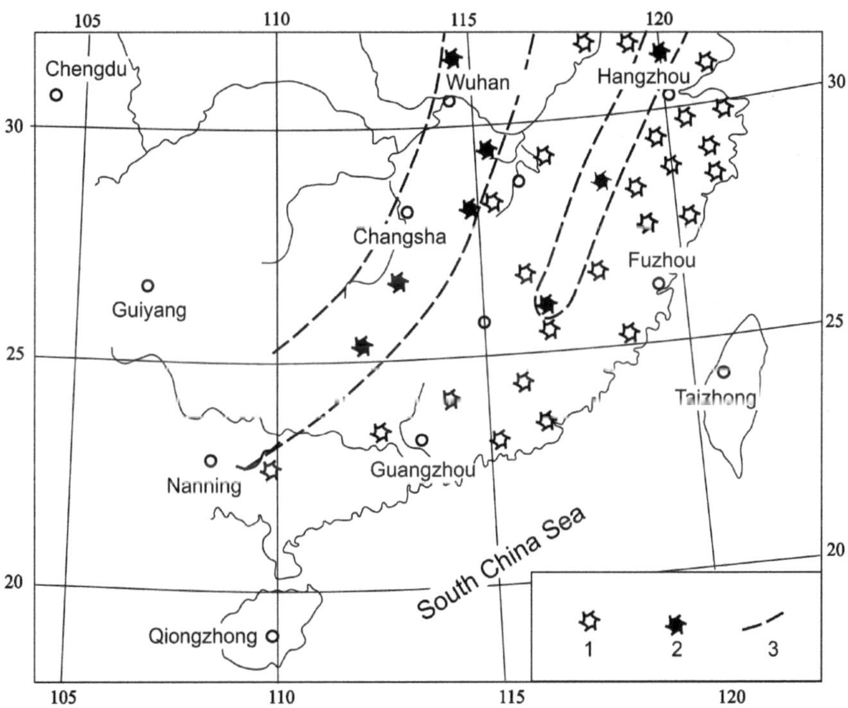

Figure 8.14. Map showing distribution of Early Cretaceous volcanic rocks (legend as in Fig. 8.12).

As discussed in Chapter 7, both remelting events reflect temporary reversals in an overall trend of decreasing crustal temperature from the Late Jurassic (Fig. 8.10) and the landward inclination of magma interfaces (MI's) are responsible for the seaward migration (younging) of granite belts in map plan view

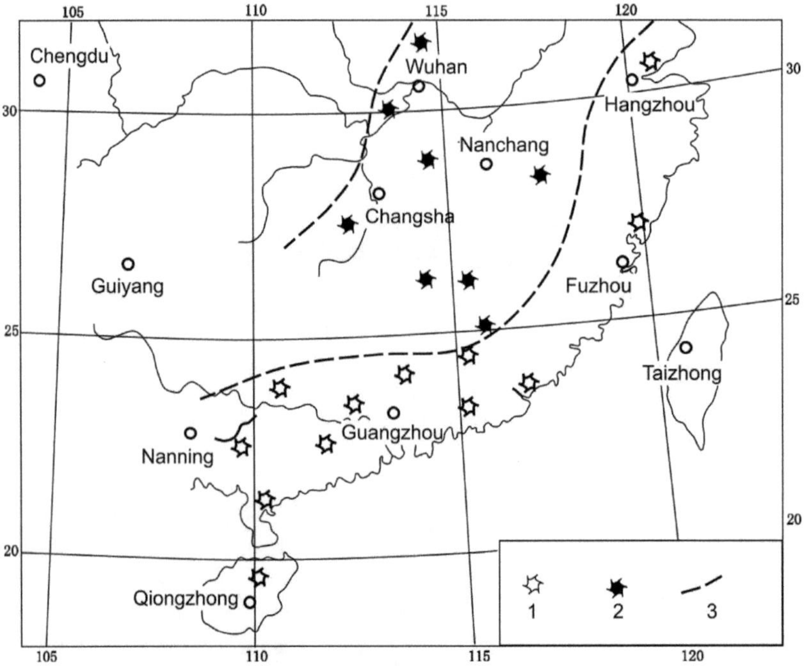

Figure 8.15. Map showing distribution of Late Cretaceous volcanic rocks (legend as in Fig. 8.12).

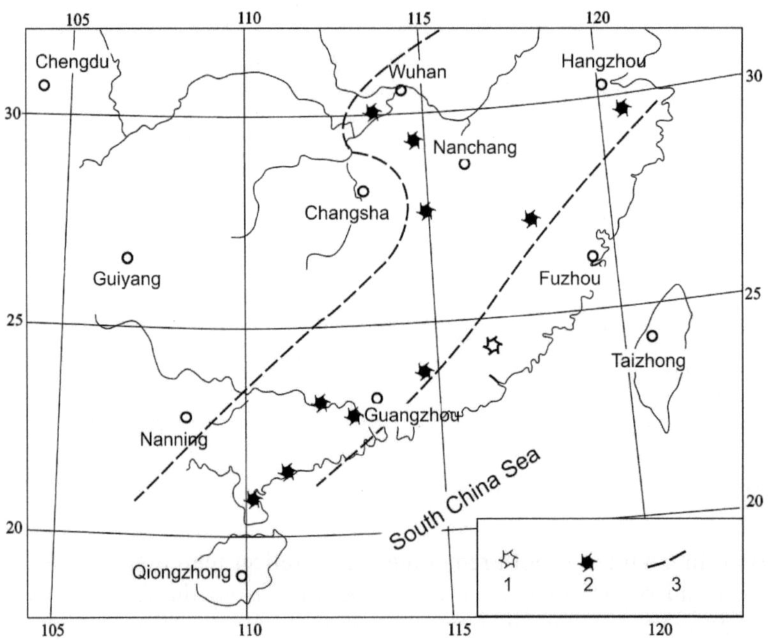

Figure 8.16. Map showing distribution of Early Tertiary volcanic rocks (legend as in Fig. 8.12).

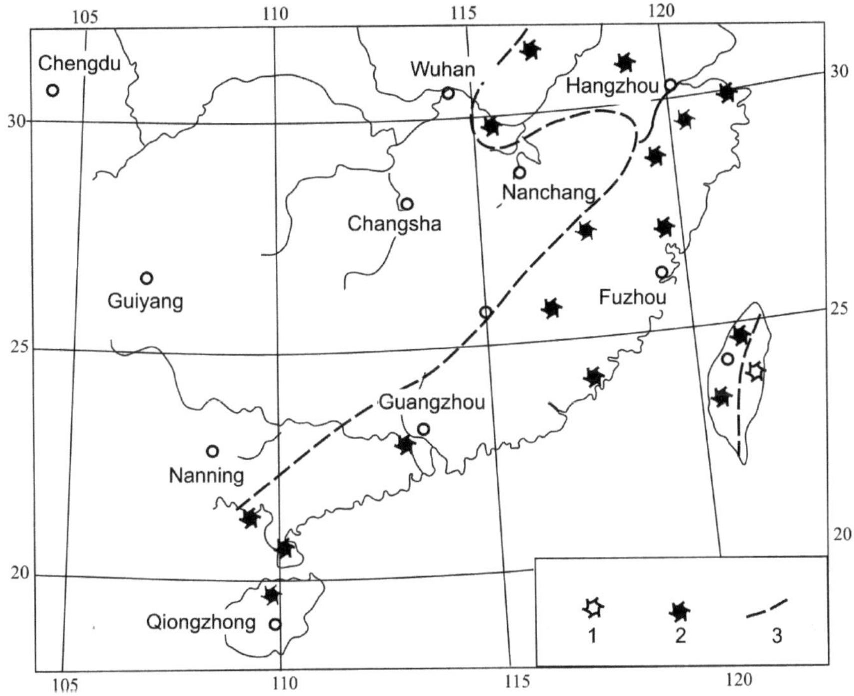

Figure 8.17. Map showing distribution of Late Tertiary volcanic rocks (legend as in Fig. 8.12).

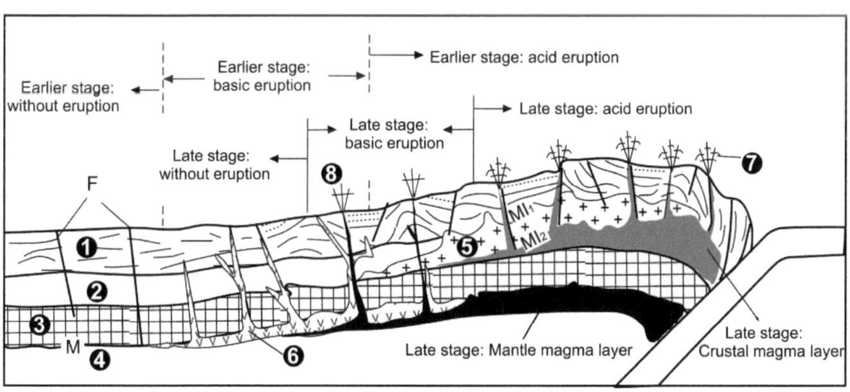

Figure 8.18 Cartoon illustrating relationship lowering MI with cooling and migration of volcanic belts in SE China. Boundary between acid-intermediate and basic volcanics suggests lateral migration of the granite melt layer coeval with volcanic activity. **1**. Upper crust; **2**. Middle crust; **3**. Lower crust; **4**. Mantle; **5**. Early stage solidified granite magma layer; **6**. Early stage solidified mantle melt layer; **7**. acid-intermediate volcanicity; **8**. basic volcanicity; **F**. Fault.

(Fig. 7.14). The migration of eruption belts also suggests depression of the MI resulting in a reduction of the areal distribution of granite magma layers with post-Late Jurassic decreasing temperatures (Fig. 8.10). The extensive Mesozoic granite magma layer is believed to have crystallised by Eocene time, allowing downward propagation of faults to mantle depths in the area of the continental margin facilitating large-scale eruptions of basalt (Fig. 8.18) associated with the development of the Cenozoic continental margin basins of the South China Sea area (Qiu 2006).

CHAPTER 9

MATERIAL AND ELEMENT CYCLING OF THE CONTINENTAL CRUST AND SUMMARY

The genetic model of sedimentary rocks (Fig. 1.1) clarifies the cycling evolution of materials in the sedimentation sphere, yet the mountains of continents are not only composed of metasedimentary-volcanic rocks, but also an abundance of granite. Weathered detritus from these rocks become basin sediments that produce a new generation of sedimentary rocks and on burial and deformation become metamorphosed or melted under appropriate conditions, which results in cycling evolution of rock materials that comprise the continental crust.

9.1. ROCK CYCLING OF CONTINENTAL MATERIAL

Transformation of the three main rocks (sedimentary, metamorphic and igneous) of the continental crust from one kind to another and the associated variation of geological environment/conditions is commonly described as 'rock cycling'. In the current model of rock cycling (Fig. 9.1), sedimentary and igneous rocks can change into metamorphic rocks by an increase in *P–T* and revert to sedimentary rock through exogenic processes. However, the rock cycling scheme does not change sedimentary–metamorphic rocks directly to igneous, because partial melting of typical crustal rocks usually produces ~30–40% granitic melt that moves from its source according to the granite intrusion model, with 60–70% of unmelted and peritectic restite material remaining in the source region as metamorphic, rather than igneous rock.

The in situ melting model of granite formation provides a logical explanation for granite formation in terms of rock cycling because the granite magma remains in the region of its source and is thus associated with its restite fraction. A revised rock cycling scheme is shown in Fig. 9.2 that provides the necessary link between sedimentary and igneous, i.e. granitic, rocks.

The revised rock cycling scheme implies that rock cycling is a reversible process, and variations of the geological environment/conditions associated with the cycle depend on the cycling direction; evolution from sedimentary rock through metamorphic to igneous rocks with increasing burial and temperature,

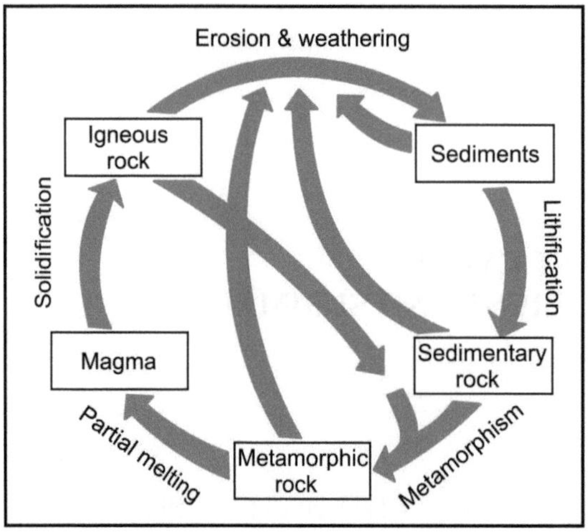

Figure 9.1. A typical rock cycling scheme (after Judson and Richardson 1995), indirectly based on an intrusion model of granite formation.

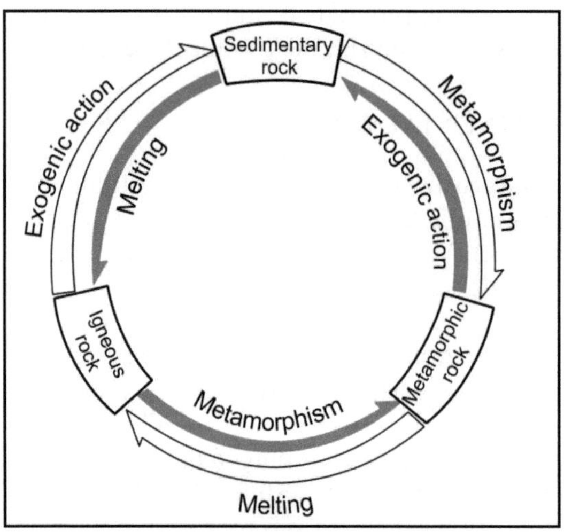

Figure 9.2. An alternative rock-cycling scheme based on an in situ melting–intercrustal convection model for granite formation.

and an opposite trend where conditions are reversed (Fig. 9.2). Direct transformation from sedimentary rock to igneous rock and the reverse process (back to sedimentary rock) should reflect relatively rapid changes in the physio-chemical conditions of the continental system that involves considerable energy input/output such as that generated by the convergence of lithospheric plates. Convergence

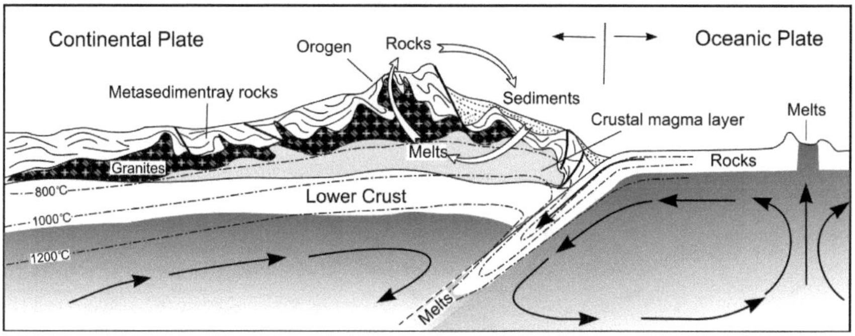

Figure 9.3. Cartoon illustrating a geodynamic interpretation for the cycling evolution of continent rock material, emphasising the relationship between the two major material (energy) cycling processes in the continental crust and the mantle.

enhances the intrinsic energy of the continental crust and leads to formation and development of a crustal magma layer, and finally results in crustal deformation (orogeny) that accelerates exogenous process of erosion, transportation and deposition. In contrast, decreasing crustal temperature results in crystallisation of the magma layer, stabilising the crust, reducing the rate and amount of erosion and deposition (Qiu 2006). The relationship between cyclic evolution of rock materials and variations in the thermodynamic environment/conditions of the continental crust caused by plate convergence is shown in Fig. 9.3.

9.2. ELEMENT CYCLING OF THE CONTINENTAL CRUST

Both energy and material flow in the Earth's interior drive the movement of the lithospheric plates. Interaction between the plates drives the flow of energy within the continental lithosphere, resulting in rock cycling (Fig. 9.3). Geological processes affecting rocks are shown in Fig. 9.4. Denudation occurs only in elevated areas of the crust with the weathering interface (WI) represented by the contact between rock and atmosphere; deposition and lithification operate in sedimentary basins or regions of low or negative relief, and the upper limit of sedimentation sphere (or hydrosphere) is defined as the accumulation interface (AI); metamorphism occurs within basement rocks, with an upper limit defined as the oxidation–reduction interface (ORI) that marks the transitional from weathered to unweathered rock; anatexis may occur in the lower part of the dominantly metasedimentary basement rocks when temperatures are high enough to intersect their solidi, with eventual formation of a convecting magma and subsequent granite layer (Fig. 9.4). Tectonic uplift exposes the new rocks (including the sedimentary, metamorphic and igneous) formed in deeper parts of the crust and the beginning of another cycle (Fig. 9.4).

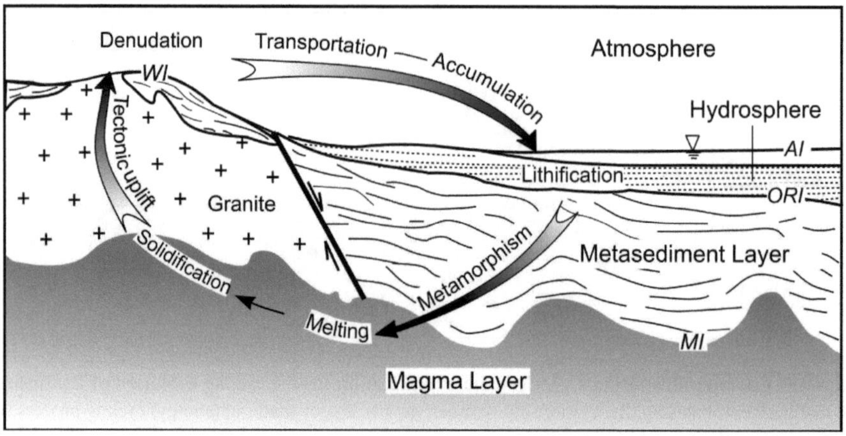

Figure 9.4. Schematic diagram to illustrate material cycling and flow paths of mid–upper continental crustal material associated with development of a granite magma layer. AI = accumulation interface; ORI = oxidation–reduction interface. See text.

Rocks are aggregations of minerals that are constructed of elements, and thus the rock cycle evolution of 'rock-flow' is paralleled by element cycling as described in Chapter 6, as represented by the four geochemical fields of elements (GFE), i.e. magmatic, hydrothermal, aqueous and gas fields in the periodic table (Fig. 8.5). Elements in the magmatic field are lithophiles that form the lattice of silicate minerals during crystallisation of magma and constitute the granite layer. Those in the hydrothermal field are mostly sulphophiles that are expelled from the magma (granite) layer during the hydrothermal stage of crystallisation and accumulate in overlying rocks to form various sulphide deposits. The hydrothermal field thus mirrors the metamorphic–sedimentary layer in space. Elements in the aqueous field, including C, N, O and halogens, act as a medium for transportation of ore-forming elements during the hydrothermal stage of magma crystallisation and they will ultimately be effluxed into the hydrosphere. The aqueous field thus mirrors the hydrosphere (or sedimentary sphere). Elements in the gas field will migrate directly or indirectly to the atmosphere after being expelled from the magma system, so that the gas field represents the atmosphere. Comparison of Figs. 9.4 and 9.5 show that the GFE reflects the texture of the upper crust, and the element-flow follows the cycling of geological processes and of rock-flow shown in Fig. 9.4.

9.3. OVERVIEW

The genetic models of sedimentary and basaltic rock formation constitute the core of both continental geology and plate tectonics, respectively, yet the absence of a model incorporating the formation of granite that occurs between

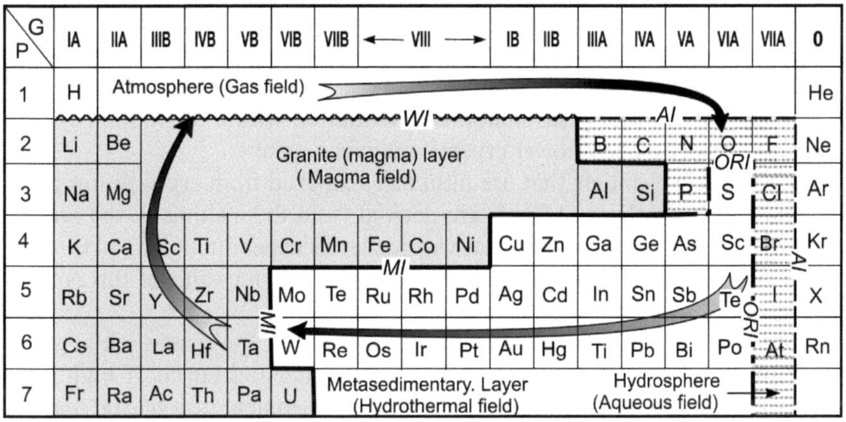

Figure 9.5. Element cycling as reflected by the geochemical field of elements (GFE). The direction of cycling is indicated by arrows, the layers (or spheres) and the various interfaces (WI, AI, ORI and MI) correspond to those in Fig. 9.4.

metasedimentary rocks of the upper crust and mafic rocks of the lower crust prevents the two geological knowledge systems from becoming a unified model. The essence of 'plutonism' is that granite is crystallised magma. The igneous origin of granite, however, does not mean that it must be intruded, i.e. from the lower to upper crust. The granite magma intrusion model regards the upper crust repository as an open system with variation of its intrinsic energy caused by heat being advected by magma intruded from below and this has resulted in a paradigm of granite genesis that is unable to explain many of the field observations of granites.

The in situ melting–intra-crustal convection model of granite genesis expounded here regards the upper crust as a relatively closed system, and formation of granite resulting in a change from order (protolith) to disorder (melt) to new order (granite) with entropy variations of crustal rocks. Convection within a crustal partial melt layer is essential for the formation of granite magma. Without convection, partial melting of rocks produces migmatites and is incapable of producing granite batholiths. The magma or melting interface (MI) divides the magma layer and the partial melting (migmatite) zone and also convection and conduction modes of heat flow. As granite is the result of crystallisation of the magma layer, it is therefore layer-like within the crust. The shape and size of individual bodies of granite reflects the geometric relationship between the paleo-magma interface (PMI) and the present-day erosion surface, rather than the volume of intrusive magma. In situ magma production eliminates the 'room problem' of granite batholiths and explains the multifarious field phenomena related to granites described above.

As a physical process, crustal melting results in reactivation of various elements that form the atomic framework of minerals, and is thus responsible for their reorganisation and redistribution within the continental crust to form granite and its associated hydrothermal mineral deposits. Variations in chemical and isotopic compositions of granites can be explained by variations in protolith (country rock) composition and differentiation within the molten system and not necessarily a deep (lower crustal) magma source.

The ore-forming elements that are ultimately expelled from crystallising granite magmas via hydrothermal fluids are derived from the melting of the granite country rock and are reprecipitated at various levels depending on T, pH, redox potential, wall rock composition etc., the potential economic value of this process being obvious.

Although possible heat sources for partial melting of the crust such as shear heating, intrusion of mafic magma, crustal thickening and burial of high radiogenic rocks, are important, the formation and development of a crustal magma layer requires constant energy (heat) input into the system. This requirement is provided by the convergence of lithospheric plates. Tectonic models explaining the relationship between variations of intrinsic energy of the continental system to plate convergence clarify material evolution within the crust and its relation to surface environmental variables with examples from the western Pacific continental margin and Tethys-Tibet: increasing temperature corresponding to periods of accelerating convergence leading to crustal melting (formation of granite magma layer) and lateral compression (folding, faulting; orogeny), and decreasing temperature corresponding to a slowing of convergence accompanied by crustal cooling, crystallisation of the magma layer and gravitational adjustment (fault-controlled basins) of rocks overlying the layer.

The formation of continental red-beds in fault-controlled basins represents an example of energy exchange between the solid crust and the sedimentation sphere. Thus, red beds can be regarded as thermally altered rocks, with gas, oil and copper deposits in the upper parts of red-bed sequences, as hydrothermal deposits.

The origin of granite, as with that of sedimentary and basaltic rocks, is not simply a petrological or geochemical problem. The various geological, geophysical, metallogenic and experimental melting observations concerning the formation of granites constitute an integrated system with an in situ melting–intra-crustal convection model for their genesis as the core, revealing the cycling movement of energy/material flow of the continental crust and its geodynamic/environmental effects, and the relationship to lithospheric plate convergence driven by the energy/material cycling of the Earth's interior. Element cycling corresponding to the geological cycle revealed by the GFE established on both in situ melting model of granite formation and the periodic table underscores the harmony and unity between the microcosm and macrocosm of the natural world.

REFERENCES

(Note: References with *et. al.* have more than 7 co-authors)

Ahern JL, Turcotte DL, Oxburgh ER (1981) On the upward migration of an intrusion. J. Geol. 89:421–432

Alsdorf D, Nelson D (1999) Tibetan satellite magnetic low: Evidence for widespread melt in the Tibetan crust? Geology 27:943–946

Anderson JL (1983) Proterozoic anorogenic granite plutonism in North America. In: Medaris JG et al. (eds) Proterozoic Geology. Geol. Soc. Am. Mem. 161:133–154

Anderson JL (1996) Status of thermobarometry in granite batholiths. Trans Roy. Soc. Edinburgh Earth Sci. 87:125–138

Arzi AA (1978) Critical phenomena in the rheology of partially melted rocks. Tectonophysics 44:173–184

Ashworth JR (ed) (1985) Migmatites. Blackie, Glasgow and London

Audren C, Triboulet C (1993) *P–T–t*-deformation paths recorded by kinzigites during diapirism in the western Variscan belt (Golfe du Morbihan, southern Brittany, France). J. Metamorph. Geol. 11:337–356

Bachi CA, Miller CF, Miller JS, Faulds JE (2001) Construction of a pluton: Evidence from an exposed cross section of the searchlight pluton, El Dorado Mountains, Nevada. Geol. Soc. Am. Bull. 113:1213–1228

Bannister S, Bryan CJ, Bibby HM (2004) Shear wave velocity variation across the Taupo Volcanic Zone, New Zealand, from receiver function inversion. Geophys. J. Int. 159:291–310

Barbarin B (1990) Granitoids: main petrogenetic classification in relation to origin and tectonic setting. Geol. J. 25:227–238

Barbarin B (1999) A review of the relationships between granitoid types, their origins and their geodynamic environments. Lithos 46:605–626

Barbarin B (2005) Mafic magmatic enclaves and mafic rocks associated with some granitoids of the central Sierra Nevada batholith, California: nature, origin and relations with the hosts. Lithos 80:155–177

Barbero L, Villaseca C, Rogers G (1995) Geochemical and isotopic disequilibrium in crustal melting: an insight from anatectic granitoids from Toledo, Spain. J. Geophys. Res. 100:15745–15765

Barboza SA, Bergantz GW, Brown M (2000) Regional granulite facies metamorphism in the Ivrea zone: is the Mafic Complex the smoking gun or a red herring? Geology 27:447–450

Barielli A, Bertini G, Buonasorte G, Cappetti G, Fiordelisi A (2000) Recent deep exploration results at the margins of the Larderello Travale geothermal system. In: Proc. World Geotherm. Congress 2000, Japan, pp 965–970

Barker F (1979) Trondhjemite: definition, environment and hypothesis of origin. In: Barker F (ed.) Trondhjemites, Dacites and Related Rocks. Elsevier, Amsterdam, pp 1–12

Barnes HL (1975) Zoning of ore deposits: types and causes. Trans. RSE 69:295–311
Barnes HL (1995) Geochemistry of hydrothermal ore deposits, 3rd edn. Wiley, New York
Bateman PC (1992) Plutonism in the central part of the Sierra Nevada batholith, California. US Geol. Surv. Prof. Paper 1483
Bateman R (1985) Progressive crystallization of a granite diapir and its relationship to stages of emplacement. J. Geol. 93:645–622
Baumgartner LP, Ferry JM (1991) A model for coupled fluid-flow and mixed-volatile mineral reactions with applications to regional metamorphism. Contrib. Mineral. Pet. 106:273–285
Beard JS, Lofgren GE (1991) Dehydration melting of water-saturated melting of basaltic and andesitic greenstones and amphibolites at 1, 3 and 6.9 kbar. J. Pet. 32:365–401
Beard JS, Abiz RJ, Lofgren GE (1993) Experimental melting of crustal xenoliths from Kilbourne Hole, New Mexico and implications for the contamination and genesis of magma. Contrib. Mineral. Pet. 115:88–102
Bergantz GW (1989) Underplating and partial melting: Implications for melt generation and extraction. Science 245:1093–1095
Bibby HM, Caldwell TG, Davey FJ, Webb TH (1995) Geophysical evidence of the structure of the Taupo Volcanic Zone and its hydrothermal circulation. J. Volc. Geotherm. Res. 68:29–58
Binns RA (1969) Hydrothermal investigations of the amphibolite–granulite facies boundary. Geol. Soc. Aust. Spec. Publ. 2:341–344
Bird P (1979) Continental delamination and the Colorado Plateau. J. Geophys. Res. 84:7561–7571
Biste M (1982) Geochemistry of south Sardinian granites compared with their tin potential. In: Evans AM (ed) Metallization Associated with Acid Magmatism. John Wiley & Sons, New York, pp 37–49
Bodorkos S, Sandiford M, Oliver NHS, Cawood PA (2002) High-*T*, low-*P* metamorphism in the Paleoproterozoic Halls Creek Orogen, northern Australia: the middle crustal response to a mantle-related transient thermal pulse. J. Metamorph. Geol. 20:217–237
Bohlen SR (1987) Pressure-temperature time paths and a tectonic model for the evolution of granulites. J. Geol. 95:617–632
Borevsky L, Milanovsky S, Yakovlev L (1995) Fluid-thermal regime in the crust-superdeep drilling data. In: Barbier E, Frye G, Iglesias E, Palmason G (eds) Proc. World Geotherm. Congress. Int. Geotherm. Assoc., Auckland, New Zealand, pp 975–981
Bowen NL (1914) The ternary system diopside-anorthite-silica. Am. J. Sci. 38:207–264
Bowen NL (1922) The reaction principle in petrogenesis. J. Geol. 30:177–198
Bowen NL (1948) The granite problems and the method of multiple prejudices. In: Gilluly J (ed) Origin of Granite. Geol. Soc. Am. Mem. 28:79–90
Brasse H, Lezaeta P, Rath V, Schwalenberg K, Soyer W, Haak V (2002) The Bolivian Altiplano conductivity anomaly. J. Geophys. Res. 107:10,1029/2001JB000391
Brearley AJ (1986) An electron microprobe study of muscovite breakdown in pelitic xenoliths during pyrometamorphism. Mineral. Mag. 357:385–397
Brearley AJ (1987a) A natural example of the disequilibrium breakdown of biotite at high temperature: TEM observations and comparison with experimental kinetics data. Mineral. Mag. 359:93–106
Brearley AJ (1987b) An experimental and kinetic study of the breakdown of aluminous biotite at 800°C: reaction microstructures and mineral chemistry. Bull. Mineral. 110:513–532
Brearley AJ, Rubie DC (1990) Effects of H_2O on the disequilibrium breakdown of muscovite + quartz. J. Pet. 31:925–956
Brown EH, McClelland WC (2000) Pluton emplacement by sheeting and vertical ballooning in part of the southeast Coast Plutonic Complex, British Columbia. Geo. Soc. Am. Bull. 112:708–719
Brown LC, Zhao W, Nelson KD, Hauck M, Alsdorf D, Ross A, Cogan M, Clark M, Liu X, Che J (1996) Bright spots, structure, and magmatism in southern Tibet from INDEPTH seismic reflection profiling. Science 274:1688–1696
Brown M (1979) The petrogenesis of the St. Malo migmatite belt, Amorican Massif, France. N. Jb. Mineral. Abh. 135:48–74

Brown M (2001) Crustal melting and granite magmatism: key issues. Phys. Chem. Earth 26:201–212
Brown M, Candela PA, Peck DL, Stephens WE, Walker RJ, Zen E-an (eds) (1996) The third Hutton symposium: The Origin of Granites and Related Rocks. Trans. Roy. Soc. Edinburgh Earth Sci. 87:1–359
Brown M, Solar GS (1998a) Shear-zone systems and melts: feedback relations and self-organization in orogenic belts. J. Struct. Geol. 20:211–227
Brown M, Solar GS (1998b) Granite ascent and emplacement during contractional deformation in convergent orogens. J. Struct. Geol. 20:1365–1393
Brown M, Solar GS (1999) The mechanism of ascent and emplacement of granite magma during transpression: a syntectonic granite paradigm. Tectonophysics 312:1–33
Brun J-P, Sokoutis D, van den Driessche J (1994) Analogue modeling of detachment fault systems and core complexes. Geology 22:319–322
Buick LS, Stevens G, Gibson RL (2004) The role of water retention in the anatexis of metapelites in the Bushveld Complex aureole, South Africa: an experimental study. J. Pet. 43:1777–1797
BGMRGd (Bureau of Geology & Mineral Resources of Guangdong Province) (1986) Regional geology of Guangdong Province. Geol. Publ. House, Beijing, pp 285–393
BGMRGx (Bureau of Geology & Mineral Resources of Guangxi Province) (1985) Regional geology of Guangxi Province. Geol. Publ. House, Beijing, pp 283–543
BGMRFj (Bureau of Geology & Mineral Resources of Fujian Province) (1985) Regional geology of Fujian Province. Geol. Publ. House, Beijing, pp 300–564
BGMRHn (Bureau of Geology & Mineral Resources of Hunan Province) (1988) Regional geology of Hunan Province. Geol. Publ. House, Beijing, pp 368–616
BGMRJx (Bureau of Geology & Mineral Resources of Jiangxi Province) (1984) Regional geology of Jiangxi Province. Geol. Publ. House, Beijing, pp 358–706
BGMRXj (Bureau of Geology & Mineral Resources of Xinjian Uygur Autonomous Region) (1993) Regional geology of Xinjiang Uygur Autonomous Region. Geol. Publ. House, Beijing
BGMRZj (Bureau of Geology & Mineral Resources of Zhejiang Province) (1989) Regional geology of Zhejiang Province. Geol. Publ. House, Beijing (in Chinese with English abstract)
Burnham CW (1997) Magmas and hydrothermal fluids. In: Barnes HL (ed) Geochemistry of Hydrothermal Ore Deposits, 3rd edn. John Wiley, New York, pp 63–123
Büsch W, Schneider G, Mehnert KR (1974) Initial melting at grain boundaries, Part II: Melting in rocks of granodioritic, quartzdioritic and tonolitic compositions. N. Jb. Mineral. Mh. H8:345–370
Cai GQ (1982) Material constituents of hydrothermal uranium deposits in South China granite. In: Papers on Granite-type Uranium Deposits. Nuclear Energy Publ. House, Beijing (in Chinese)
Cai QZ, Zheng SF (1997) The formation and evolution of Wuguishan granite, Guangdong Province, Guangdong. Geology 12:33–39 (in Chinese)
Campos TFC, Neiva AMR, Nardi LVS (2002) Geochemistry of the Rio Espinharas hybrid complex, northeastern Brazil. Lithos 64:131–153
Cameli GM, Dini I, Liotta D (1993) Upper crustal structure of the Larderello geothermal field as a feature of post-collisional extension tectonics (southern Tuscany, Italy). Tectonophysics 224:413–423
Candella PA, Holland HD (1986) A mass transfer model for copper and molybdenum in magmatic hydrothermal systems: the origin of porphyry-type ore deposits. Econ. Geol. 81:1–19
Carl C, Wendt I (1993) Radiometrische Datierung der Fichtelgebirgs-granite. Z. Geol. Wiss. 21:49–72
Castro A, Corretgé LG, El-Biad M, El-Hmidi H, Fernández C, Patiño Douce AE (2000) Experimental constraints on Hercynian anatexis in the Iberian massif, Spain J. Pet. 41:1471–1488
Chamberlain CP, Sonder LJ (1990) Heat-producing elements and the thermal and baric patterns of metamorphic belts. Science 250:763–769
Chamberlin RT, Link TA (1927) The theory of laterally spreading batholiths. J. Geol. 35:319–352
Chang CF, Pan YS, Zheng XL, Zhang XM (1982) Geological structures of Qing-zhang Plateau. Science Press, Beijing (in Chinese)

Chappell BW, White AJR (1974) Two contrasting granite types. Pacific Geol. 8:173–174

Chappell BW, White AJR (1976) Plutonic rocks of the Lachlan Mobile Zone 25th Int. Geol. Congr. Sydney, Excursion Guide 13C

Chen GD (1985) Methods of metallotectonic research. Geol. Publ. House, Beijing (in Chinese)

Chen GN (1987) Studies on the Mesozoic tectogeneses in the Diwa region of southeast China, Geotecton. Metallog. 11:65–73 (in Chinese)

Chen GN (1992) On genesis of the Mesozoic granite in southeast China and its relation to plate subduction. In: Progress in Geology of China-contributions to the 29th International Geological Congress. Geol. Publ. House, Beijing, pp 3–6

Chen GN (1997) Pondering over some problems in the study of continental geology. Geol. J. Chinese Univ. 4:342–347 (in Chinese with English abstract)

Chen GN (1998a) Geochemical field of the elements and its geo-implications. Geochimica 27:566–574 (in Chinese)

Chen GN (1998b) Advances in the study of genesis and metallogeny of granite: a brief introduction of the in-situ melting hypothesis and geochemical field of the elements. Adv. Earth Sci. 13:140–144 (in Chinese)

Chen GN (2005) On rock genesis and rock cycling. http://www.paper.edu.cn

Chen GN, Grapes RH (2003) In-situ melting model for granite formation: evidence from SE China. Int. Geol. Rev. 45:611–622

Chen GN, Cao JJ, Zhang K (1996) In-situ melting and the geochemical field of elements. Geol. Publ. House, Beijing, pp 1–98 (in Chinese)

Chen GN, Grapes R, Zhang K (2003) A model for Mesozoic crustal melting and tectonic deformation in Southeast China. Int. Geol. Rev. 45:948–955

Chen L, Booker JR, Jones AG, Wu N, Unsworth MJ, Wei W, Tan H (1996) Electrically conductive crust in southern Tibet from INDEPTH magnetotelluric surveying. Science 274:1694–1696

Chen YC et al (1989) The geology of nonferrous and rare metal deposits related to Mesozoic granitoids in Nanling region. Geol. Publ. House, Beijing (in Chinese with English abstract)

Chen ZX, Li SZ, Zhu JG (1989) The study on the geology and mineralization of the Xihuashan and Hongling. In: Collection of Scientific Research Reports on Geology and Mineral Resources of Nanling. China University of Geoscience Press, Wuhan (in Chinese with English abstract)

Chmielowski J, Zandt G, Haberland C (1999) The central Andean Altiplano-Puna magma body. J. Geophys. Res. Lett. 26:783–786

Cho M, Fawcett JJ (1986) A kinetic study of clinochlore and its high temperature equivalent forsterite-cordierite-spinel at 2 kbar water pressure. Am. Mineral. 71:68–77

Choudhuri A, Winkler HGF (1967) Anthopyllit und Hornblende in einigen metamorphen Reaktionen. Contrib. Mineral. Pet. 14:293–315

Clemens JD (1988) Observations on the origin and ascent mechanisms of granitic magmas. J. Geol. Soc. Lond. 155:843–851

Clemens JD (1990) The granulite-granite connection. In: Vielzeuf D, Vidal PH (eds) Granulites and Crustal Evolution. NATO ASII Ser C, 311:25–36, Kluwer Acad., Norwell, MA

Clemens JD (2005) Granites and granitic magmas: strange phenomena and new perspectives on some old problems. Proc. Geol. Assoc. 116:9–16

Clemens JD, Mawer CK (1992) Granite magma transport by fracture propagation. Tectonophysics 204:339–360

Clemens JD, Vielzeuf D (1987) Constraints on melting and magma production in the crust. Earth Planet. Sci. Lett. 86:287–306

Clemens JD, Wall VJ (1981) Origin and crystallization of some peraluminous (S-type) granite magmas. Can. Mineral. 19:111–131

Cloos H (1923) Das Batholiten Problem. Fortschr Geol Pal (1)80pp

Cole JW (1990) Structural control and origin of volcanism in the Taupo Volcanic Zone, N. Z. Bull. Volcanol. 52:445–459

Collins WJ, Richards SR, Healy BE, Ellison PI (2000) Origin of heterogeneous mafic enclaves by two-stage hybridization in magma conduits (dykes) below and in granite magma chambers. Trans. Roy. Soc. Edinburgh Earth Sci. 91:27–45

Connolly JAD, Podladchikov YY (2004) Fluid flow in compressive tectonic settings: implications for midcrustal seismic reflectors and downward fluid migration. J. Geophys. Res. 109: B04201
Conrad WK, Nicholls IA, Wall VJ (1988) Water-saturated and undersaturated melting at 10kbar: evidence for the origin of silicic magmas in the Taupo Volcanic Zone, New Zealand, and other occurrences. J. Pet. 29:765–803
CGQXP Editorial Committee (1983) Contribution to the geology of the Qinghai-Xizang (Tibet) Plateau 3: Paleontology and stratigraphy. Geol. Publ. House, Beijing (in Chinese with English abstract)
Crunden AR (1998) On the emplacement of tabular granites. J. Geol. Soc. Lond. 155:853–862
Crunden AR, McCaffrey KJW (2001) Growth of plutons by floor subsidence: implications for rates of emplacement, intrusion spacing and melt-extraction mechanisms. Phys. Chem. Earth 26:303–315
Currie CA, Wang K, Hyndman RD, He J (2004) The thermal effects of steady-state slab-driven mantle flow above a subducting plate: the Cascadia subduction zone and backarc. Earth Planet. Sci. Lett. 223:35–48
Currie KL, Pajari Jr. GE (1981) Anatectic peraluminous granites from the Carmanville area, northwestern Newfoundland. Can. Mineral. 19:147–161
Daly RA (1914) Igneous rocks and the depths of the Earth. McGraw-Hill, New York
Dana JD (1873) On some results of the Earth's contraction from cooling, including a discussion on the origin of mountains and the nature of the Earth's interior. Am. J. Sci. 5:423–443
Davidson C, Schmid SM, Hollister LS (1994) Role of melt during deformation in the deep crust. Terra Nov. 7:133–142
Davies GR, Tommasini S (2000) Isotopic disequilibrium during rapid crustal anatexis: implications for petrogenetic studies of magmatic processes. Chem. Geol. 162:169–191
Davies JH, von Blanckenburg F (1995) Slab breakoff: A model of lithosphere detachment and its test in the magmatism and deformation of collision orogens. Earth Planet. Sci. Lett. 129:85–102
Davies JH, Stevenson DJ (1992) Physical model of source region of subduction zone volcanics. J. Geophys. Res. 97:2037–2070
Davison I, McCarthy M, Powell D, Torres HHF, Santos CA (1995) Laminar flow on shear zones: The Pernambuco Shear Zone, NE Brazil. J. Struct. Geol. 17:149–161
DeBoer CB, Dekkers MJ (2001) Unusual thermomagnetic behaviour of hematites: neoformation of highly magnetic spinel phase on heating in air. Geophys. J. Int. 144:481–494
Dekkers MJ (1990) Magnetic properties of natural goethite. Magnetic behaviour and properties of minerals originating from goethite dehydration during thermal demagnetization. Geophys. J. Int. 103:233–250
D'Lemos RS, Brown M, Strachan RA (1992) Granite magma generation, ascent and emplacement within a transpressional orogen. J. Geol. Soc. Lond. 149:487–490
Dewey JF (1988) Extensional collapse of orogens. Tectonics 7:1123–1139
De Yoreo JJ et al. (1989a) The Acadean thermal history of western Maine. J. Metamorph. Geol. 7:169–190
De Yoreo JJ, Lux DR, Guidotti CV (1989b) The role of crustal anatexis and magma migation in the thermal evolution of regions of thickened continental crust. In: Daly JS, Cliff RA, Yardley BWD (eds) Evolution of Metamorphic Belts. Spl. Publ. Geol. Soc. Lond. 43:187–202
De Yoreo, JJ, Lux DR, Guidotti CV (1991) Thermal modeling in low pressure/high temperature metamorphic belts. Tectonophysics 188:209–238
Dickinson WR (1978) Plate tectonic evolution of north Pacific rim. In: Uyeda S, Murphy RW, Kobayashi K (eds) Geodynamics of the Western Pacific. Japan Sci. Soc. Press, Tokyo, pp 1–20
Didier J (1973) Granites and their Enclaves: The bearing of enclaves on the origin of granites. Developments in Petrology 3. Elsevier, Amsterdam
Dietz RS (1961) Continent and ocean basin evolution by spreading of the ocean floor. Nature 190:854–857
Ding K (1986) Discussion of vapor–phase transport of Hg in the process of ore–formation. Ann Rpts Inst Geochem Acad Sinica Guizhou. Peoples Publ. House, Guizhou:116–117
Dixon JM (1975) Finite strain and progressive deformation in models of diapiric structures. Tectonophysics 28:89–124

Dontsova EI, Milovsky AV, Melnikov FP (1972) Fractionation of oxygen isotopes during the process of pegmatite body formation, Russia. Geochemistry 5:562–567 (in Russian with English abstract)

Dorais MJ, Whitney JA, Roden MF (1990) Origin of mafic enclaves in the Dinkey Creek pluton, central Sierra nevada, California. J. Pet. 31:853–881

Droop GTR, Clemens JD, Dalrymple DJ (2003) Processes and conditions during contact anatexis, melt escape and restite formation: the Huntly gabbro complex, NE Scotland. J. Pet. 44:995–1029

Du JE, Ma CH, Wei L (1996) Geochemical characteristics of Changkeng Au–Ag deposit, Guangdong. Geol. Guangdong 11:49–59 (in Chinese)

Du LT, et al. (1982) Colloquia on granite type uranium deposits. Nuclear Energy Publ. House, Beijing (in Chinese)

Eaton GP (1980) Regional geophysics, Cenozoic tectonics and geologic resources of the Basin and Range Province and adjoining regions. In: Newman GM, Goode HP (eds) Basin and Range Symposium. Rocky Mountain Assoc. Geol., Denver, CO pp 11–13

Eberz GW, Nicholls IA, Maas R (1990) The Nd- and Sr-isotopic composition of I-type microgranitoid enclaves and theor host rocks from the Swifts Creek pluton, southeast Australia. Chem. Geol. 85:119–134

EGCAS (Exploration Group of Chinese Academy of Sciences) (1981) Igneous activities and metamorphism of Tibet. Science Press, Beijing

Elburg MA (1996) Evidence of isotopic equilibration between microgranitoid enclaves and host granodiorite, Warburton Granodiorite, Lachlan Fold Belt, Australia. Lithos 38:1–22

Elkins-Tanton LT, Grove TL, Donnelly-Nolan J (2001) Hot shallow melting under the Cascades volcanic arc. Geology 29:631–634

Emmermann R (1975) Petrologic significance of rare earths distribution in granites. Contrib. Mineral. Pet. 52:267–283

Emmermann R, Lauterjung J (1997) The German continental deep drilling program KTB: overview and major results. J. Geophys. Res. 102, 18:179–18, 201

Emmons WH (1924) Primary downward changes in ore deposits. Trans. Am. Inst. Min. Met. Eng. 70:964–992

Emmons WH (1926) Relations of metallifeous load systems to igneous intrusions. Trans. Am. Inst. Min. Met. Eng. 74:29–70

England PC (1993) Convective removal of the thermal boundary layer of thickened continental lithosphere: a brief summary of causes and consequences with special reference to the Cenozoic tectonics of the Tibetan Plateau and surrounding regions. Tectonophysics 223:67–73

England, PC, Thompson AB (1984) Pressure–temperature–time paths of regional metamorphism. I. Heat transfer during evolution of regions of thickened continental crust. J. Pet. 25:894–928

England PC, Thompson AB (1986) Some thermal and tectonic models for crustal melting in continental collision zones. In: Coward MP, Ries AC (eds) Collision Tectonics. Geol. Soc. Lond. Spec. Publ. 19:83–94

Escuder V, Indares A, Arenas R (2000) *P-T* paths derived from garnet growth zoning in an extensional setting: An example from the Tormes gneiss dome (Iberian massif, Spain). J. Pet. 41:1489–1515

Etheridge MA, Wall VJ, Cox SF, Vernon RH (1984) High fluid pressures during regional metamorphism and deformation: Implications for mass transport and deformation mechanisms. J. Geophys. Res. 89:4344–4358

Faure G (1977) Principles of Isotope Geology. John Wiley and Sons, New York

Feng ZZ, Yang YQ, Bao ZD (1999) Lithofacies palaeogeography of the Carboniferous in South China. J. Palaeogeogr. 1:75–86 (in Chinese with English abstract)

Fialko Y, Simons M, Khazan Y (2001) Finite source modeling of magmatic unrest in Socorro, New Mexico, and Long Valley, California. Geophys. J. Int. 146:191–200

Flood RH, Vernon RH (1978) The Cooma granodiorite, Australia: an example of crustal anatesis? Geology 6:81–84

Fountain JC, Hodge DS, Shore RP (1989) Melt segregation in anatectic granites: A thermo-mechanical model. J. Volcan. Geotherm. Res. 39:279–296

Frost BR, Barnes CG, Collins WJ, Arculus RJ, Ellis DJ, Frost CD (2001) A geochemical classification for granitic rocks. J. Pet. 42:2033–2048

Frost BR, Bucher K (1994) Is water responsible for geophysical anomalies in the deep continental crust?: A petrological perspective. Tectonophysics 231:293–309
Fu YL, Ren JH, Chen TY, Liu ZG (1989) $^{40}Ar/^{39}Ar$ isotopic dating of migmatitic granite and granulite from the Jingjiang area, Fujian. Geol. Rev. 55:552–557 (in Chinese with English abstract)
Fyfe WS, Price NJ, Thompson AB (1978) Fluids in the Earth's crust. Elsevier, the Netherlands
García-Casco A, Haissen F, Castro A, El-Hmidi H, Torres-Roldán RL, Millán G (2003) Synthesis of staurolite in melting experiments of a natural metapelite: consequences for the phase relations in low-temperature pelitic migmatites. J. Pet. 44:1727–1757
Gardien V, Thompson AB, Grujic G, Ulmer P (1995) Experimental melting of biotite + plagioclase + quartz ± muscovite assemblages and implications for crustal melting. J. Geophys. Res. 100:15581–15591
Gardien V, Thompson AB, Ulmer P (2000) Melting of biotite + plagioclase + quartz gneisses: the role of H_2O in the stability of amphibole. J. Pet. 41:651–666
Gavrilenko P, Gueguen Y (1989) Percolation in the crust. Terra Res. 1:63–68
Gerbi CC, Johnson SE, Koons PO (2006) Controls on low-pressure anatexis. J. Metamorph. Geol. 24:107–118
Gerdes A, Wörner G, Henk A (2000) Post collisional granite generation and HT-LP metamorphism by radiogenic heating: the Variscan South Bohemian Batholith. J. Geol. Soc. Lond. 157:577–587
Gianelli G, Ruggieri G (2000) Contact metamorphism in the Larderello geothermal system. In: Proc. World Geotherm. Congress 2000, Japan, pp 1163–1168
Gibson RL, Bickle MJ (1994) Thermobarometric constraints on the conditions of metamorphism in the Canigoi massif, Pyrenees: implications for Hercynian geothermal gradients. J. Geol. Soc. Lond. 151:987–997
Gilbert MC, Helz RT, Popp RK, Spear FS (1982) Experimental studies of amphibole stability. In: Velben DR, Ribbe PH (eds). Amphiboles: Petrology and Experimental Phase Relations. Rev. Mineral. 9B:229–346, Min. Soc. Am.
Gilluly J (ed) (1948) The origin of granite. Geol. Soc. Am. Mem. 28:139pp
Gleeson SA, Yardley BED, Boyce AJ, Fallick AE, Munz LA (2000) From basin to basement: The movement of surface fluids into the crust. J. Geochem. Explor. 69:527–531
Gomes MEP, Neiva AMR (2005) Geochemistry of granitoids and their minerals from Rebordelo-Agrochao area, northern Portugal. Lithos 81:235–254
Graessner T, Schenk V (2001) An exposed Hercynian deep crustal section in the Sila Massif of northern Calabria: mineral chemistry, petrology and a P–T path of granulite-facies metapelitic migmatites and metabasites. J. Pet. 32:931–961
Graham IJ (1987) Petrography and origin of metasedimentary xenoliths in lavas from Tongariro Volcanic Centre. NZ. J. Geol. Geophys. 30:139–157
Grant JA (2004) Liquid compositions from low-pressure experimental melting of pelitic rock from Morton Pass, Wyoming, USA. J. Metamorph. Geol. 22:65–78
Grapes RH (1986) Melting and thermal reconstitution of pelitic xenoliths, Wehr volcano, East Eifel, Germany. J. Pet. 27:343–396
Grapes, RH (1995) Uplift and exhumation of Alpine schist, Southern Alps, New Zealand: thermobarometric constraints. NZ J. Geol. Geophys. 38:525–533
Greenfield JE, Clarke GL, Bland M, Clark DC (1996) In-situ migmatite and hybrid diatexite at Mt. Stafford, central Australia. J. Metamorph. Geol. 14:413–426
Greenfield JE, Clark GL, White RW (1998) A sequence of partial melting reactions at Mt. Stafford, central Australia. J. Metamorph. Geol. 16:363–378
GRGN (Granite Research Group of Nanling) (1989) Geology of granitoids of Nanling region and their petrogenesis and mineralization. Geol. Publ. House, Beijing (in Chinese with English abstract)
GSF (Geological Survey of Fujian Province) (1982) Characteristics of the Mesozoic metamorphic belt along the coast of Fujian. Fujian Geol. 2:29–51 (in Chinese)
Guan SC (1988) Guan Sicong's corpus geology. Geol. Publ. House, Beijing (in Chinese)
Guo LZ, Shi YS, Ma RS (1980) Geotectonic framework and crustal evolution of the southeast China. In: Scientific Papers on Geology for International Exchange (I). Science Press, Beijing, pp 109–116

Haapala I, Rämö OT (1992) Tectonic setting and origin of Proterozoic rapakivi granites of southeastern Fennoscandia. Trans. Roy. Soc. Edinburgh Earth Sci. 83:165–171

Harley SL (1989) The origins of granulites: a metamorphic perspective. Geol. Mag. 126:215–247

Harris C, Faure K, Diamon RE (1997) Oxygen and hydrogen geochemistry of S- and I-type granitoids: the Cape granite suite, South Africa. Chem. Geol. 143:95–114

Harris N, Ayres M (1998) The implications of Sr-isotope disequilibrium for rates of prograde metamorphism an melt extraction in anatectic terrains, In: Treloar PJ, O'Brien PJ (eds) What Drives Metamorphism and Metamorphic Reactions? Geol. Soc. Lond. Spec. Publ. 138:171–182

Harris NBW, Inger S (1992) Trace element modeling of pelitic-derived granites. Contrib. Mineral. Pet. 110:46–56

Harris NBW, Vance D, Ayres MW (2000) From sediment to granite: timescales of anatexis in the upper crust. Chem. Geol. 162:155–167

Harrison TM, Lovera OM, Grove M (1997) New insights into the origin of two contrasting Himalayan granite belts. Geology 25:899–902

Hecht L, Vigneresse JL, Morteani G (1997) Constraints on the zonation of the granite complexes in the Fichtelgebirge (Germany and the Czech Republic): evidence from a gravity and geochemical study. Geol. Rundsch. 86, Suppl.:S93–S109

Henry DJ, Guidotti CV, Thomson JA (2005) The Ti-saturation surface for low-to-medium pressure metapelitic biotite: implications for geothermometry and Ti-substitution mechanisms. Am. Mineral. 90:316–328

Hess HH (1962) History of ocean basins. In: Engel AEJ, James HL, Leonard BF (eds) Petrological Studies: A Volume in Honour of A.F. Buddington. Geological Society of America, New York, pp 104–115

Hilde TW (1977) Evolution of the western Pacific and its margin. Tectonophysics 38:14–165

Holmes A (1926) The origin of igneous rocks. Sci. Prog. 11:67–73

Holness MB (1993) Temperature and pressure dependence on quartz-aqueous fluid dihedral angles: the control of absorbed H_2O on the permeability of quartzites. Earth Planet. Lett. 117:363–377

Holtz F, Johannes W (1991) Genesis of peralumonous granites: I. Experimental investigation of melt compositions at 3 and 5 kb and various H_2O activities. J. Pet. 32:935–958

Holyoke CW, Rushmer T (2002) An experimental study of grain scale melt segregation mechanisms in two common crustal rock types. J. Metamorph. Geol. 20:493–512

Hosking WT (1951) Primary ore deposition in Cornwall. Roy. Geol. Soc. Cornwall Trans. 18:309–356

Hosking WT, Shrimpton GJ (eds) (1964) Present Views of Some Aspects of the Geology of Cornwall and Devon. Royal Geol. Soc. Cornwall, Penzance

Houseman GA, McKenzie DP, Molnar P (1981) Convective instability of a thickened boundary layer and its relevance for the thermal evolution of continental convergent belts, J. Geophys. Res. 86:6115–6132

Hua R (1999) On the Yanshanian metallogenic explosion in east China. J. Geosci. China 1:8–12

Huang JQ, Ren JS, Jiang CF, Zhang ZK, Qin DY (1985) The geotectonic evolution of China. Science Press, Beijing, pp 61–66

Huang WL, Wyllie PJ (1973) Melting relations of muscovite-granite to 35 kbar as a model for fusion of metamorphosed subducted oceanic sediments. Contrib. Mineral. Pet. 42:1–14

Huang WL, Wyllie PJ (1981) Phase relationships of S-type granite with H_2O to 35 kbar: Muscovite granite from Harney Peak, South Dakota. J. Geophys. Res. 86:1015–1029

Huerta AD, Royden LH, Hodges KV (1998) The thermal structure of collisional orogens as a response to accretion, erosion, and radiogenic heat production. J. Geophys. Res. 103:15, 287–15,302

Huppert HE, Sparks RSJ (1988a) The generation of granitic magmas by intrusion of basalt into continental crust. J. Pet. 29:599–624

Huppert HE, Sparks RSJ (1988b) The fluid dynamics of crustal melting by injection of basaltic sills. Trans. Roy. Soc. Edinburgh Earth Sci. 79:237–243

Hutton DWH, Dempster TJ, Brown PE, Becker SD (1990) A new mechanism of granite emplacement: intrusion in active extensional shear zones. Nature 343:452–455

Hyndman RD, Lewis TS (1999) Geophysical consequences of the Cordillera-craton thermal transition in S.W. Canada. Tectonophysics 306:397–422
IGCAGS (Institute of Geology, Chinese Academy of Geological Sciences) (1987) Metallogenic map of endogenic ore deposits of China. Cartographic Publ. House, Beijing
Ingham M (2005) Deep electrical structure of the Central Volcanic Region and Taupo Volcanic Zone, New Zealand. Earth Planet. Space 57:591–603
Ishihara S (1977) The magnetite-series and ilmenite-series granitic rocks. Mining Geol. 8:293–305
Jamieson RA, Beaumont C, Fullsack P, Lee B (1998) Barrovian regional metamorphism: where's the heat? In: Treloar PJ, O'Brien PJ (eds) What Drives Metamorphism and Metamorphic Reactions? Geol. Soc. Lond. Spec. Publ. 138:23–51
Jenkins DM, Chernosky JV Jr (1986) Phase equilibria and crystallochemical properties of Mg-chlorite. Am. Mineral. 71:924–936
Jin JF, Huang GR (1991) Uranium deposits. Nuclear Energy Publ. House, Beijing (in Chinese)
Johannes W (1984) Beginning of melting in the granite system Qz-Or-Ab-An-H_2O. Contrib. Mineral. Pet. 86:264–303
Johannes W, Holz F (1996) Petrogenesis and experimental petrology of granitic rocks. Springer-Verlag, Berlin Heidelberg
Johnson TE, Brown M, Solar GS (2003) Low-pressure subsolidus phase equilibria in the MnNCKFMASH system: constraints of regional metamorphism in western Maine, northern Appalachians. Am. Mineral. 88:624–638
Jones KA, Brown M 1989 The metamorphic evolution of the Southern Brittany metamorphic belt, France. In: Daly JS, Cliff RA, Yardley BWD (eds) Evolution of Metamorphic Belts. Geol. Soc. Lond. Spec. Publ. 43:187–202
Judson S, Richardson SM (1995) Earth—an Introduction to Geologic Change. Prentice–Hall Inc., Englewood Cliffs, NJ
Kaczor SM, Hanson GN, Peterman ZE (1988) Disequilibrium melting of granite at the contact of a basic plug: a geochemical and petrological study. J. Geol. 96:61–78
Kawate S, Arima M (1998) Petrogenesis of the Tanzawa plutonic complex, central Japan: Exposed felsic middle crust of the Izu-Bonin-Mariana arc. Isl. Arc 7:342–358
Kennedy WQ (1933) Trends of differentiation in basaltic magmas. Am. J. Sci. 25:239–256
Kerrick DM (ed) (1991) Contact Metamorphism. Rev. Mineral. 26, Min. Soc. Am.
Kifle K (1992) High temperature-low pressure, water-saturated disequilibrium melting experiments of quartzofeldspathic rock compositions. Unpubl. Ph.D. Thesis, School of Earth Sciences, Victoria University of Wellington, New Zealand
Kind R, Ni J, Zhao W, Wu J, Yuan X, Zhao L, Sandovi E, Reese C, Nabelek J, Hearn T (1996) Evidence from earthquake data for a partially molten crustal layer in southern Tibet. Science 274:1692–1694
Kistler RW, Peterman ZE (1973) Variations in Sr, Rb, K, Na and initial Sr^{87}/Sr^{86} in Mesozoic granitic rocks and intruded wall rocks in central California. Geol. Soc. Am. Bull. 84: 3489–3512
Knesel KM, Davidson JP (1996) Isotopic disequilibrium during melting and implications for crustal contamination of magmas. Geology 24:243–246
Koester E, Pawley AR, Fernandes LAD, Porcher CC, Soliani E Jr (2002) Experimental melting of cordierite gneiss and the petrogenesis of Syntranscurrent peraluminous granites in southern Brazil. J. Pet. 43:1595–1616
Kolzlovsky Y (1984) The world's deepest well. Sci. Am. 251:106–112
Kremenetsky AA (compiler) (1991) Models and cross-sections of the earth's crust based on superdeep drilling data of the USSR. Ministry of Geology of the USSR, Moscow
Kremenetsky AA, Ovchinnikov LN (1986) The Precambrian continental crust: its structure, composition and evolution as revealed by deep drilling in the U.S.S.R. Precamb. Res. 33:11–43
Lachenbruch AH, Sass JH (1978) Models of an extensing lithosphere and heat flow in the Basin and Range province. Geol. Soc. Am. Mem. 152:209–250
Lai YQ (1981) Ore bearing property of the Devonian System in Guangdong. Geol. Rev. 27(1):8–15

Lambert IB, Wyllie PJ (1972) Melting of gabbro (quartz eclogite) with excess water to 36 kilobars, with geological applications. J. Geol. 80:693–708

Lapierre H, Jahn BM, Yu YW (1997) Mesozoic felsic arc magmatism and continental olivine tholeiites in Zhejiang Province and their relationship with the tectonic activity in SE China. Tectonophysics 274:321–338

Laporte D, Rapaille C, Provost A (1997) Wetting angles, equilibrium melt geometry, and the permeability threshold of partially molten crustal protoliths. In: Bouchez JL, Hutton D, Stephens WE (eds) Granite: From Melt Segregation to Emplacement Fabrics. Kluwer Academic Publishers, the Netherlands, pp 31–54

Laube N, Springer J (1998) Crustal melting by ponding of mafic magmas: A numerical model. J. Volc. Geotherm. Res. 81:19–35

Le Bel L, Li YD, Sheng JF (1984) Granite evolution of the Xihuashan-Dangping tungsten-bearing system. Tschermaks Mineral. Petrogr. Mitt. 33:149–167

Le Breton N, Thompson AB (1988) Fluid-absent (dehydration) melting of biotite in metapelites in the early stages of crustal anatexis. Contrib. Mineral. Pet. 99:226–237

Ledru P, Courrioux G, Dallain C, Lardeaux JM, Montel JM, Vanderhaeghe O, Vitel G (2001) The Velay dome (French Massif Central): melt generation and granite emplacement during orogenic evolution. Tectonophysics 342:207–237

Le Fort P, Cuney M, Deniel C, France-Lancord C, Sheppard S, Upreti BN, Vidal P (1987) Crustal generation of Himalayan leucogranite. Tectonophysics 134:39–57

Lewis TS, Hyndman RD, Flück P (2003) Heatflow, heat generation, and crustal temperature in the northern Canadian Cordillera: thermal controls of tectonics. J. Geophys. Res. 108:2316

Li F, Pu WM (2000) Geology of Cooper deposits in the red beds of West Yunnan Province. Yunnan University Press, Kunming (in Chinese)

Li H, Liu J, Wei L (1993) Chronological study of fluid-inclusions from hydrothermal mineral deposits and its geological implications. Geol. Publ. House, Beijing (in Chinese)

Li JL (1992) A study of lithosphere structure and evolution of Southeast China continent. Chinese Sci. Tech. Press, Beijing (in Chinese)

Li SB, et al. (1975) Rare element mineralization series in endogenic ore–forming processes related to granite. In: Symposium on All–China Rare Element Geology. Science Press, Beijing (in Chinese)

Li SG (1973) Conspectus of geomechanics. Geol. Publ. House, Beijing (in Chinese)

Li SH, Unsworth MJ, Brooker JR, Wei WB, Tan HD, Jones AG (2003) Partial melt or acqueous fluid in the mid-crust of southern Tibet? Constraints from INDEPTH magnetotelluric data. Geophys. J. Int. 153:289–304

Li XW, Zhou XM (1999) Late Mesozoic subduction zone of southeastern China. Geol. J. China Univ. 5:164–169

Li YS (1982) Discussion on metallogenic epoch and origin of South China granite type uranium deposits. In: Granite Type Uranium Deposits. Nuclear Energy Publ. House, Beijing (in Chinese)

Lin XM (2005) Study of the mechanism of hypothermal deposits: an example of major quartz reef tungsten deposits in north Guangdong. Unpubl. Ph.D. Thesis, Department of Earth Sciences, Sun Yat-sen University, Guangzhou, China

Little TA, Holcombe RJ, Ilg BR (2002) Kinematics of oblique collision and ramping inferred from microstructures and strain in middle crustal rocks, central Southern Alps, N. Z. J. Struc. Geol. 24:219–239

Liu GD (1985) Some progress in studies of the crust and the upper mantle. J. Seism. 1–2:1–14 (in Chinese)

Liu GH (1984) The characteristics and genesis of the granitoid belts in southern Xizang (Tibet). In: Li GC, Mercier JL (eds) Sino-French Cooperative Investigation in Himalayas. Geol. Publ. House, Beijing, pp 239–272 (in Chinese with English abstract)

Liu YJ, Cao LM, Li ZL, Wang HN, Zhu XT, Zhang JR (1984) Element Geochemistry. Science Press, Beijing (in Chinese)

Liu YJ, et al. (1984) Element geochemistry. Science Press, Beijing (in Chinese)
Liu YM et al. (1975) Characteristics of spatial distribution of endogenic REE deposits in China. In: Symposium on All-China Rare Element Geology. Science Press, Beijing (in Chinese)
Liu YS, Gao S, Jin SY, Hu SH, Sun M, Zhao ZB, Feng JL (2001) Geochemistry of lower crustal xenoliths from Neogene Hannuoba basalt, North China craton: implications for petrogenesis and lower crustal composition. Geochim. Cosmochim. Acta 65:2589–2604
Liu YS, Li Y, Pan JH, Zhu XR (1985) Melting characteristics of granitoids in South China. Colloq. Int. Geol. Exch. 5, Geol. Publ. House, Beijing (in Chinese)
Liu ZQ (1980) Metallogenic prognosis of vein type tungsten deposits. Science Press, Beijing (in Chinese)
Loosveld RJH, Etheridge MA (1990) A model for low pressure facies metamorphism during crustal thickening. J. Metamorph. Geol. 8:257–267
López S, Castro A (2001) Determination of the fluid-absent solidus and supersolidus phase relationships of MORB-derived amphibolites in the range 4–14 kbar. Am. Mineral. 86:1396–1403
Lowenstern JB (1994) Dissolved volatile concentrations in an ore-forming magma. Geology 22:893–896
Lu HZ, Shi JX, Yu CM (1974) Determination of the temperature of ore formation and diagenesis of a deposit from Southern China. Geochemistry 3:145–156
Luth WC, Jahns RH, Tuttle OF (1964) The granite system at pressures of 4 to 10 kbars. J. Geophys. Res. 69:759–773
Lux DR, DeYoreo JJ, Guidotti CV, Decker ER (1986) Role of plutonism in low-pressure metamorphic belt formation, Nature 323:794–797
Lyell C (1868) Principles of geology, vol.II, 10th edn. John Murray, London, England
Ma HM (2006) Studies of the genesis of Shangqi composite granite body and associated migmatite, eastern Guangdong. Unpubl Ph.D. Thesis, Department of Earth Sciences, Sun Yat-sen University, Guangzhou, China
Maas R, Nicholls IA, Legg C (1977) Igneous and metamorphic enclaves in the S-type Deddick Granodiorite, Lachlan Fold Belt, SE Australia: petrographic, geochemical and Nd-Sr isotopic evidence fro crustal melting and magma mixing. J. Pet. 38:815–841
McCaffey KJW, Petford N (1997) Are granitic intrusions scale invariant? J. Geol. Soc. Lond. 154:1–4
McKee E, Rytuba JJ, Xu K (1988) Geochronology of the Xihuashan composite granite body and tungsten mineralization, Jiangxi Province, South China. Econ. Geol. 82:218–223
Mao JW, Li XF, Zhang RH, Wang YT, He Y, Zhang ZH (2005) Mantle-derived fluid-related ore-forming system. China Land Publ. House, Beijing (in Chinese)
Marsh BD (1982) On the mechanics of igneous diapirism, stoping and zone melting. Am. J. Sci. 282:808–855
Maruéjol P, Cuney M, Turpin L (1990) Magmatic and hydrothermal R.E.E. fractionation in the Xiahuashan granites (SE China). Contrib. Mineral. Pet. 104:668–680
Mattauer M (1980) Les deformations des materaux de l'ecorce terrestre Deuxieme edition. Corrigeo Janvier Hermann, Paris
Mehnert KR (1968) Migmatites and the origins of granitic rocks. Elsevier, New York
Mehnert KR, Büsch W, Schneider G (1973) Initial melting at grain boundaries of quartz and feldspar in gneisses and granulites. N. Jb. Mineral. Mh:11–183
Meng LY (1993) Granite and mineralization. Science Press, Beijing (in Chinese)
Meyerhoff AA, Taner I, Morris AEL, Agocs WB, Kamen-Kaye M, Bhat MI, Smoot NC, Choi DR (1996) Surge tectonics: a new hypothesis of global dynamics. Series: Solid Earth Sciences Library, Vol.9. Springer-Verlag
Miller CF, Watson EB, Harrison TM (1988) Perspectives of the source, segregation and transport of granitoid magmas. Trans. Roy. Soc. Edinburgh Earth Sci. 79:135–156
Miller RB, Paterson SR (1999) In defence of magmatic diapirs. J. Struct. Geol. 21:1162–1173
Minear JW, Toksöz MN (1970) Thermal regime of a downgoing slab and the new global tectonics. J. Geophys. Res. 75:1397–1419
Mo Z, Ye P, Pan W, Wang S, Zhang J, Gao B, Liu J, et al. (1980) Geology of Nanling granite. Geol. Publ. House, Beijing, pp 288–289 (in Chinese with English abstract)

Möller P, 25 co-authors (1997) Paleo- and recent fluids in the upper continental crust – Results from the German Continental deep drilling program (KTB). J. Geophys. Res. 102:18,245–18,256

Molnar P, England PC (1990) Temperatures, heat flux, and frictional stress near major thrust faults. J. Geophys. Res. 95:4833–4856

Molnar P, Houseman GA, Conrad CP (1998) Raleigh-Taylor instability and convective thinning of mechanically thickened lithosphere: effects of non-linear viscosity decreasing exponentially with depth and of horizontal shortening of the layer. Geophys. J. Int. 133:568–584

Montel JM, Marignac C, Barbey P, Pichavant M (1992) Thermobarometry and granite genesis: the Hercynian low-*P* high-*T* Velay anatectic dome (French Massif Central). J. Metamorph. Geol. 10:1–15

Montel JM, Vielzeuf D (1997) Partial melting of metagreywackes, Part II. Compositions of minerals and melts. Contrib. Mineral. Pet. 128:176–196

Nabelek PI, Liu M (1999) Leucogranites in the Black Hills, South Dakota: the consequence of shear heating during continental collision. Geology 27:523–526

Nabelek PI, Liu M (2004) Petrologic and thermal constraints on the origin of leucogranites in collisional orogens. Trans. Roy. Soc. Edinburgh Earth Sci. 95:73–85

Nabelek PI, Liu M, Sirbescu M-L (2001) Thermo-rheological, shear heating model for leucogranite generation, metamorphism, and deformation during the Proterozoic Trans-Hudson orogeny, Black Hills, South Dakota. Tectonophysics 342:371–388

Nair R, Chacko T (2002) Fluid-absent melting of high-grade semi-pelites: *P-T* constraints on orthopyroxene formation and implications for granulite genesis. J. Pet. 43:2121–2142

Nanling Tungsten Deposits Research Group (1985) Tungsten deposits in South China. Metall. Indust. Publ. House, Beijing (in Chinese)

Nelson KD, et al. (1996) Partially molten middle crust beneath southern Tibet: synthesis of project INDEPTH results. Science 274:1684–1688

Niemi AN, Courtney TH (1983) Settling in solid-liquid systems with specific application to liquid phase sintering. Acta. Metall. 9:1393–1401

Norlander BN, Whitney DL, Tessier C Vanderhaeghe O (2002) Partial melting and decompression of the Thor-Odin dome, Shuswap metamorphic core complex, Canadian Cordillera. Lithos 61:103–125

Northrup CJ, Royden LH, Burchfield BC (1995) Motion of the Pacific plate to Eurasia and its potential relation to Cenozoic extension along the eastern margin of Eurasia. Geology 23:719–722

O'Connor JT (1965) A classification of quartz-rich igneous rock based on feldspar ratios. US Geol. Surv. Prof. Paper 525B:B79–B84

Ogawa Y, Bibby HM, Caldwell TG, Takakura S, Matsushima N, Bennie SL, Tosha T, et al. (1999) Wide-band magnetotelluric measurements across the Taupo volcanic zone: preliminary results. Geophys. Res. Lett. 26:3673–3676

O'Neil JR, Chappell BW (1977) Oxygen and hydrogen isotope relations in the Berridale batholith. J. Geol. Soc. Lond. 133:559–571

Owens TJ, Zandt G (1997) Implications of crustal property variations fro models of Tibetan plateau evolution. Nature 387:37–43

Oxburgh ER, Turcotte DL (1970) Thermal structure of island arcs. Geol. Soc. Am. Bull. 81:1665–1688

Park CF, MacDiarmid RA (1975) Ore deposits, 3rd ed. W.H. Freeman and Co., San Francisco

Paterson SR, Schmidt KL (1999) Is there a close spatial relationship between faults and plutons? J. Struct. Geol. 21:1131–1142

Patiño-Douce AE (1995) Experimental generation of hybrid silicic melts by reaction of high-Al basalt with metamorphic rocks. J. Geophys. Res. 100:15623–15639

Patiño-Douce AE (1997) Generation of metaluminous A-type granites by low-pressure melting of calc-alkaline granitoids. Geology 25:743–746

Patiño-Douce AE, Beard JS (1995) Dehydration melting of biotite gneiss and quartz amphibolite from 3 to 15 kbar. J. Pet. 36:707–738

Patiño-Douce AE, Beard JS (1996) Effects of *P*, $f(O_2)$ and Mg/Fe ratio on dehydration melting of model greywacke. J. Pet. 37:999–1024

Patiño-Douce AE, Harris N (1998) Experimental constraints on Himalayan anatexis. J. Pet. 39:689–710
Patiño-Douce AE, Humphreys ED, Johnston AD (1990) Anatexis and metamorphism in tectonically thickened continental crust exemplified by the Sevier hinterland, western North America. Earth Planet. Sci. Lett. 97:290–315
Patiño-Douce AE, Johnston AD (1991) Phase equilibria and melt productivity in the pelitic system: implications for the origin of peraluminous granitoids and aluminous granulites. Contrib. Mineral. Pet. 107:202–218
Pattison DRM (1992) Stability of andalusite, sillimanite and the Al_2SiO_5 triple point: constraints from the Ballachulish aureole, Scotland. J. Geol. 100:423–446
Pattison DRM, Chacko T, Farquhar J, McFarlane CRM (2003) Temperatures of granulite-facies metamorphism: constraints from experimental phase equilibria and thermobarometry corrected from retrograde exchange. J. Pet. 44:867–900
Peacock SM, Rushmer T, Thompson AB (1994) Partial melting of subducting oceanic crust. Earth Planet. Sci. Lett. 121:227–244
Peng Z (2006) Inhomogeneous mafic microgranular enclaves in granites and their genesis. Unpubl. Ph.D. Thesis, Department of Earth Sciences, Sun Yat-sen University, Guangzhou, China (in Chinese with English abstract)
Petford N, Clemens JD (2000) Granites are not diapiric! Geol. Today (Sep–Oct):180–184
Petford N, Crunden AR, McCaffrey KJW, Vigneresse J.L (2000) Granite magma formation, transport and emplacement in the Earth's crust. Nature 408:669–673
Petford N, Gallagher K (2001) Partial melting of mafic (amphibolitic) lower crust by periodic influx of basalt magma. Earth Planet. Sci. Lett. 193:483–499
Petrini K, Podladchikov Y (2000) Lithospheric pressure-time relationship in compressive orogens of thickened crust. J. Metamorph. Geol. 18:67–78
Pickering J, Johnston A (1998) Fluid-absent melting behavior of a two-mica metapelite: experimental constraints on the origin of the Black Hills granite. J. Pet. 39:1787–1804
Pitcher WS (1993) The nature and origin of granite. Chapman & Hall, London
Piwinskii WS (1968) Experimental studies of igneous rock series, central Sierra Nevada batholith, California. J. Geol. 76:548–570
Piwinskii WS (1973a) Experimental studies of igneous rock series, central Sierra Nevada batholith, California: Part II. N. Jb. Mineral. Mh:193–215
Piwinskii WS (1973b) Experimental studies of granitoids from the central and southern Coast Ranges, California. Tschermaks Mineral. Petrog. Mitt. 20:107–130
Piwinskii WS, Wyllie PJ (1968) Experimental studies of igneous rock series: a zoned pluton in the Wallowa batholith, Oregon. J. Geol. 76:205–234
Pons J, Barbey P, Dupuis D, Léger JM (1995) Mechanism of pluton emplacement and structural evolution of a 2.1 Ga juvenile continental crust: the Birimian of southwest Niger. Precamb. Res. 70:281–301
Powell R, White RW (2001) Granulite facies processes: constraints from partial melting equilibria calculated. Abs. Eleventh Ann V.M. Goldschmidt Conference 2001; 3043.pdf
Prodehl C, Giese P (1990) Seismic investigation around the EGT in central Europe. In: Freeman R, et al. (eds) The European Geotraverse: Integrative Studies, Europ. Science Foundation, Strasbourg, pp 77–98
Puxeddu M (1984) Structure and Late Cenozoic evolution of the upper lithosphere in southwest Tuscany (Italy). Tectonophysics 101:357–382
Qiu Y (2006) Mechanism for the Cenozoic tecto-geomorphologic evolution of the SE China continental margin. Unpubl Ph.D. Thesis, Department of Earth Sciences, Sun Yat-sen University, Guangzhou, China
Qiu YM, Groves DI, McNaughton RJ, Phillips GN (2002) Nature, age and tectonic setting of granitoid-hosted, orogenic gold deposits of the Jiaodong Peninsula, eastern North China craton, China. Mineralium Deposita 37:283–305
Railsback LB (2003) An earth scientist's periodic table of the elements and ions. Geology 31:737–740
Ramsey JG (1989) Emplacement mechanics of a granite diapir: The Chidamora batholith, Zimbabwe. J. Struct. Geol. 11:191–209

Rapp RP, Watson EB (1995) Dehydration melting of metabasalt at 8–32 kbar: implications for continental growth and crust-mantle recycling. J. Pet. 36:891–931
Rapp RP, Watson EB, Miller CF (1991) Partial melting of amphibolite/eclogite and the origin of Archean trondhjemites and tonalite. Precamb. Res. 51:1–25
Read HH (1948) Granites and granites. Geol. Soc. Am. Mem. 28:1–19
Read HH (1957) The Granite controversy. Thomas Murby, London
Ribe NM (1987) Theory of melt segregation—a review. J. Volc. Geotherm. Res. 33:241–253
Richards SW, Collins WJ (2002) The Cooma metamorphic complex, a low-*P*, high-*T* (*LPHT*) regional aureole beneath the Murrumbidgee batholith. J. Metamorph. Geol. 20:119–134
Robertson JK, Wyllie PJ (1971) Experimental studies on rocks from the Deboullie Stock, northern Maine, including melting relations in a water-deficient environment. J. Geol. 79:549–571
Rosler HJ, Lange H (1972) Geochemical tables. Elsevier, Amsterdam
Rubie DC (1986) The catalysis of mineral reactions by water and restrictions on the presence of aqueous fluid during metamorphism. Mineral. Mag. 50:399–415
Rubie DC, Brearley AJ (1987) Metastable melting during the breakdown of muscovite + quartz at 1 kbar. Bull. Mineral. 110:533–549
Rudnick RL, Fountain DM (1995) Nature and composition of the continental crust: a lower crustal perspective. Rev. Geophys. 33:267–309
Rudnick RL, Gao S (2003) Composition of the continental crust. In: Treatise on Geochemistry 3, Elsevier Ltd, Amsterdam, pp 1–64
Rudnick RL, Presper T (1990) Geochemistry of intermediate-to high-pressure granulites. In: Vielzeuf D, Vidal PH (eds) Granulites and Crustal Evolution. NATO ASII Ser C, 311:523–550, Kluwer Acad., Norwell, MA
Rushmer T (1991) Partial melting of two amphibolites: contrasting experimental results under fluid-absent conditions. Contrib. Mineral. Pet. 107:41–59
Sallet R, Moritz R, Fontignie D (2000) Fluorite $^{87}Sr/^{86}Sr$ and REE constraints on fluid-melt relations, crystallization time span and bulk Dsr of evolved high-silica granites, Tabulerio granites, Santa Catarina, Brazil. Chem. Geol. 164:81–92
Sandiford M, Hand M (1998) Australian Proterozoic high-temperature, low-pressure metamorphism in the conductive limit. In: Treloar P, O'Brien P (eds) What Drives Metamorphism and Metamorphic Reactions? Geol. Soc. Lond. Spec. Publ. 138:103–114
Sandiford M, Hand M, McLaren S (1998) High geothermal gradient metamorphism during thermal subsidence. Earth Planet. Sci. Lett. 163:149–165
Sandiford M, Powell R (1986) Deep crustal metamorphism during continental extension: modern and ancient examples. Earth Planet Sci. Lett. 79:151–158
Sawyer EW (1998) Formation and evolution of granites magmas during crustal reworking: the significance of diatexites. J. Pet. 39:1147–1167
Schilling F, Partzsch G (2001) Quantifying partial melt fraction in the crust beneath the central Andes and the Tibetan Plateau. Phys. Chem. Earth 26:239–246
Schilling FR, Partzsch GM, Braase H, Schwarz G (1997) Partial melting below the magmatic arc in the central Andes deduced from geoelectromagnetic field experiments and laboratory data. Phys. Earth Planet. Inter. 103:17–32
Schmidt MW, Poli S (2004) Magmatic epidote. In: Liebscher A, Franz G (eds) Epidotes. Rev. Mineral. Geochem. 56:399–430
Schmidt MW, Thompson AB (1996) Epidote in calc-alkaline magmas: An experimental study of stability, phase relationships, and the role of epidote in magmatic evolution. Am. Mineral. 81:462–474
Schmitz M, Heinsohn W-D, Schilling FR (1997) Seismic, gravity and petrological evidence for partial melt beaneath the thickened Central Andean crust (21–23°S). Tectonophysics 270:313–326
Schödlbauer S, Hecht L, Höhndorf A, Morteani G (1997) Enclaves in the S-type granites of the Kösseine massif (Fichtelgebirge, Germany): implications for the origin of granites. Geol. Rundsch. 86, Suppl.: S125–140

SEGCA (Scientific Exploration Group, Chinese Academy) (1981) Magmatism and metamorphism of Tibet. Science Press, Beijing (in Chinese)
SEGCA (Scientific Exploration Group, Chinese Academy) (1982) Geological structures of Qinghai-Xizhang Plateau. Science Press, Beijing (in Chinese)
Sen C, Dunn T (1994) Dehydration melting of a basaltic composition amphibolite at 1.5 and 2.0 GPa: implications for the origin of adakites. Contrib. Mineral. Pet. 117:394–409
Sengor AMC (1984) The Cimmeride orogenic system and tectonic Eurasia. Geol. Soc. Am. Special paper 195
Shaw HR (1980) The fracture mechanism of magma transport from the mantle to the surface. In: Hargraves RB (ed) Physics of Magmatic Processes. Princeton University Press, Princeton, NJ, pp 201–264
Sherburn S, Bannister S, Bibby H (2003) Seismic velocity structure of the central Taupo Volcanic Zone, New Zealand, from local earthquake tomography. J. Volcan. Geotherm. Res. 122:69–88
Shibata K, Matsumoto T, Yanagi T, Hamamoto R (1978) Isotopic ages and stratigraphic control of Mesozoic igneous rocks in Japan. In: AAPG Studies in Geology 6, Contribution to the Geology Timescales, pp 143–164
Sieger R, van der Laan L, Wyllie PJ (1993) Experimental interaction of granitic and basaltic magmas and implications for mafic enclaves. J. Pet. 34:491–517
Silva MMVG, Nevia AMR, Whitehouse MJ (2000) Geochemistry of enclaves and host granites from the Neslas area, central Portugal. Lithos 50:153–170
Singh J, Johannes W (1996) Dehydration melting of tonalities. Part II. Composition of melts and solids. Contrib. Mineral. Pet. 125:26–44
Sisson TW, Ratajski K, Hankins WB, Glazner AF (2005) Voluminous granitic magmas from common basaltic sources. Contrib. Mineral. Pet. 148:635–661
Skjerlie KP, Johnston AD (1993) Fluid-absent melting behaviour of an F-rich tonalitic gneiss at mid-crustal pressures: implications for the generation of anorogenic granites. J. Pet. 34:785–815
Skjerlie KP, Johnston AD (1996) Vapour-absent melting from 10 to 20 kbar of crustal rocks that contain multiple hydrous phases: implications for anatexis in the deep to very deep continental crust and active continental margins. J. Pet. 37:661–691
Skjerlie KP, Patiño-Douce AE (1995) Anatexis of interlayered amphibolite and pelite at 10 kbar: effect of diffusion of major components on phase relations and melt fraction. Contrib. Mineral. Pet. 122:62–78
Skjerlie KP, Patiño-Douce AE, Johnston AD (1993) Fluid-absent melting of a layered crustal protolith: implications for the generation of anatectic granites. Contrib. Mineral. Pet. 114:365–378
Slemmons DB, Engdahl ER, Zoback MD, Blackwell DD (1991) Neotectonics of North America. Geol. Soc. Am. Decade Map 1:498pp
Smith RB, Nagy WC, Julander KAS, Viveiros JJ, Barker CA, Gants DJ (1989) Geophysical and tectonic framework of the eastern Basin and Range-Colorado Plateau-Rocky Mountain transition. In: Pakiser LC, Mooney WD (eds) Geophysical Framework of the Continental United States. Geol. Soc. Am. Mem. 172:205–233
Soengkono S (1995) A magnetic model for deep plutonic bodies beneath the central Taupo Volcanic Zone, North Island, N. Z. J. Volcan. Geotherm. Res. 68:193–207
Solar GS, Pressley RA, Brown M, Tucker RD (1998) Granite ascent in convergent orogenic belts: testing a model. Geology 26:711–714
Spear FS (1981) An experimental study of hornblende stability and compositional variability in amphibolite. Am. J. Sci. 281:697–734
Spear FS, Kohn MJ, Cheney JT (1999) P-T paths from anatectic pelites. Contrib. Mineral. Pet. 134:17–32
Spera FJ (1980) Aspects of magma transport. In: Hargraves RB (ed) Physics of Magmatic Processes. Princeton University Press, Princeton, NJ, pp 263–323
Springer M (1999) Interpretation of heat-flow density in the central Andes. Tectonophysics 306:377–395

Springer M, Förster A (1998) Heat-flow density across the central Andean subduction zone. Tectonophysics 291:123–129
Springer W, Seck A (1997) Partial fusion of basic granulite at 5 to 15 kbar: implications for the origin of TTG magmas. Contrib. Mineral. Pet. 127:30–45
Spurr JE (1907) A theory of ore deposition. Econ. Geol. 2:781–785
Steiner A (1958) Petrogenetic implications of the 1954 Ngauruhoe lava and its xenoliths. N. Z. J. Geol. Geophys. 1:325–363
Stephens G, Clemens JD, Droop GTR (1997) Melt production during granulite facies anatexis: experimental data from 'primitive' metasedimentary protoliths. Contrib. Mineral. Pet. 128:352–370
Stephens WE, Halliday AN (1979) Compositional variation in the Galloway plutons. In: Atherton IP, Tarney J (eds) Origin of Granite Batholiths–Geochemical Evidence: 9–11
Stern CR, Wyllie PJ (1975) Water-saturated and -undersaturated melting relations of a granite to 35 kilobars. Earth Planet. Sci. Lett. 18:163–167
Stern TA (1985) A back arc basin formed within continental lithosphere: the Central Volcanic Region of New Zealand. Tectonophysics 112:385–409
Stern TA (1987) Asymmetric back-arc spreading, heat flux and structure associated with the Central Volcanic Region of New Zealand. Earth Planet. Sci. Lett. 85:265–276
Stern TA, Davey FJ (1987) A seismic investigation of crustal and upper mantle structure within the Central Volcanic Region of New Zealand. N. Z. J. Geol. Geophys. 30:217–231
Stober I, Bucher K (1999) Origin of salinity of deep groundwater in crystalline rocks. Terra Nov. 11:181–185
Stober I, Bucher K (2004) Fluid-sinks in Earth's crust. Geofluids 4:143–151
Streckeisen AL (1976) Classification and nomenclature of igneous rocks (final report of enquiry). N. Jb. Mineral. Abh. 134:1–4
Streckeisen AL, Le Maitre RW (1979) A chemical approximation to the modal QAPF classification of the igneous rocks. N. Jb. Mineral. Abh. 136:169–206
Stüwe K, Sandiford M (1994) Contribution of deviatoric stresses to metamorphic P-T paths: An example appropriate to low-P, high-T metamorphism. J. Metamorph. Geol. 12:445–454
Stüwe K, Sandiford M, Powell R (1993) Episodic metamorphism and deformation in low-pressure, high-temperature terranes. Geology 21:829–832
Sun G, Shi M, Zhang HL, Hu XW (1989) Petrology, geochemistry and mineralization of the Dajishan granite. In: Collection of Scientific Research Reports on Geology and Mineral Resources of Nanling, China. Univ. Geosci. Press, Wuhan (in Chinese with English abstract)
Suyehiro K, Takahashi N, Ariie Y, Yokoi Y, Hino R, Shinohara M, Kanazawa T, et al (1996) Continental crust, crustal underplating, and low-Q upper mantle beneath an oceanic island arc. Science 272:390–392
Taylor Jr HP (1974) The application of oxygen and hydrogen isotope studies to problems of hydrothermal alteration and ore deposition. Econ. Geol. 69:843–883
Taylor Jr HP (1978) Oxygen and hydrogen isotope studies of plutonic granitic rocks. Earth Planet. Sci. Lett. 38:177–210
Teyssier C, Whitney DL (2002) Gneiss domes and orogeny. Geology 30:1139–1142
Thompson AB (1982) Dehydration melting of pelitic rocks and the generation of H_2O-undersaturated granitic liquids. Am. J. Sci. 282:1567–1595
Thompson AB (1999) Some time–space relationships for crustal melting and granite intrusion at various depths. In: Castro A, Fernandez C, Vigneresse JL (eds) Understanding Granites: Integrating New and Classical Techniques. Geol. Soc. Lond. Spec. Publ. 168:7–25
Thompson AB (2001a) Clockwise *P–T* paths for crustal melting and H_2O recycling in granite source regions and migmatite terrains. Lithos 56:33–34
Thompson AB (2001b) P–T paths, H_2O recycling, and depth of crystallisation for crustal melts. Phys. Chem. Earth 26:231–237

Thompson AB, Algor JR (1977) Model systems for anatexis in pelitic rocks: I. Theory of melting reactions in the systems $KAlO_2$-$NaAlO_2$-Al_2O_3-SiO_2-H_2O. Contrib. Mineral. Pet. 63:247–269
Thompson AB, Connolly AD (1995) Melting of the continental crust: some thermal and petrological constraints on anatexis in continental collision zones and other tectonic settings. J. Geophys. Res. 100:15,565–15,579
Thompson AB, England PC (1984) Pressure-temperature-time paths of regional metamorphism, II. Their influence and interpretation using mineral assemblages in metamorphic rocks. J. Pet. 25:929–955
Thompson AB, Tracy RJ (1979) Model systems for anatexis of pelitic rocks. II. Facies series melting and reactions in the system CaO-$KAlO_2$-$NaAlO_2$-Al_2O_3-SiO_2-H_2O. Contrib. Mineral. Pet. 70:429–438
Tod EW, Maas R, Nicholls IA (2001) Geochemical investigations of microgranitoid enclaves in the S-type Cowra Granodiorite, Lachlan Fold belt, SE Australia. Lithos 40:321–339
Tommasini S, Davies GR (1997) Isotope disequilibrium during anatexis: a case study of contact melting, Sierra Nevada, California. Earth Planet. Sci. Lett. 148:273–285
Tracy RJ (1978) High grade metamorphic reactions and partial melting in pelitic schist, west-central Massachussetts. Am. J. Sci. 278:150–178
Turcotte DL, Schubert B (1982) Geodynamics–applications of continuum physics to geological problems. John Wiley & Sons, New York, pp 188–197
Turner P (1980) Continental red beds. Developments in sedimentology. Elsevier, Amsterdam
Tuttle OF, Bowen NL (1958) Origin of granite in the light of experimental studies in the system $NaAlSi_3O_8$–$KAlSi_3O_8$–SiO_2–H_2O. Geol. Soc. Am. Mem. 74:1–153
Unsworth MJ, Jones AG, Wei W, Marquis G, Gokarn SG, Spratt JE & INDEPTH-MT team (2005) Crustal rheology of the Himalaya from magnetotelluric data. Nature 438:78–81
Upton P, Koons PO, Chamberlain CP (1995) Penetration of deformation-driven meteoric water into ductile rocks: Isotopic and model observations from the Southern Alps, New Zealand. N. Z. J. Geol. Geophys. 38:535–543
Vanderhaeghe O (2001) Melt segregation, pervasive melt migration and magma mobility in the continental crust: the structural record from pores to orogens. Phys. Chem. Earth 26:213–223
Vanderhaeghe O, Burg J-P, Tessier C (1999) Exhumation of migmatites in two collapsed orogens: Canadian Cordillera and French Variscides. In: Ring U, Lister MT, Willett SD (eds) Exhumation Processes: Normal Faulting, Ductile Flow and Erosion. Geol. Soc. Lond. Spec. Publ. 154:181–204
Van der Molen I, Paterson MS (1979) Experimental deformation of partially melted granite. Contrib. Mineral. Pet. 70:299–318
Van Houten FB (1973) Origin of red beds. A review 1961–1972. Annu. Rev. Earth Planet. Sci. 1:39–61
Vernon RH (1984) Microgranular enclaves in granites—globules of hybrid magma quenched in a plutonic environment. Nature 309:438–439
Vernon RH, Etheridge MA, Wall VJ (1988) Shape and microstructure of microgranitoid enclaves: indicators of magma mingling and flow. Lithos 22:1–11
Vielzeuf D, Holloway JR (1988) Experimental determination of the fluid-absent melting relations in the pelitic system: Consequences for crustal differentiation. Contrib. Mineral. Pet. 98:257–276
Vielzeuf D, Montel JM (1994) Partial melting of metagreywackes. Part l. Fluid-absent experiments and phase relationships. Contrib. Mineral. Pet. 117:375–393
Vielzeuf D, Schmidt MW (2001) Melting relations in hydrous systems revisited: applications to metapelites, metagreywackes and metabasalts. Contrib. Mineral. Pet. 141:251–267
Vigneresse JL (1990) Use and misuse of geophysical data to determine the shape at depth of granite intrusions. Geol. J. 25:249–260
Villaseca C, Downes H, Pin C, Barbero L (1999) Nature and composition of the lower continental crust in central Spain and the granulite–granite linkage: inferences from granulitic xenoliths. J. Pet. 40:1465–1496

Wada H, Harayama S, Yamaguchi Y (2004) Mafic enclaves densely concentrated in the upper part of a vertically zoned felsic magma chamber: the Kurobegawa granitic pluton, Hida Mountain Range, central Japan. Geol. Soc. Am. Bull. 116:788–780

Walker TR (1976) Diagenetic origin of continental red beds. In: Falke H (ed) The Continental Permian in Central, West, and South Europe. D. Reidel Pub Com Dordrecht, the Netherlands pp 240–282

Wall VJ, Clemens JD, Clarke DB (1987) Models for granitoid and source compositions. J. Geol. 95:731–749

Walton M (1960) Granite problems. Science 131:635–645

Walther JV, Orville PM (1982) Volatile production and transport in regional metamorphism. Contrib. Mineral. Pet. 79:252–257

Wang CZ (1985) Geological characteristics of granite type uranium deposits of China. Nuclear Energy Publ. House, Beijing (in Chinese)

Wang LK, Huang ZL (2000) Experiments of the liquid immiscibility of Li-F granite. Sci. Press, Beijing (in Chinese)

Wang NW (1983) Development of the Mesozoic formations in the lakes region, north Tibet and its plate tectonic implications. In: CGQXP Editorial Committee Contribution to the Geology of the Qinghai-Xizang (Tibet) Plateau 8: Geol. Publ. House, Beijing (in Chinese with English abstract)

Wang NW (1984) Qingzangindia paleocontinent and its welding to Cathaysia. In: Li GC, Mercier JL (eds) Sino-French cooperative investigation in Himalayas. Geol. Publ. House, Beijing, pp 39–62 (in Chinese with English abstract)

Wang R, Shen W, Xu S, Xu K, Lai M (1998) Oxygen isotope study of the Suzhou granite. Chin. J. Geochem. 17:128–134

Wang SY et al. (1990) The Mangchang stratabound type tin polymetallic deposit. China Univ. Geosci. Press, Wuhan (in Chinese)

Wang YM, Zhu JA, Yu QH (1988) Geology of lead zinc deposits in Hunan. Geol. Publ. House, Beijing (in Chinese)

Wang YT (1982) South China granite and uranium mineralization. In: Papers on Granite-type Uranium Deposits. Nuclear Energy Publ. House, Beijing (in Chinese)

Wannamaker PE (1994) Fluids in the Earth's crust: Electromagnetic inferences on existence and distribution. U.S. Geol. Surv. Open File Rep. 94:162–257

Warner M (2004) Free water and seismic reflectivity in the lower continental crust. J. Geophys. Eng. 1:88–101

Watanabe T, Langseth MG, Anderson RN (1977) Heat flow in back-arc basins of the western Pacific. In: Taiwani M, Pittman WCIII (eds) Island Arcs, Deep Sea Trenches and Back-Arc Basins. Maurice Ewing Ser 1, Am. Geophys. Union, Washington DC, pp 137–161

Watson EB, Brenan JM (1987) Fluids in the lithosphere, 1. Experimentally-determining wetting characteristics of CO_2-H_2O fluids and their implications fro fluid transport, host-rock physical properties, and fluid inclusion formation. Earth Planet. Sci. Lett. 85:497–515

Watson S, McKenzie D (1991) Melt generation by plumes: A study of Hawaiian volcanism. J. Pet. 32:501–537

Weibe RA, Smith D, Sturm M, King EM, Seckler MS (1997) Enclaves in the Cadillac Mountain Granite (coastal Maine): samples of hybrid magma from the base of the chamber. J. Pet. 38:393–423

Weibel R, Grobety B (1999) Pseudomorphous transformation of goethite needles into hematite in sediments of the Triassic Skagerrak Formation, Denmark. Clay Miner. 34:657–660

Weinberg RF, Podladchikov YV (1994) Diapiric ascent of magmas through power law crusts and mantle. J. Geophys. Res. 99:9543–9559

Wen GZ, Guo LX, Ding CG (1984) A preliminary study on mineralizing zonation of Xianghualing Sn–Pb–Zn polymetal deposit, Linwu County, Hunan Province. Geol. Hunan (in Chinese)

Weng SJ, Kong QS, Huang H (1987) Late Mesozoic volcanism of Zhejiang, Fujiang, Jiangxi and Guangdong. Geol. Publ. House, Beijing (in Chinese with English abstract)

White RW, Pomroy NE, Powell R (2005) An in situ metatexite–diatexite transition in upper amphibolite facies rocks from Broken Hill, Australia. J. Metamorph. Geol. 23:579–602

White RW, Powell R, Clarke GL (2003) Prograde metamorphic assemblage evolution during partial melting of metasedimentary rocks at low pressures: migmatites from Mt. Stafford, central Australia. J. Pet. 44:1937–1960

Wickham KT, Taylor HP Jr (1985) Stable isotopic evidence for large-scale seawater infiltration in a regional metamorphic terrane: the Trois Seigneurs Massif, Pyrenees, France. Contrib. Mineral. Pet. 91:122–137

Wickham SM (1987) Crustal anatexis and granite petrogenesis during low-pressure regional metamorphism: the Trois Seigneurs Massif, Pyrenees, France. J. Pet. 28:127–169

Wickham SM, Oxburgh ER (1985) Continental rifts as a setting for regional metamorphism. Nature 318:330–333

Wilson CJN, Houghton BF, McWilliams MO, Lanphere MA, Weaver SD, Briggs RM (1995) Volcanic and structural evolution of the Taupo Volcanic Zone, New Zealand: a review. J. Volcan. Geotherm. Res. 68:1–28

Wimmenauer W, Bryhni I (2002) Towards a unified nomenclature of metamorphic petrology: 6. Migmatites and related rocks. IUGS Subcommission for the Nomenclature of Metamorphic Rocks (SCMR). 18/08/2004 SCMR_MIG.doc

Winkler HGF (1976) Petrogenesis of metamorphic rocks. Springer-Verlag, New York

Winther KT, Newton RC (1991) Experimental melting of hydrous low-K tholeiite: evidence on the origin of Archean cratons. Bull. Geol. Soc. Den. 39:213–228

Wolf MB, Wyllie PJ (1994) Dehydration melting of amphibolite at 10 kbar: the effects of temperature and time. Contrib. Mineral. Pet. 115:369–383

Wood BJ, Walther JV (1986) Fluid-flow during metamorphism and its implications for fluid-rock ratios: In: Walther JV, Wood BJ (eds) Fluid-Rock Interactions during Metamorphism. Springer, New York Berlin Heidelberg Tokyo, pp 89–108

Wu P, Yang ZQ (1978) The Cretaceous–Early Tertiary lithofacies, paleogeography and ore–bearing characteristics in the Central SE China. Geol. Publ. House, Beijing (in Chinese)

Wu YL, et al. (1987) Geology of the Xihuashan tungsten ore field. Geol. Publ. House, Beijing (in Chinese)

Wu YL, Mei YW, Liu CP (1987) Geology of the Xihuashan tungsten ore field. Geol. Publ. House, Beijing (in Chinese)

Wyllie PJ (1977) Crustal anatexis. Tectonophysics 43:41–71

Xia HY, Liang SY (1991) Genetic series of granite related to tungsten, tin and rare metals in South China. Science Press, Beijing (in Chinese)

Xiang MM (1989) The geochemistry of Hg in country rocks of Guizhou Province. In: Geology of Guizhou Mercury Deposits. Geol. Publ. House, Beijing (in Chinese)

Xiao XC, Li TD, Li GC, et al. (1988) An introduction to tectonic evolution of lithosphere in Himalaya Mountains. Geol. Publ. House, Beijing (in Chinese)

Xie J, Gok R, Ni J, Aoki Y (2004) Lateral variations of crustal seismic attenuation along the INDEPTH profiles in Tibet from *Lg Q* inversion. J. Geophys. Res. 109:1–17

Xu GM (1987) Basining period of the Mesocenozoic continental basins of China. In: Guan (ed) Development, Sedimentation and Oil/Gas-bearing Rocks of the Meso-Cenozoic Continental Basins of China. Pet. Ind. Press, Beijing, pp 19–32

Xu K, et al. (1983) On genetic series of granites, as exemplified by the Mesozoic granites of South China. Acta Geologica. Sinica 2 (in Chinese with English abstract)

Xue H, et al. (1992) Compositional gradient in Mesozoic volcanic rocks in southeastern China continent. In: Lu Z, Tao K (eds) Colloqui of Geology and Mineral Resources of Southeastern Coastal Volcanic Rocks of China. Geol. Publ. House, Beijing (in Chinese with English abstract)

Yan JP, et al. (1989) Geology of mercury deposits of Guizhou Province, China. Geol. Publ. House, Beijing (in Chinese)

Yan MZ, Wu XL, Li CY (1980) Metallogenic systems of tungsten in southeastern China and their mineralization characteristics. In: Ishihara S, Takenouchi S (eds) Granite Magmatism and Related Mineralization. Min. Geol. Jpn., Special Issue 8:215–221. Soc. Min. Geol. Japan, Tokyo

Yan Z (1985) Granites of Shanxi, Xian. Xian Jiaotong Univ. Press, Xian (in Chinese)

Yang JH, Wu FU, Chung SL, Wilde SA, Chu MF (2004) Multiple sources for the origin of granites: Geochemical and Nd/Sr isotopic evidence from the Gudaoling granite and its mafic enclaves, northeast China. Geochim. Cosmochim. Acta 68:4469–4483

Yang ZY, Chen YQ, Wang HZ (1989) The geology of China. China Univ. Geosci. Press, Beijing, pp 249–256

Yardley BWD (1981) Effect of cooling on the water content and mechanical behaviour of metamorphosed rocks. Geology 9:405–408

Yardley BWD, Valley JW (1994) How wet is the Earth's crust? Nature 371:205–206

Yardley BWD, Valley JW (1997) The petrologic case for a dry lower crust. J. Geophys. Res. 102:12,173–12,185

Yin JH, Ruan HH, Xie JY, Feng ZZ, Yu YW (1991) Mesozoic volcanic cycles, volcanic structures and its ore-controlling significance of SE continent of China. Geol. Publ. House, Beijing pp 3–132

Yin JS, Ruan HH, Xia JY, Feng ZZ, Yu YW (1991) Mesozoic volcanic cycles, volcanic structures and its ore-controlling significance of southeast continent of China. Geol. Publ. House, Beijing (in Chinese with English abstract)

Yin ZX, Lai MH, Xiong SB, Liu HB, Teng JW, Kong XR (1995) Crustal structure and velocity distribution from deep seismic sounding along the profile of Lianxian-Gangkou in south China. Chinese J. Geophys. 42:383–392 (in Chinese with English abstract)

Young DA (2003) Mind over magma—the story of igneous petrology. Princeton University Press, Princeton Oxford

Yu CW, Luo TC, Bao ZY, Hu YZ, Liang YC, Wei XJ (1987) Regional geochemistry of Nanling Region. Geol. Publ. House, Beijing pp 157–226 (in Chinese with English abstract)

Yuan X, Sobolev SV, Kind R, Oncken O, Bock G, Asch G, Schurr B, Graeber F, et al. (2000) Subduction and collision processes in the central Andes constrained by converted seismic phases. Nature 408:958–961

Zandt G, Ammon CJ (1995) Continental crust composition constrained by measurements of crustal Poisson's ratio. Nature 374:152–154

Zeitler PK, et al. (2001) Crustal reworking at Nanga Parbat, Pakistan: Metamorphic consequences of thermal-mechanical coupling facilitated by erosion. Tectonics 20:712–728

Zen E (1988) Thermal modelling of stepwise anatexis in a thrust-thickened sialic crust. Trans. Roy. Soc. Edinburgh Earth Sci. 79:223–235

Zeng Y, Liu J, Zhu Y (2002) Short-chain carboxylates in high-temperature ore fluids of W-Sn deposits in south China. Geochem. J. 36:219–234

Zhang DQ, Sun GY (1988) Granites of Eastern China. China Univ. Geosci. Press, Wuhan (in Chinese)

Zhang HL, Sun GA, Huang HX, Cai JH (1986) The study on the petrography, geochemistry and metallization of the Dengfuxian muticomponent granite. In: Yuchang Institute of Geological and Mineralogical Research. Contrib. Geol. Miner. Res. Nanling area 2:123–140 (in Chinese)

Zhang LG (1989) Petrogenic and minerogenic theories and prospecting. Beijing Univ. Tech. Press, Beijing (in Chinese with English abstract)

Zhang ZH, Zhang BT (1992) On the uranium-bearing granites and their related uranium deposits in South China. Nuclear Energy Publ. House, Beijing (in Chinese)

Zhu X, et al. (1983) The geology of Dexing porphyry copper ore field. Geol. Publ. House, Beijing (in Chinese)

Zhu Y, Shi Y (1990) Shear heating and partial melting of granite: thermal structure of overthrust terranes in the Greater Himalaya. Chin. J. Geophys. 33:341–351

Zhu YL, Li CZ, Lin YZ (1981) Tungsten deposits in the southern Jiangxi Province. Jiangxi People's Publ. House, Nanchang (in Chinese)

Zhuang WM, Chen GN, Lin XM, Peng ZL, Ma HM (2006) Composition of oxygen isotopes and discussion on the formation of the Changkeng Au-Ag deposit, Guangdong Province. J. Jilin. Univ. Earth Sci. 36:521–526

Zhou, XM, Li WX (2000) Origin of Late Mesozoic rocks in southeastern China: implications for lithospheric subduction and underplating of mafic magmas. Tectonophysics 326:269–287

APPENDIX 1

MAP OF SE CHINA SHOWING PROVINCES

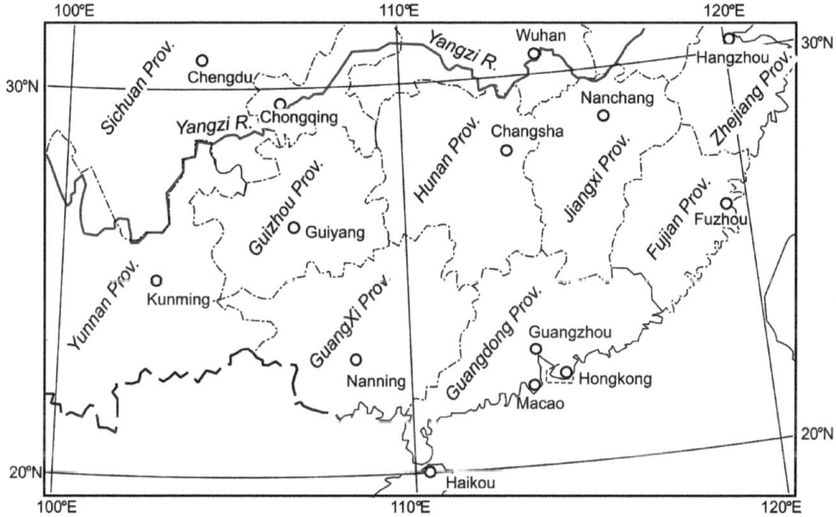

APPENDIX 2

RESULTS OF EXPERIMENTAL ROCK MELTING

A. Mica-bearing metasedimentary compositions

1. *Quartz–plagioclase-rich layer* with Qz, Pl, minor Bt, Gt, Ap, Ilm, in amphibolite grade Alpine schist, southern Alps, New Zealand (Kifle 1992).
Experimental conditions: Powdered starting material; 625–800°C; 3kb; fO_2 ~NNO buffer; H_2O-added; run-times 120–336 h.
Crystalline products: Opx, Cd, Os, Oamp, Sp, Mt.
Melt composition: Granite to granodiorite with increasing *T* (Fig. A2a).

2. *Quartz-rich micaschist* with Qz (55%), Pl (19%), Ksp (2%), Ms (7%), Bt (15%), Sil (2%), from Black Hills, South Dakota, USA (Pickering and Johnston 1998).
Experimental conditions: Powdered starting material; no added H_2O; fO_2 < QFM buffer; 10 kb; 812–975°C; run-times 3–17 days.
Crystalline products: Ksp, Sil, Gt, Bt.
Melt composition: Granite (leucogranite) (Fig. A2a).

3. *Muscovite schist* with Qz (43%), Pl (18%), Ms (22%), Bt (2%), Gt (5%), from hanging wall of the Main Central Thrust of the Himalayas (Patiño-Douce and Harris 1998).
Experimental conditions: Powdered starting material; no added H_2O and 1–4% added H_2O; 700–900°C; 6–10 kb; fO_2 QFM–QFM-2; run-times 8–16 days.
Crystalline products: Bt, Pl (An_{15-26}), Ksp (Or_{63-80}), Gt, Sil, Trm, (St), (Sp).
Melt composition: Granite (leucogranite) (no added H_2O); granite, adamellite, trondhjemite (H_2O-added) (Fig. A2a).

4. Metagreywacke sandstone with *neometamorphic* Qz, Ab, Ms, Ch, Pp, Stp, Ep, Tn, Cc, Py; *relic (detrital)* Qz, Pl, Ksp, Ms, Ch, Pp, Ep, Gt, Tn, Ap, Zr, Ilm, Rt, Pyr, graphite, from Taupo volcanic zone, New Zealand (Kifle 1992).
Experimental conditions: Lightly crushed rock; H_2O-added; 650–850°C; 3 kb; fO_2 ~NNO buffer conditions; run-times 19–888 h.
Crystalline products: Cd, Os, Opx, Sp, Mt.
Melt Composition: Granite–adamellite–granodiorite with increasing *T* (Fig. A2a).

	1	2	3	4	5
SiO2	79.07	77.14	75.28	70.87	70.04
TiO2	0.29	0.50	0.36	0.43	0.38
Al2O3	10.19	11.20	14.29	14.13	14.92
FeO	2.31	3.39	2.40	2.89	3.65
MnO	0.03		0.13	0.06	0.09
MgO	0.95	1.23	0.68	0.95	1.39
CaO	1.09	0.75	0.94	1.33	1.79
Na2O	2.84	1.42	2.77	3.97	2.65
K2O	1.23	2.44	2.40	2.62	3.18
P2O5	0.09			0.11	
F		0.11			
H2O	1.70	0.92		2.07	1.90
Total	99.79	99.68	99.23	99.43	99.99

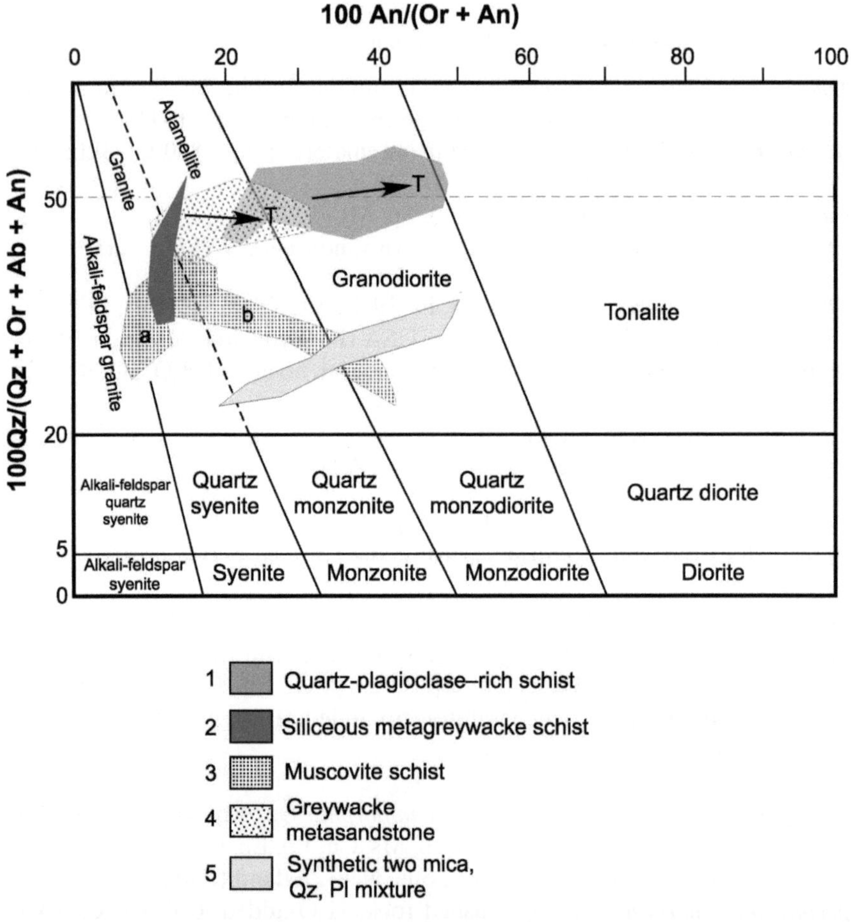

Figure A2a. Normative Qz, Or, Ab, An—ratio plot showing compositions of granite melts derived from experimental melting of mica-bearing meta-sedimentary rocks (Nos. A.1–5), with bulk compositions listed in the table. **a** and **b** for composition A.3 indicate H_2O-added and no H_2O-added, respectively. Arrows indicate direction of increasing *T*.

5. *Synthetic mixture* of Qz (44.8%), Pl (26.8%), Ms (15.3%), Bt (13%), (Gardien et al. 1995).
Experimental conditions: Powdered starting material; no added H_2O; 750–950°C; 10 kb; fO_2 ~NNO; run-times 122–333 h.
Crystalline products: Gt, Bt, Opx, Sil, (Ksp).
Melt composition: Adamellite–granodiorite (Fig. A2a).

6. *Biotite schist* (metagreywacke) with Qz (40%), Pl (32%), Bt (25%), minor Trm, Ap, Mz, Zr, Py, from 'Série Cévenole' Ardèche, France (Vielzeuf and Montel 1994).
Experimental conditions: Powdered and glass starting materials; no added H_2O; 780–1000°C; 1–10 kb; fO_2 ~NNO and $MnO\text{-}Mn_3O_4$ buffer conditions; run-times 120–607 h.
Crystalline products: Ksp, Opx, Gt/Cd/Sp, Mt, Ilm, Ap, Rt, Po related by way of the melt-producing reactions,

$$Qz + Pl + Bt = Opx + Gt/Cd/Sp + Ksp + L$$

$$Qz + Pl + Bt = Gt + Ksp + L$$

Melt composition: Granite (Fig. A2b).

7. *Augen gneiss* with Ksp megacrysts and matrix of Qz (42%), Pl (20%), Ksp (8%), Bt (10%), Ms (20%) from the Iberian Massif, Spain (Castro et al. 2000).
Experimental conditions: Powdered starting material; 2 and 4wt % added H_2O and no added H_2O; 675–900°C; 2–15 kb, fO_2 between QFM and QFM-2; run-times 73–304 h.
Crystalline products: Bt, Pl, Ksp, Cd, Opx, AS, Gt (>10kb) in fluid-present experiments; Bt, Pl, Ksp, Cd, AS, Gt (>10 kb), Sp in fluid-absent experiments.
Melt composition: Granite (Fig. A2b).

8. Synthetic mixture of Qz (45%), Pl (28.5%), Bt (26.8%) (Gardien et al. 1995).
Experimental conditions: Powdered starting material; no added H_2O; 750–950°C; 10 kb; fO_2 ~NNO; run-times 122–333 h.
Crystalline products: Pl, Gt, Bt, Opx, Sil, (Ksp), Rt.
Melt composition: Adamellite–granodiorite (Fig. A2b).

9. *Semischistose greywacke sandstone* with *neometamorphic* Qz, Ab, Ms, Bt, Ep, accessory Ch, Tn, Zr; *relic (detrital)* Qz, Ab, Ep, Tn Ap, Ilm, from greenschist facies (biotite zone) rocks, central southern Alps, New Zealand (Kifle 1992).
Experimental Conditions: Lightly crushed starting material; H_2O-added; 625–850°C; 1–3 kb; fO_2 ~NNO buffer; run-times 144–360 h.
Crystalline products: Ca-Pl, Opx, Oamp, Os, Sp, Mt.
Melt Composition: Adamellite–granodiorite (Fig. A2b).

	6	7	8	9	10
SiO2	69.99	69.10	68.43	68.08	67.03
TiO2	0.70	0.52	0.81	0.58	0.79
Al2O3	12.96	15.20	12.45	14.45	16.26
FeO	4.42	3.77	6.36	4.01	5.5
MnO	0.06	0.04	0.12	0.05	0.14
MgO	2.36	1.51	2.54	1.48	1.88
CaO	1.67	1.28	1.77	2.55	1.06
Na2O	2.95	3.06	2.52	3.53	1.36
K2O	2.41	3.81	3.18	2.41	3.39
P2O5	0.20			0.12	
F	0.09				
H2O	1.43	1.30	1.82	1.83	
Total	99.24	99.60	100.00	99.09	97.41

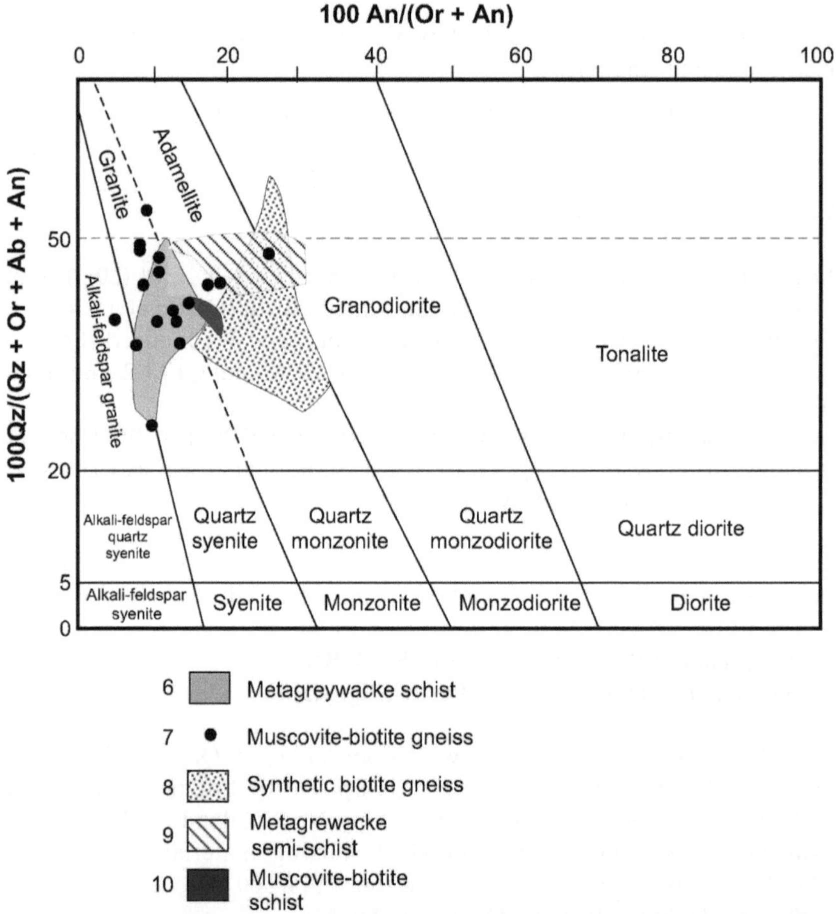

Figure A2b. Normative Qz, Or, Ab, An—ratio plot showing compositions of granite melts derived from experimental melting of mica-bearing meta-sedimentary rocks (Nos. A.6–10), with bulk compositions listed in the table.

10. *Muscovite–biotite schist* with Qz (38%), Pl (11%), Ms (29%), Bt (13%), Gt (6%), Trm + Ky + St (4%), from hanging wall of the Main Central Thrust of the Himalayas (Patiño-Douce and Harris 1998).
Experimental conditions: Powdered starting material; no added H_2O and 1–4% added H_2O; 700–900°C; 6–10 kb; fO_2 QFM–QFM-2; run-times 8–16 days.
Crystalline products: Bt, Pl, Ksp, Gt, Sil, Trm, (St), (Sp). Compared with biotite compositions in starting material, newly-formed biotite is more aluminous, and has higher Mg/Fe, Ti and F.
Melt composition: Granite (leucogranite) (Fig. A2b).

11. *Cordierite gneiss* with Qz (35%). Pl (23%), Ksp (2%), Bt (25%), Cd (9%), Gt (3%), retrograde Ms (2%), accessory (1%) Ap, Zr, Sp, Py, from mega-enclave (~2km) in migmatite, Sul-rio-grandense Shield, southern Brazil (Koester et al. 2002).
Experimental conditions: Powdered starting material; H_2O-added and H_2O-absent; 700–900°C; 5, 19, 15 kb; fO_2 between QFM and QFM-2 buffer conditions and higher; run-times 61–290 h.
Crystalline products: Opx, Mt (5 kb); Gt, AS, Ksp (10 and 15 kb).
Melt Composition: Granite–adamellite (Fig. A2c).

12. *Banded biotite gneiss* with Qz (24%), Pl (42%), Ksp (8%), Bt (25%), Ep (9%), Tn, Ap, Zr, Ilm, from the Lepontine nappes, Ticino Valley, Switzerland (Gardien et al. 2000).
Experimental conditions: Powdered starting material; 2% and 4% H_2O-added and H_2O-absent; 800–900°C; 10, 15, 20 kb; fO_2 between QFM and QFM-1.2; run-times 73–232 h.
Crystalline products: Ksp, Pl, Bt, Hb, Gt, (Cpx, Ep, Rt at 20 kb).
Melt composition: Adamellite–granodiorite (10 kb); granodiorite (15 kb); adamellite–quartz monzonite (20 kb) (Fig. A2c).

13. *Pelitic gneiss* with Qz, Pl, Ms, Bt, Ky, Gt, ± St, ± Ch (retrograde) from the Cariño gneisses, Cabo Ortegal complex, Galicia, NW Spain (Vielzeuf and Holloway 1988).
Experimental conditions: Powdered starting material; 750–1250°C; 7,10 and 12 kb; fO_2 unbuffered; no added H_2O; run-times 14 h to 7 days for experiments <1000°C, 20 min–24 hr for experiments >1000°C.
Crystalline products: Qz, (Ksp), Sill, Gt, Sp by way of the melt-producing reactions (at 10 kb) up to 875°C,

$$\mathrm{Ms + Bt + Pl + Qz = AS + L}$$

$$\mathrm{Ms + Pl + Qz = Bt + AS + (Ksp) + L}$$

$$\mathrm{Bt + AS + Pl + Qz = Gt + (Ksp) + L}$$

	11	12	13	14	15	16
SiO2	65.80	64.59	64.35	63.6	62.55	61.89
TiO2	0.78	0.67	0.82	2.5	0.99	0.76
Al2O3	14.29	16.40	18.13	12.3	18.41	17.24
FeO	5.87	4.59	6.26	7.6	6.73	8.02
MnO	0.12	0.05	0.09	0.1	0.15	0.06
MgO	2.97	1.37	2.44	4.7	1.74	2.81
CaO	1.61	3.85	1.52	2.1	1.19	0.09
Na2O	1.43	3.58	1.66	2	1.22	0.80
K2O	3.16	3.41	2.56	3.6	3.37	5.93
P2O5	0.30	0.21			0.18	
F				0.1		
H2O	1.89		2.15	1.6	2.83	
Total	98.22	98.72	99.98	100.2	99.36	97.60

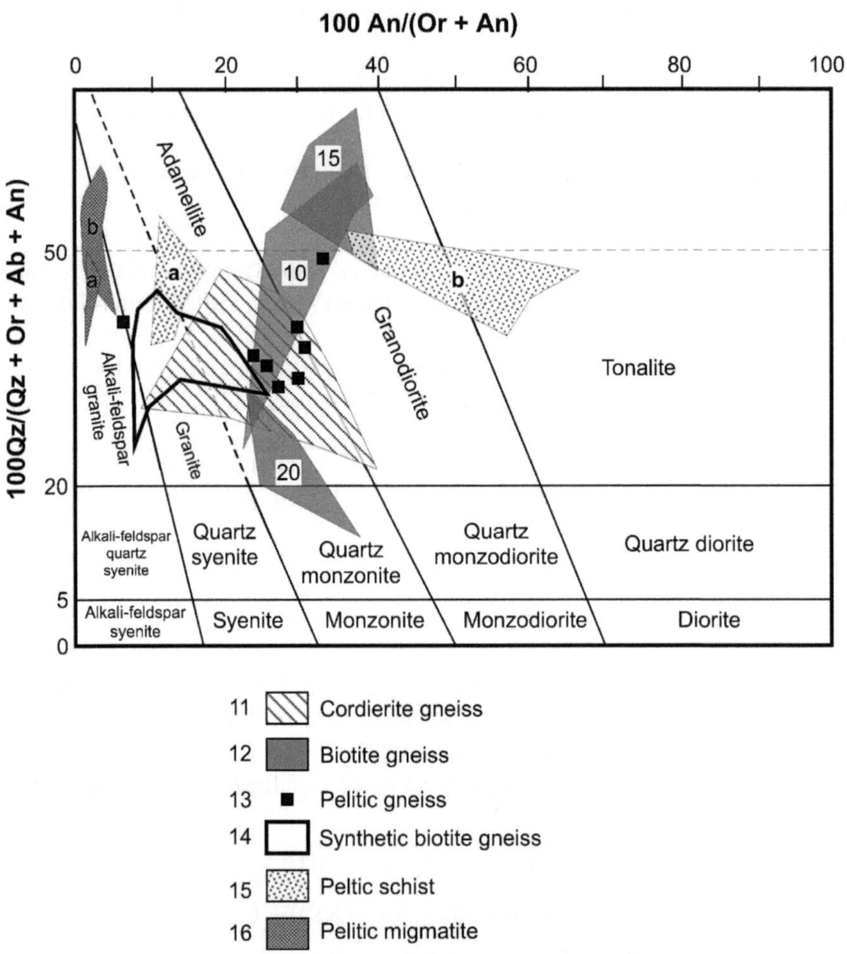

Figure A2c. Normative Qz, Or, Ab, An—ratio plot showing compositions of granite melts derived from experimental melting of mica-bearing meta-sedimentary rocks (Nos. A.11–16), with bulk compositions listed in the table. **a** and **b** for compositions A.15,16 indicate H_2O-added and no H_2O-added, respectively. Composition fields for 10, 15 and 20 kb indicated for No.A.12.

and above 1000°C,

$$Gt = Sp + (Qz) + L$$

$$Sp + (Qz) = Sil + L$$

Melt composition: Adamellite–granodiorite and granite (Fig. A2c).

14. *Synthetic biotite gneiss* with Qz (34.3%), Pl (26.5%), Bt (37.2%), Ilm (2.0%), (Patiño-Douce and Beard 1995).
Experimental conditions: Powdered and glass staring compositions; no added H_2O; 875–1000°C; 3–15 kb; fO_2 between QFM-2–+ 3 buffer; run-times 5–21 days.
Crystalline products: Pl, Ksp, Opx, Cpx (>15 kb), Bt, Gt, Ilm, Rt (>10 kb).
Melt composition: Adamellite–granodiorite (Fig. A2c).

15. *Metapelite* with Qz (40.6%), Pl (14.1%) Ms (13.2%), Bt (18.2%), Ky + Sill (8.3%), Gt (4.9%), graphite; *accessory* Ap, Tr,, Zr, Ilm, Pyr; *retrograde* Ch, from Isle of Pines, western Cuba (García-Casco et al. 2003).
Experimental conditions: Powdered starting material; 10 wt %. added H_2O and no added H_2O; 650–850°C; 4–14 kb; fO_2 between QFM and QFM-2 buffer conditions; run-times 70–339 h.
Crystalline phases: Bt, Gt, St in fluid-present experiments; Ksp, ± Pl, Bt, ± AS, ± Gt, ± St, ± Sp in fluid-absent experiments.
Melt composition: Adamellite, granodiorite, tonalite in fluid-present experiments; granite in fluid-absent experiments (Fig. A2c).

16. *Pelitic migmatitic schlieren* with Qz (24%), Ksp (38%), Bt (11%), Cd (25%), Ilm (1%), Mt (1%), Sp (hercynite; trace) (**a**), from Morton Pass, Wyoming, USA (Grant 2004). Two additional compositions derived by; adding 12.8% Qz (**b**), and adding 22.6% Sil (**c**) to composition (a).
Experimental conditions: Powdered starting material; both added H_2O and no H_2O added; 700–840°C; 1–3.5 kb (mainly at 2 kb); fO_2 = QFM; run-times 6–13 days.
Crystalline products:
H_2O-added. At T near beginning of melting:

$$Qz + Ksp + Cd + Sil + Sp + V = L$$

$$Qz + Ksp + Cd + Sil + Sp + Ilm + V = Bt + L$$

$$Qz + Bt + V = Cd + Opx + Ilm + L$$

H_2O-absent. From ~800°C:

$$Qz + Bt = Ksp + Cd + Opx + Ilm + L$$

Melt composition: Granite–alkali-feldspar granite (Fig. A2c).

17. *Pelitic hornfels* with Qz (27.1%), Pl (10.3%), Ms (8.4%), Bt (22.6%), Cd (30.6%) and (1.0%); *trace* Ilm, Ap, Mz, Zr, from contact aureole beneath the Bushveld Intrusion, South Africa (Buick et al. 2004).
Experimental conditions: Powdered starting material; no added H_2O, but with 1.2% H_2O^+; 700–1000°C; 3 kb; fO_2 unbuffered; run-times 138–232 h.
Crystalline products: Pl, Ksp, Bt, Cd, Sil, Opx, Sp, Ilm.
Melt composition: Granite (Fig. A2d).

18. *Metasiltstone (argillite)* with *neometamorphic* Qz, Ab, Ms, Ch, Pp, Pr, Ep, Tn, Cc, Py; *relic (detrital)* Qz, Pl, Ksp, Ilm, Ap, Zr, from Taupo volcanic zone, North island, New Zealand (Kifle 1992).
Experimental conditions: Lightly crushed starting material; 675–850°C; 3 kb; fO_2 ~NNO buffer; run-times 85–744 h.
Crystalline products: Pl, Bt, Cd, ± Sil, ± Opx, Os, Sp, Mt.
Melt composition: Adamellite–granodiorite (Fig. A2d).

19. *Metapelite* with Qz (31%), Pl (4%), Sil (19%), Ms (10%); Bt (30%); Gt (5%); Ilm (1%), from northern Idaho, USA (Patiño-Douce and Johnston 1991).
Experimental conditions: Powdered starting material; no added H_2O; 825–1075°C; 7, 10 and 13 kb; fO_2 about 1 to 1.5 log units above NNO buffer; run-times 48–360 h.
Crystalline products: Qz, AS, ± Bt, ± Gt; ± Ilm; ± Sp; ± Rt.
Melt composition: Alkali-feldspar granite (Fig. A2d).

20. *Pelitic hornfels* with Qz (30.9%), Pl (10.8%), Ms (29.5%), Ch (27.8%), Ilm (1.0%), Zr (trace), from contact aureole beneath the Bushveld Intrusion, South Africa (Buick et al. 2004).
Experimental conditions: Powdered starting material; no added H_2O, but with 4.4% H_2O^+; 700–1000°C; 3 kb; fO_2 unbuffered; run-times 138–232 h.
Crystalline products: Cd, Sp, Ilm, (Pl, Ksp, Bt, Sil below 750°C)
Melt composition: Adamellite (Fig. A2d).

21. *Mica-rich layer* with Qz, Pl, Ms, Bt, Gt, Ep, Ilm, Rt, Ap, in amphibolite grade Alpine schist, southern Alps, New Zealand (Kifle 1992).
Experimental conditions: Lightly crushed starting material; H_2O-added; 650–850°C; 3 kb; *f*O2 ~NNO buffer; run-times 120–384 h.
Crystalline products: Ksp, Cd, Os, Sil, ± Co, Sp, Mt.
Melt composition: Adamellite–granite with increasing *T* (arrow) (Fig. A2d).

B. Dacite/basalt-micaceous quartzofeldspathic mixtures/quartz amphibolite

1. *Dacite*, central volcanic zone, North Island, New Zealand (Conrad et al. 1988).
Experimental conditions: Glass starting material; added H_2O and CO_2 to give XH_2O in vapour phase of 0.75, 0.50, 0.25; 675–975°C; 10 kb; run-times 24–384 h.
Crystalline products: Qz, Pl, Hb, Bi, (Ksp), Cpx, Opx, Ap, Ilm, Rt.
Melt compositions: Adamellite–granodiorite–tonalite with increasing *T* (Fig. B2a).

	17	18	19	20	21
SiO2	59.79	58.81	57.36	57.18	50.36
TiO2	0.84	0.82	1.26	0.77	1.25
Al2O3	20.43	18.22	23.24	20.02	22.83
FeO	9.37	6.72	8.59	9.83	7.66
MnO	0.13	0.08	0.17	0.10	0.13
MgO	3.58	2.19	2.72	3.35	2.98
CaO	0.64	1.73	0.40	0.61	1.65
Na2O	1.36	1.74	0.48	1.36	2.71
K2O	3.04	4.16	3.63	2.59	5.45
P2O5		0.17			0.36
F			0.11		
H2O	1.11	3.79	1.69	4.33	4.33
Total	100.29	98.43	99.65	100.14	99.71

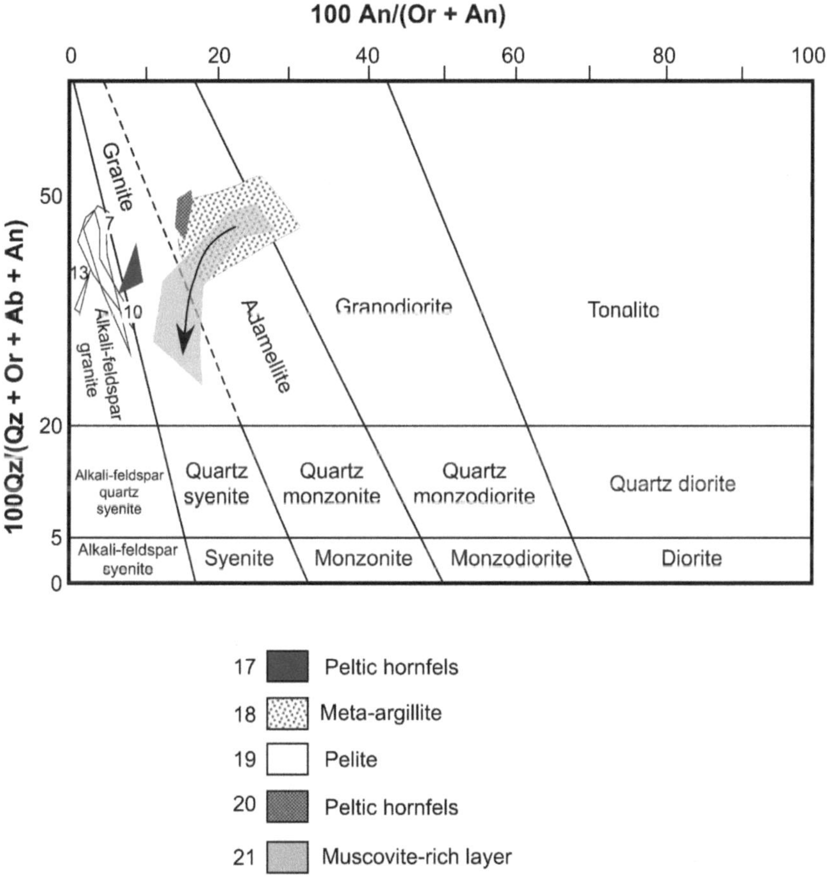

Figure A2d. Normative Qz, Or, Ab, An—ratio plot showing compositions of granite melts derived from experimental melting of mica-bearing meta-sedimentary rocks (Nos. A.17–21), with bulk compositions listed in the table. Melt composition fields at 7, 10 and 13 kb are shown for A.19. Arrow for composition A.21 indicates increasing *T*.

	1	2	3	4	5	6	7
SiO2	65.20	65.00	63.74	60.40	59.03	55.7	52.7
TiO2	0.60	0.73	0.78	1.70	1.12	1.7	1.1
Al2O3	15.70	16.50	16.05	11.30	15.12	15.7	21.3
FeO	5.00	7.00	5.62	7.90	9.17	8.0	8.5
MnO	0.10	0.10	0.12	0.20	0.16	0.2	0.2
MgO	2.50	1.90	2.41	6.70	2.48	7.8	6.9
CaO	5.30	3.50	3.14	7.60	5.31	7.0	6.1
Na2O	3.60	3.40	4.22	1.90	3.28	2.1	1.3
K2O	1.70	1.90	2.15	0.70	1.64	1.9	1.9
P2O5	0.10	0.15	0.14		0.20		
F							
H2O			1.88	1.40			
Total	99.80	100.18	100.25	99.80	97.51	100.1	100.0

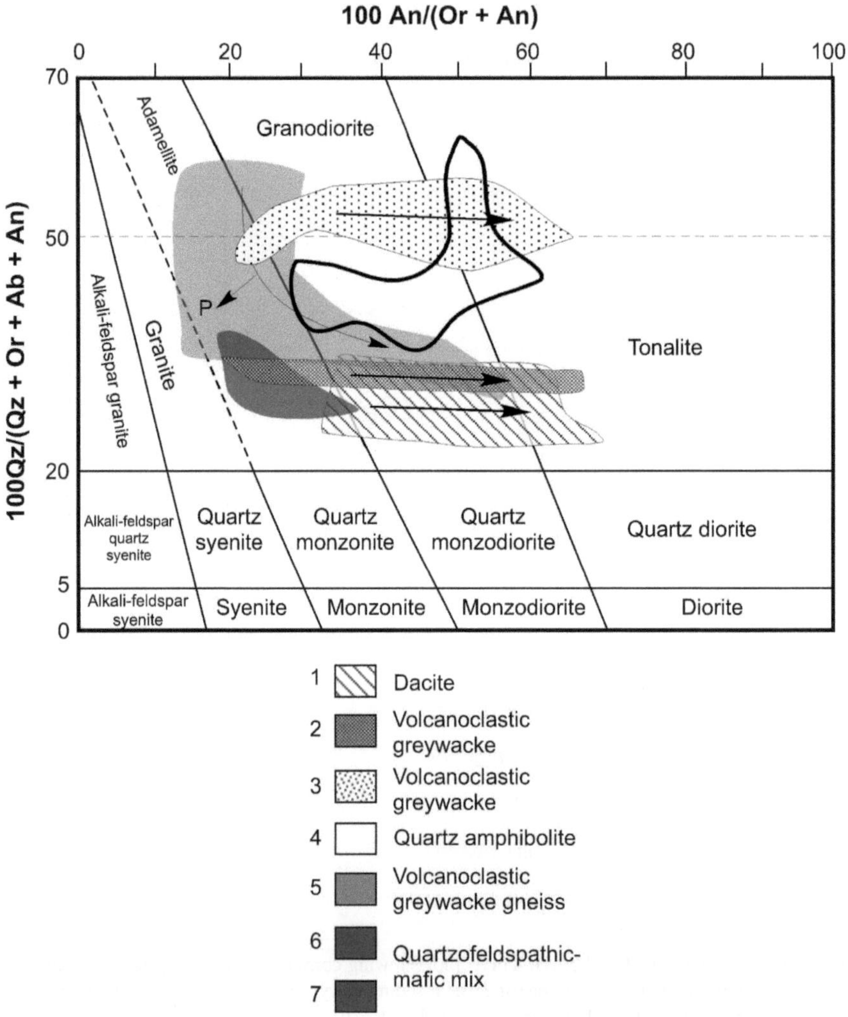

Figure B2a. Normative Qz, Or, Ab, An—ratio plot showing compositions of granite melts derived from experimental melting of dacite, volcanogenic greywacke, synthetic basalt-meta-sedimentary mix, and quartz amphibolite (Nos. B.1–7), with bulk compositions listed in the table. Values below anhydrous totals of Analyses 6 and 7 are H_2O values. Arrows indicate increasing T and P = increasing P.

2. *Prehnite-pumpellyite facies volcanoclastic greywacke*, central volcanic zone, North Island, New Zealand (Conrad et al. 1988).
Experimental conditions: Glass starting material; added H_2O and CO_2 to give XH_2O in vapour phase of 0.75, 0.50, 0.25; 675–975°C; 10 kb; run-times 24–384 h.
Crystalline products: Qz, Pl, Amp (ortho and Ca-amphibole), Gt, Ap, Ilm.
Melt compositions: Adamellite–granodiorite–tonalite with increasing *T* (Fig. B2a).

3. *Prehnite-pumpellyite facies greywacke* with *detrital* Qz, Pl (An_{0-55}), Ksp, Cpx, Opx, Hb, Ms, Ep, rare Trm, Ap, Tn, Zr, Ilm, Rt; *authigenic* (matrix) Ab, Ms, Ch, Ep, Pp, Pr, Tn, Cc, Py from western part of central volcanic zone of North Island, New Zealand (Kifle 1992).
Experimental conditions: Lightly crushed starting material; added H_2O; fO_2 QFM–NNO; 650–850°C; 3 kb; run-times 2–5 weeks.
Crystalline products: Qz, Pl, Amp (hornblende and cummingtonite), Opx, Bt, Mt.
Melt compositions: Granite–adamellite–granodiorite with increasing *T* (Fig. B2a).

4. Synthetic *quartz amphibolite* with Qz (24.5%), Pl (19.6%), Hb (63.9%), Ilm (2%) (Patiño-Douce and Beard 1995).
Experimental conditions: Glass starting material; no added H_2O; 840–1000°C; 3–15 kb; fO_2 ΔQFM-1.4–2.7; run-times 7–21 days.
Crystalline phases: Qz, Pl (An_{41-59}), Hb, Opx, Cpx, Gt, Ilm, Mt, Rt.
Melt compositions: Granodiorite–adamellite (Fig. B2a).

5. *Volcanoclastic gneiss* with Qz, Pl (An_{18}), Bt (16%), Hb (15%), Ep (13%), from the western gneiss region, Norway (Skjerlie and Johnston 1996).
Experimental conditions: Powdered starting material; No H_2O-added; 850–1050°C; 10–20 kb; fO_2 ~ 1 log unit <QFM buffer conditions; run-times 97–663 h.
Crystalline products: Gt, ± Hb, Cpx, Opx, Ksp through the melt-producing reactions (at 10 kb),

$$Qz + Bt + Ep = Hb + Gt + Ksp + L$$

and

$$Hb + Pl + Qz = Cpx + Opx \pm Gt + L$$

at higher temperatures (875–925°C).
Melt composition: Granite–adamellite with increasing *T* (Fig. B2a).

6. Synthetic *high alumina-olivine tholeiite* (50%) + synthetic *biotite gneiss* (50%). Biotite gneiss synthesised from a mixture of Qz (34%), Pl (27%), Bt (37%), Ilm (2%) (Patiño-Douce 1995).
Experimental conditions: Glass starting material; no added H_2O; 5–15 kb; 1000°C; fO_2<QFM; run-times 141–201 h.
Crystalline products: Pl, Opx, Cpx, Gt, Sp, Ilm, Rt.
Melt composition: Adamellite (Fig. B2a).

7. Synthetic *high-alumina olivine tholeiite* (50%) + natural *metapelite* (50%). Metapelite with Qz (31%), Pl (4%), Ms (10%), Bt (30%), Gt (5%), Sil (19%), Ilm (1%) from northern Idaho, USA (Patiño-Douce 1995).
Experimental conditions: Glass starting material; no added H_2O; 5–15 kb; 1000°C; fO_2<QFM; run-times 141–201 h.
Crystalline products: Pl, Opx, Cpx, Gt, Sp, Ilm, Rt.
Melt composition: Adamellite (Fig. B2a).

C. Mafic rocks

1. *Greenschist facies basaltic andesite* with Qz, Ab, Ep, Act (40%), Tn, Fe-Ox, (relic Cpx), from the Smartville Complex, northern California, USA (Beard and Lofgren 1991).
Experimental conditions: Powdered starting material; H_2O-added and H_2O-absent; 800–1000°C; 1, 3 and 6.9 kb; fO_2 externally unbuffered; run-times 72–312 h.
Crystalline products: *H_2O-absent* Pl, Opx, Cpx, Mt, ± Ilm; *H_2O-added* Hb, Mt, ± Cpx, ± Pl, ± Ilm.
Melt composition: Granodiorite–tonalite (trondhjemite) (Fig. C2a).

2. *Greenschist facies andesite* with Qz, Ab, Act (35%), Tn, Fe-Ox, from the Smartville Complex, northern California, USA (Beard and Lofgren 1991).
Experimental conditions: Powdered starting material; H_2O-added and H_2O-absent; 850–1000°C; 1, 3 and 6.9 kb; fO_2 externally unbuffered; run-times 72–138 h.
Crystalline products: *H_2O-absent* Pl, Opx, Cpx, Mt, ± Ilm; *H_2O-added* Hb, Mt, ± Cpx, ± Pl, ± Ilm.
Melt composition: Granodiorite–tonalite(trondhjemite)–quartz-diorite (Fig. C2a).

3. *Hornblende-quartz gabbro* (= medium-K [1.19 wt % K_2O] *basaltic andesite*) with Hb, Pl, Bt, minor Qz, Fe-Ti Ox; trace Ksp, Ap, Tn, Zr, S; alteration phases, Ch, Ser (= sericite), Carb, from sill, Sierra Nevada batholith, California, USA (Sisson et al. 2005).
Experimental conditions: Powdered samples; no added H_2O; 825–925°C; 7 kb; fO_2 ΔNNO-1.0–3.5; run-times 4 days.
Crystalline products: Pl, Hb (tschermakitic–magnesiohornblende), Opx, Cpx, Ap; Qz at lowest *T* and highest fO_2. Under reduced conditions FeS ± ilmenite are stable and are replaced by Ti-magnetite or Ti-magnetite/ilmeno-hematite at intermediate and high-oxidation conditions.
Melt composition: Adamellite–granodiorite–quartz monzodiorite (Fig. C2a).

4. *Upper greenschist facies tholeiitic basalt* with Qz, Pl, Act (46%), Fe-Ox, from Smartville Complex, northern California, USA (Beard and Lofgren 1991).
Experimental conditions: Powdered starting material; H_2O added and not added; 800–1000°C; 1, 3 and 6.9 kb; fO_2 externally unbuffered; run-times 72–137 h.
Crystalline products: *H_2O-absent*: Pl, Opx, Cpx, Mt, ± Ilm; *H_2O-added*: Hb, Mt, ± Cpx, ± Pl, ± Ilm.
Melt composition: Granodiorite–tonalite (Fig. C2a).

	1	2	3	4	5	6
SiO2	57.02	55.11	53.95	52.47	51.69	51.39
TiO2	0.60	1.66	1.03	1.74	1.00	1.55
Al2O3	15.39	14.94	17.85	15.29	16.31	15.82
FeO	8.01	11.28	8.11	11.79	8.70	12.23
MnO	0.17	0.21	0.15	0.22	0.11	0.26
MgO	5.52	4.01	5.43	5.29	7.51	4.42
CaO	9.20	6.07	9.33	9.21	8.90	8.95
Na2O	2.54	4.29	2.74	2.55	3.09	3.30
K2O	0.44	0.03	1.19	0.16	0.26	0.37
P2O5	0.18	0.30	0.21	0.29	0.10	0.30
H2O	1.56	0.47		0.49	1.36	0.61
Total	100.63	98.37	99.99	99.50	99.03	99.20

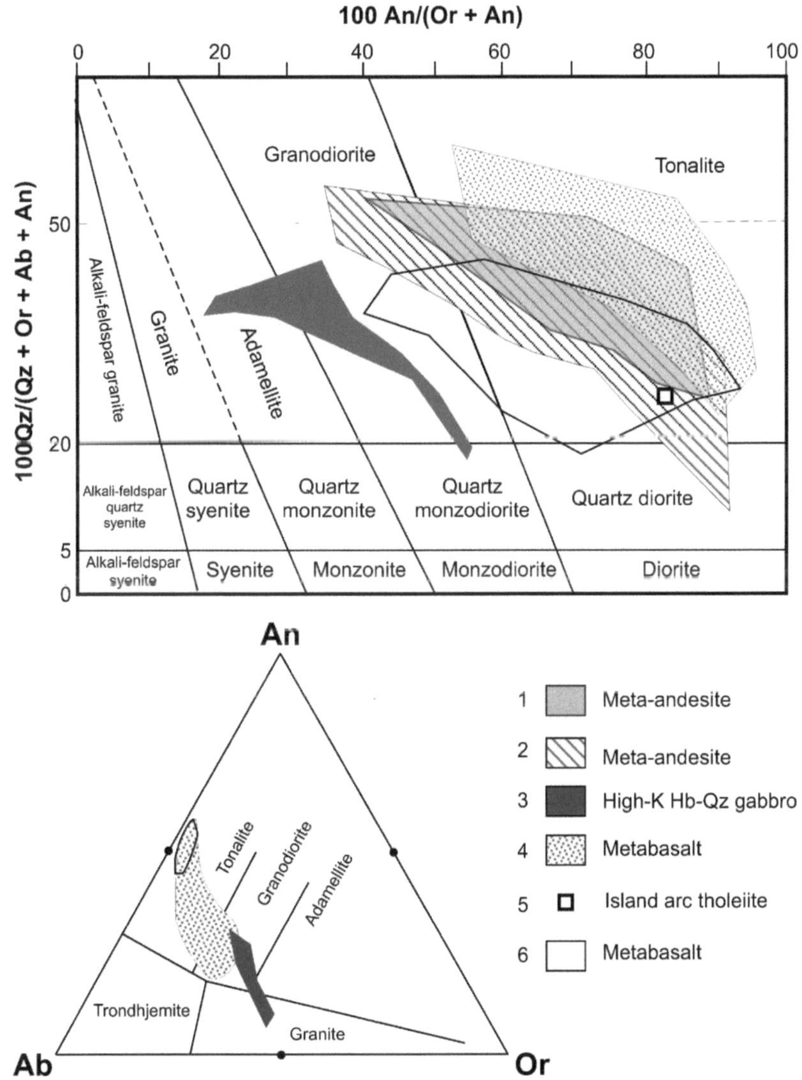

Figure C2a. Normative Qz, Or, Ab, An—ratio plot showing compositions of granite melts derived from experimental melting of mafic amphibolite and granulite (Nos. C.1–6), with bulk compositions listed in the table. Note that all compositions are not plotted in the An-Or-Ab diagram because of overlap.

5. *Amphibolitic island arc tholeiite* with Act-Hb (44%), Pl (32%), Qz (17%), Cum (5%), Bt (1%), Ilm (1%), Ep, from Hidaka metamorphic belt, Hokkaido, Japan (Rushmer 1991).
Experimental conditions: Powdered starting material; no added H_2O; 640–950°C; 8 kb; fO_2 ~QFM; run-times 1–10 days.
Crystalline products: Pl, Cpx, Opx, Ilm.
Melt Compositions: Tonalite (Fig. C2a).

6. *Metabasaltic hornfels* with Pl, Hb (34%), Fe-Ox, (relic Cpx), from Smartville Complex, northern California, USA (Beard and Lofgren 1991).
Experimental conditions: Powdered starting material; H_2O added and H_2O-absent; 800–1000°C; 1, 3 and 6.9 kb; fO_2 externally unbuffered; run-times 72–304 h.
Crystalline products: *H_2O-absent*: Pl, Opx, Cpx, Mt, ± Ilm; *H_2O-added*: Hb, Mt, ± Cpx, ± Pl, ± Ilm.
Melt composition: Granodiorite–tonalite (trondhjemite) (Fig. C2a).

7. *Hornblende gabbro* (= medium-K [1.01 wt % K_2O] high-alumina basalt) with Hb, Pl, Bt, minor Qz, Fe-Ti Ox; trace Ksp, Ap, Tn, Zr, S; alteration phases are Ch, Ser (= sericite), Carb, from sills, Sierra Nevada batholith, California, USA (Sisson et al. 2005).
Experimental conditions: Powdered samples; no added H_2O; 825–975°C; 7 kb; fO_2 ΔNNO-1.3–4.0.
Crystalline products: Pl, Hb (hastingsitic), trace Opx, Ap; depending on fO_2, Bt at or >875°C, Cpx at or >900°C. Under reduced conditions FeS ± ilmenite are stable and are replaced by Ti-magnetite or Ti-magnetite/ilmeno–hematite at intermediate and high-oxidation conditions.
Melt composition: Adamellite through quartz monzonite, quartz monzodiorite to monzodiorite with increasing *T* (Fig. C2b).

8. *Amphibolitic alkali basalt pillow lava* with Pl, Hb, Fe-Ti oxides, from Kalmath Mountains, California, U.S.A (Rapp et al. 1991; Rapp and Watson 1995).
Experimental conditions: Powdered and glass starting material; no H_2O added; 900–1150°C; 8–32 kb; fO_2 ~QFM buffer; run-times 2–28 days.
Crystallne products: Opx (8 kb), Cpx, Gt (>8 kb), Ilm (8 kb), Rt (>8 kb).
Melt composition: Wide variation; adamellite to tonalite, and quartz monzonite to monzodiorite with increasing *T* (Fig. C2b).

9. *Hornblende-quartz gabbro* (= medium-K [1.19 wt % K_2O] basaltic andesite) with Hb, Pl, Bt, minor Qz, Fe-Ti Ox; trace Ksp, Ap, Tn, Zr, S; alteration phases are Ch, Ser (= sericite), Carb, from sills, Sierra Nevada batholith, California, U.S.A (Sisson et al. 2005).
Experimental conditions: Powdered samples; no added H_2O; 825–975°C; 7 kb; fO_2 ΔNNO-1.3–4.0.

	7	8	9	10	11	12	13	14
SiO2	51.32	51.19	50.74	49.20	49.14	49.04	48.8	48.79
TiO2	1.29	1.18	1.25	1.19	1.61	1.27	0.4	0.54
Al2O3	19.37	16.62	19.71	15.60	16.00	16.37	14.6	17.13
FeO	8.82	11.32	9.69	13.60	10.94	9.18	8.4	8.03
MnO	0.17	0.23	0.15	0.15	7.17	0.18	0.2	0.22
MgO	4.38	6.59	4.52	6.89	0.22	7.45	10.7	10.86
CaO	8.98	5.49	7.39	10.80	10.70	10.81	14.3	11.66
Na2O	4.29	4.33	3.76	2.35	3.29	3.42	1.0	1.54
K2O	1.01	0.82	2.32	0.26	0.09	0.44	0.1	0.05
P2O5	0.38		0.46			0.16		0.12
H2O		1.53				1.77		0.44
Total	100.01	99.30	99.99	100.04	99.16	100.09	98.5	99.38

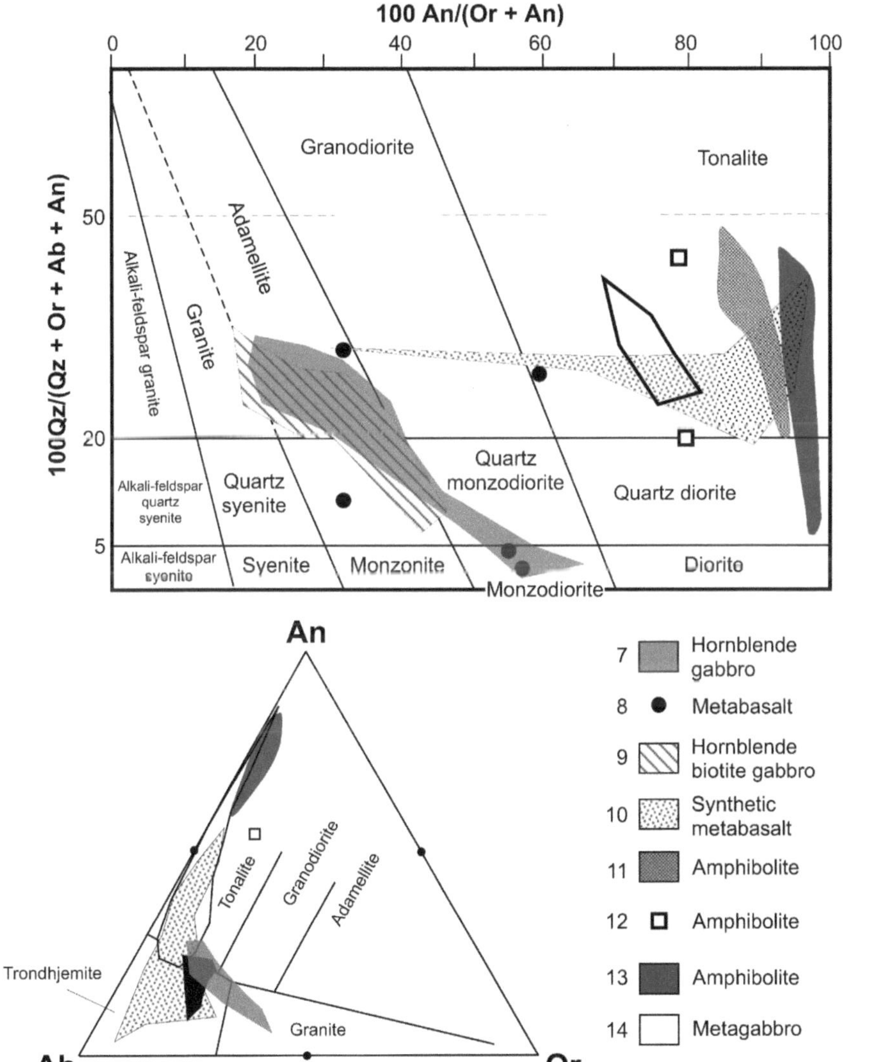

Figure C2b. Normative Qz, Or, Ab, An—ratio plot showing compositions of granite melts derived from experimental melting of mafic amphibolite and granulite (Nos. C.7–14), with bulk compositions listed in the table. Note that all compositions are not plotted in the An-Or-Ab diagram because of overlap.

Crystalline products: Pl, Hb (tschermakitic–magnesiohornblende), Opx, Cpx, Ap; Qz at lowest *T* and highest fO_2. Under reduced conditions FeS ± ilmenite are stable and are replaced by Ti-magnetite or Ti-magnetite/ilmeno-hematite at intermediate and high-oxidation conditions.
Melt composition: Granite–adamellite to quartz monzonite with increasing *T* (Fig. C2b).

10. *Synthetic average Archean tholeiite* (Winther and Newton 1991).
Experimental conditions: Powdered starting material: H_2O added; 750–1100°C; 5–30 kb; fO_2 QFM buffer; run-times 1–10 days.
Crystalline products: Qz, Pl, Hb, Cpx, ± Opx, Gt, (Ky), Mt, Ilm.
Melt composition: Tonalite–trondhjemite (Fig. C2b).

11. *Amphibolite* (MORB) with Pl (46%); Hb (49%); Ilm (5%), from the Acebuches metamorphic belt, SW Spain (López and Castro 2001).
Experimental conditions: Powdered starting material; no added H_2O; 800–950°C; 4–14 kb; run-times 215–261 h.
Crystalline products: Cpx, Opx, ± Gt, ± Ep through the melting reactions,

$$(P < 10 \text{ kb})\ \text{Hb} + \text{Pl} = \text{Cpx} + \text{L}$$

$$(P = 10 \text{ kb})\ \text{Hb} + \text{Pl} = \text{Gt} + \text{Opx} + \text{L}$$

$$(P > 10 \text{ kb})\ \text{Hb} + \text{Pl} = \text{Ep} \pm \text{Gt} \pm \text{Cpx} + \text{L}$$

Melt composition: Tonalite (Fig. C2b).

12. *Amphibolitic alkali metabasalt* with Qz (8.4%), Pl (36.4%), Mg-Hb (54.4%), Ilm (0.8%), from the Ivrea Zone, Italy (Rushmer 1991).
Experimental conditions: Powdered starting material; no added H_2O; 640–1000°C; 8 kb; fO_2 ~QFM; run-times 1–10 days.
Crystalline products: Pl (An_{60}), Cpx, Opx, Tn, Ilm.
Melt Composition: Tonalite (Fig. C2b).

13. *Low K-amphibolite* with Pl (An_{90}; 32.5%); Hb (67.4%); trace Tn, Fe-Ti oxides from western Foothills Metamorphic Belt, Sierra Nevada, California, USA (Wolf and Wyllie 1994).
Experimental conditions: Powdered starting material; no added H_2O; 750–1000°C; 10 kb; fO_2 ~0.5–1.0 log units above NNO buffer; run-times 1–9 days.
Crystalline products: Cpx, Opx, Al-Hb, Ca-Hb (probably metastable), Gt.
Melt composition: Tonalite–quartz diorite (Fig. C2b).

14. *Metagabbro (olivine tholeiite)* with Pl (51.5%), Cpx (24.5%), Opx (14.5%), Gt (1.5%), Hb (7.5%), Fe-oxides (0.5%), occurring as a xenolith in alkaline tuffs, Eifel, Germany (Springer and Seck 1997).
Experimental conditions: Powdered starting material; H_2O and CO_2 added to give $H_2O/(H_2O + CO_2) = 0.5$ and 0.75; 850–1200°C; 5–15 kb; fO_2 ~QFM–NNO; run-times 7–194 h.

Crystalline products: Pl, Cpx, Opx dominant at 5 kb; Gt, Cpx, Opx dominant at 15 kb.
Melt compositions: Trondhjemite through tonalite to diorite with increasing *T* (Fig. C2b).

15. *Amphibolite (high Al-basalt)* with Pl, Hb, Fe-Ti oxides, from Ammonoosuc Volcanics, Vermont, USA (Rapp et al. 1991; Rapp and Watson 1995).
Experimental conditions: Powdered and glass starting material; no H_2O added; 900–1150°C; 8–32 kb; fO_2 ~MW and QFM buffers; run-times 2–28 days.
Crystalline products: Hb, Pl, Opx, Cpx, Gt, Ol, Ilm, Rt (>8 kb).
Melt composition: Trondhjemite–tonalite (Fig. C2c).

16. *Amphibolite (oceanic tholeiite)* with Qz, Pl, Hb, Tn, from Nunatak Fjord, Alaska, USA. (Rapp et al. 1991; Rapp and Watson 1995).
Experimental conditions: Powdered and glass starting material; no H_2O added; 900–1150°C; 8–32 kb; fO_2 ~MW and QFM buffers; run-times 2–28 days.
Crystalline products: Hb, Pl, Opx, Cpx, Gt, Ol, Ilm, Mt.
Melt composition: Trondhjemite–tonalite (Fig. C2c).

17. *Amphibolitic pillow lava* from Wind River Ranges, Wyoming, USA. (Rapp et al. 1991; Rapp and Watson 1995).
Experimental conditions: Powdered and glass starting material; no H_2O added; 900–1150°C; 8–32 kb; fO_2 ~MW and QFM buffers; run-times 2–28 days.
Crystalline products: Hb, Pl, Opx, Cpx, Gt, Mt.
Melt composition: Trondhjemite–tonalite (Fig. C2c).

18. *Amphibolitic metabasalt (quartz tholeiite)* with Pl (40%), Cpx (32%), Opx (8%), Gt (8%), Hb (7%), Fe-oxides (5%), occurring as a xenolith in alkaline tuffs, Eifel, Germany (Springer and Seck 1997).
Experimental conditions: Powdered starting material; H_2O and CO_2 added to give $H_2O/(H_2O + CO_2) = 0.5$ and 0.75; 850–1200°C; 5–15 kb; fO_2 ~QFM–NNO; run-times 7–194 h.
Crystalline products: Pl, Cpx, ±Opx, ±Gt, ±Sp.
Melt compositions: Tonalite–quartz diorite (Fig. C2c).

19. Amphibolite with Hb (76.3%), Pl (20.5%), Qz (2.3%), Tn (0.9%), Gt (<0.1%), from British Columbia, Canada (Sen and Dunn 1994).
Experimental conditions: Powdered starting material; no H_2O added; 850–1150°C; 15 and 20 kb; fO_2 0.5 log units <QFM buffer and between QFM and HM buffer conditions; run-times 24–216 h.
Crystalline products: Cpx, Gt.
Melt composition: Granite, through granodiorite/trondhjemite to tonalite with increasing *T* (Fig. C2c).

	15	16	17	18	19	20
SiO2	48.60	48.30	47.60	46.92	46.88	45.97
TiO2	2.06	0.72	1.19	2.02	1.22	1.58
Al2O3	17.03	15.30	14.18	14.37	15.00	14.81
FeO	10.69	10.70	13.77	13.82	13.09	11.29
MnO	0.21	0.19	0.19	0.17	0.26	0.21
MgO	6.07	8.40	6.86	7.32	8.25	8.11
CaO	9.66	12.60	10.99	12.14	11.28	13.16
Na2O	3.30	2.27	2.56	1.80	2.51	2.10
K2O	0.21	0.08	0.19	0.17	0.80	0.10
P2O5				0.24		0.35
H2O	0.70	0.98	1.65	0.37	1.50	1.14
Total	98.53	99.54	99.18	99.34	100.79	98.82

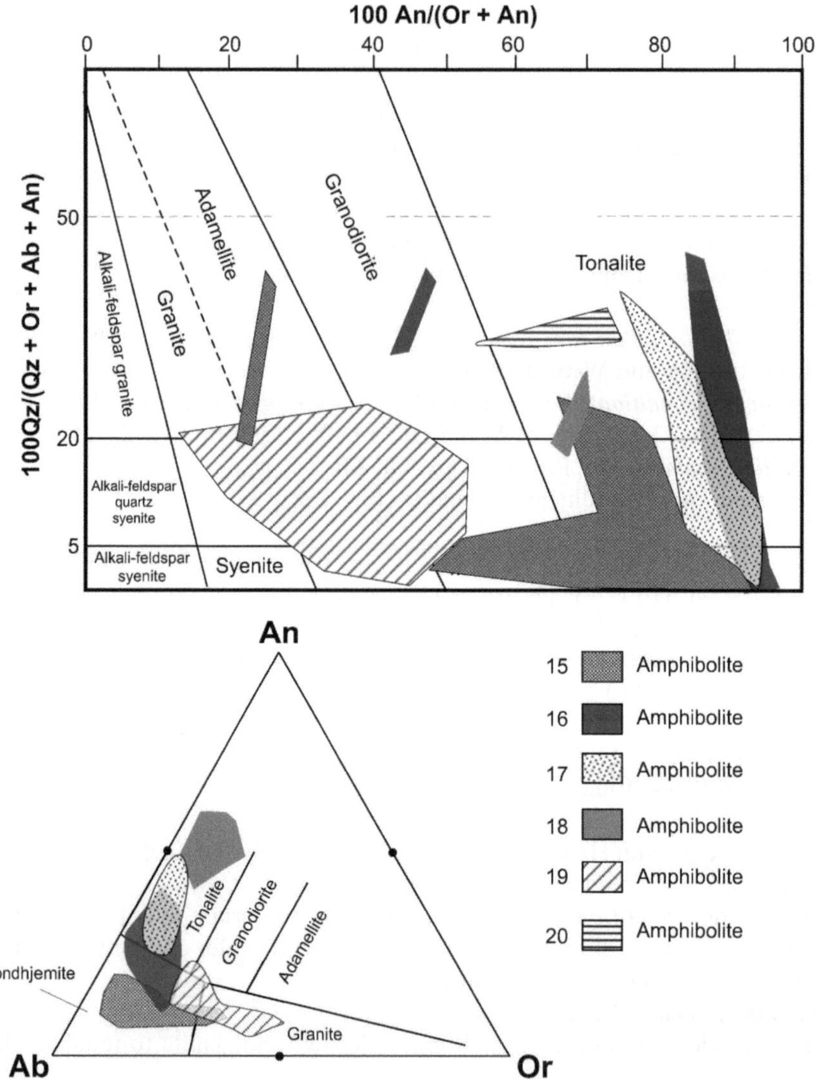

Figure C2c. Normative Qz, Or, Ab, An—ratio plot showing compositions of granite melts derived from experimental melting of mafic amphibolite and granulite (Nos. C.15–20), with bulk compositions listed in the table. Note that all compositions are not plotted in the An-Or-Ab diagram because of overlap.

20. *Amphibolitic metabasalt (olivine tholeiite)* with Pl (32%), Cpx (33%), Opx (1.5%), Gt (6.5%), Hb (23.5%), Fe-oxides (1.5%), occurring as a xenolith in alkaline tuffs, Eifel, Germany (Springer and Seck 1997).
Experimental conditions: Powdered starting material; H_2O and CO_2 added to give $H_2O/(H_2O + CO_2) = 0.5$ and 0.75; 850–1200°C; 5–15 kb; fO_2 ~QFM–NNO; run-times 7–194 h.
Crystalline products: Cpx, ±Pl, ±Opx, ±Hb, ±Gt, ±Sp.
Melt compositions: Tonalite (Fig. C2c).

D. Granitoids

1. *Granodiorite gneiss* with Qz, Anorthoclase, Ksp, Gt, Opx, Sp, Ilm as a xenolith from Kilbourne Hole, New Mexico, USA (Beard et al. 1993).
Experimental conditions: Powdered starting material; no added H_2O and H_2O added; 900–1000°C; 6.9 kb; fO_2 unbuffered; run-times 4–7 days.
Crystalline products: Same as starting minerals but with disappearance of Ksp at 1000°C.
Melt composition: K-feldspar granite and granodiorite with increasing *T* (Fig. D2a).

2. *Granodiorite* with Qz (32%), Pl (30%), Ksp (24%), Hb (6%), Bt (7%); accessories (1%), from Sierra Nevada batholith, California, USA (Patiño-Douce 1997).
Experimental conditions: Powdered starting material; no added H_2O; 950°C; 4 and 8 kb; fO_2 1 log unit below QFM; run-times 7 days.
Crystalline products: ± Pl, Ksp, Bt, Opx, Cpx, Ilm.
Melt compositions: Alkali-feldspar granite (4 kb; ~20% melt fraction), adamellite (8 kb) (<15% melt fraction) (Fig. D2a).

3. *F-rich tonalitic gneiss* with Qz (30%), Pl (48.2%), Bt (19.4%), Hb (1.8%), Ep (0.2%), Ap (0.4%), Zr (0.04%) from Swaziland, South Africa (Skjerlie and Johnston 1993).
Experimental conditions: Powdered starting material; no added H_2O; 875–1075°C; 6, 10, 14 kb; fO_2 ~1–3 log units >NNO buffer; run-times161–502 h.
Crystalline products: Opx, Gt, Fe-Ti oxides.
Melt composition: Adamellite–granodiorite (Fig. D2a).

4. *Granodiorite gneiss* with Qz, Anorthoclase, Ksp, Opx, Ilm, Py as a xenolith from Kilbourne Hole, New Mexico, USA (Beard et al. 1993).
Experimental conditions: Powdered starting material; no added H_2O and H_2O added; 900–1000°C; 6.9 kb; fO_2 unbuffered; run-times 4–7 days.
Crystalline products: Same as starting minerals but with disappearance of Mt and Ilm at 950 and 1000°C, respectively.
Melt composition: Granite (Fig. D2a).

5. *Synthetic tonalite* with Qz (30%), Pl (An_{45}; 50%), Bt (XMg_{50}; 20%) (Singh and Johannes 1996).
Experimental conditions: Fine-grained mineral mixtures; 750–900°C; 2–15 kb; fO_2 NNO and close to CoCoO buffers; run-times 4–12 days.

	1	2	3	4	5	6	7
SiO2	71.51	69.06	68.05	67.91	66.63	61.48	57.49
TiO2	0.04	0.28	0.48	0.08		0.9	0.68
Al2O3	15.23	16.81	15.19	15.74	15.75	17.07	15.57
FeO	1.45	1.39	4.67	1.11	4.76	5.92	7.04
MnO	0.05	0.04	0.06	0.05		0.11	0.14
MgO	0.32	0.75	1.73	0.37	2.6	2.68	3.32
CaO	2.21	3.85	2.93	2.53	4.66	5.39	8.29
Na2O	3.82	4.2	4.47	3.65	2.99	3.88	3.55
K2O	3.76	3.61	2.05	5.36	2.27	2.57	1.39
P2O5	0.04		0.15	0.07			0.15
F			0.1				
H2O	0.2			0.3			0.5
Total	98.63	99.99	99.88	97.17	99.66	100.00	98.12
		99.4				98.83	

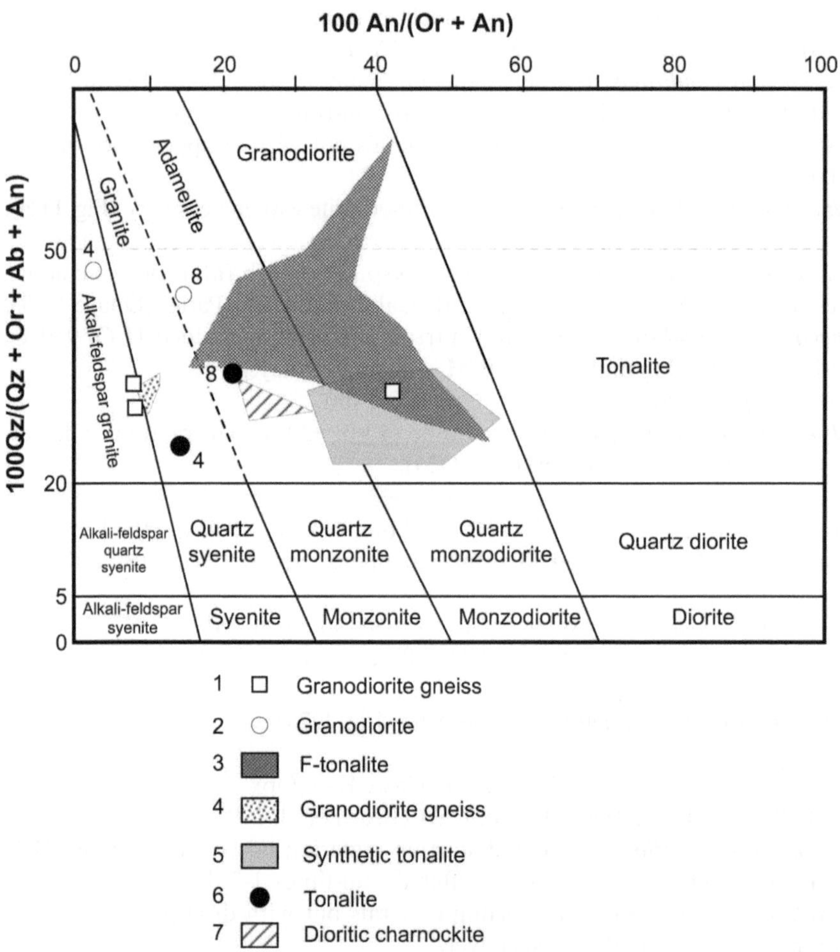

Figure D2a. Normative Qz, Or, Ab, An—ratio plot showing compositions of granite melts derived from experimental melting of granitic rocks (Nos. D.1–7), with bulk compositions listed in the table. Values below recalculated anhydrous totals of Analyses 2 and 6 represent hydrous totals. Experiments at 4 and 8kb indicated.

Crystalline products: Qz, Pl, ± Ksp, Opx, ± Cpx, ± Hb, ± Bt, ± Gt, ± Ilm
Melt composition: Adamellite–granodiorite with increasing *T* (Fig. D2a).

6. *Tonalite* with Qz (20%), Pl (45%), Ksp (7%), Hb (13%), Bt (13%); accessories (2%), from Sierra Nevada batholith, California, USA (Patiño-Douce 1997).
Experimental conditions: Powdered starting material; no added H_2O; 950°C; 4 and 8 kb; fO_2 1 log unit below QFM; run-times 7 days.
Crystalline products: ± Pl, ± Qz, Bt, Opx, Cpx, Ilm, Rt (only at 8 kb).
Melt composition: Adamellite (4 kb; ~40% melt fraction), granite (8 kb; ~30% melt fraction) (Fig. D2a).

7. *Dioritic 'charnockite' gneiss* with Qz, Anorthoclase, Ksp, Cpx, Opx, Ilm, Mt, Ap as a xenolith from Kilbourne Hole, New Mexico, USA (Beard et al. 1993).
Experimental conditions: Powdered starting material; no added H_2O and H_2O added; 900–1000°C; 6.9 kb; fO_2 unbuffered; run-times 4–7 days.
Crystalline products: Same as starting minerals but with disappearance of Ksp at 900°C and Mt at 950°C.
Melt composition: Adamellite (Fig. D2a).

INDEX

MIX
Papier aus verantwortungsvollen Quellen
Paper from responsible sources
FSC® C105338

If you have any concerns about our products,
you can contact us on
ProductSafety@springernature.com

In case Publisher is established outside the EU,
the EU authorized representative is:
Springer Nature Customer Service Center GmbH
Europaplatz 3, 69115 Heidelberg, Germany

Printed by Libri Plureos GmbH
in Hamburg, Germany